Nutritional Neurosciences

Series Editor

Mohamed Essa, Sultan Qaboos University
Qaboos, Oman

This book series aims to publish volumes focusing on both basic and clinical research in the field of nutritional neuroscience with a focus on delineating the effect of nutrition on brain function and behavior. The books will examine the role of different nutrients, food agents and supplements (both macro and micro) on brain health, neurodevelopment, neurochemistry, and behaviour. The books will examine the influence of diet, including phytochemicals, antioxidants, dietary supplements, food additives, and other nutrients on the physiology and metabolism of neurons, neurotransmitters and their receptors, cognition, behavior, and hormonal regulations.

The books will also cover the influence of nutrients and dietary supplements on the management of neurological disorders. It details the mechanism of action of phytonutrients on signaling pathways linked with protein folding, aggregation, and neuroinflammation. The books published in the series will be useful for neuroscientists, nutritionists, neurologists, psychiatrists, and those interested in preventive medicine.

Mullaicharam Bhupathyraaj
K. Reeta Vijayarani
Muralikrishnan Dhanasekaran
Kalander Ameer • Kamal Dua
Editors

Neuroprotective Marine Bioactives

Ocean-Derived Compounds for Targeting CNS Disorders

Springer

Editors
Mullaicharam Bhupathyraaj
College of Pharmacy
National University of Science & Technology
Muscat, Oman

Muralikrishnan Dhanasekaran
Harrison College of Pharmacy
Auburn University
Auburn, AL, USA

Kamal Dua
NICM Health Research Institute
and School of Science
Western Sydney University
Sydney, NSW, Australia

K. Reeta Vijayarani
Department of Pharmaceutics
Surya School of Pharmacy
Vikravandi, Villupuram, Tamil Nadu, India

Kalander Ameer
Faculty of Medicine
AIMST University
Bedong-Kedah, Malaysia

ISSN 2730-6712 ISSN 2730-6720 (electronic)
Nutritional Neurosciences
ISBN 978-981-95-5790-5 ISBN 978-981-95-5791-2 (eBook)
https://doi.org/10.1007/978-981-95-5791-2

This Springer imprint is published by the registered company Springer Nature Singapore Pte Ltd.
The registered company address is: 152 Beach Road, #21-01/04 Gateway East, Singapore 189721, Singapore

Preface

A Perspective Indication, Modern Score, and Perspective Necessity for Marine-Derived Natural Bioactive Compounds in Neuroscience

The animal and human central nervous system (CNS), composed of the brain and spinal cord, intuitively and coherently orchestrates the complex symphony of cognitive, sensory, and motor functions essential to the quality and longevity of life. Yet this intricate central network is vulnerable to an expanding spectrum of diseases/disorders, ranging from neurodegenerative and/or neurological diseases such as Alzheimer's and Parkinson's disease to traumatic injuries, seizures, sleep pathologies, neuroinfections, neuropsychiatric syndromes, and primary brain malignancies. The global prevalence of CNS disorders continues to rise drastically, driven by longer life expectancy, environmental stressors, and modern lifestyle factors. Despite progress in understanding the etiology, pathophysiology, and molecular pathways in neuroscience, current therapeutic strategies remain largely symptomatic, with limited ability to halt or reverse the neurodegeneration. This imminent and contemporary reality underscores an urgent need for novel natural bioactives or synthetic agents capable of addressing the multifactorial nature of CNS pathology.

Perspective Indication

Amid this critical healthcare landscape, marine-derived natural bioactives have emerged as a positively stirring frontier in neuroscience to diagnose, prevent, or treat various neurological and neurodegenerative diseases (Table 1). Covering nearly 70% of the Earth's surface, the oceans harbor extraordinary marine biodiversity, where organisms ranging from microalgae and corals to sponges, mollusks, and marine microbes produce structurally unique and biologically potent bioactives molecules with excellent pharmacodynamic and pharmacokinetic properties, with few or minimal adverse effects and hypersensitivity reactions. In recent years, such marine bioactives have demonstrated neuroprotective, antioxidant, anti-apoptotic, mitochondrial energy enhancing, anti-excitotoxic, anti-inflammatory, regenerative, and antimicrobial properties, aligning closely with the complex therapeutic

Table 1 Neurological and neurodegenerative diseases and their important aspects

CNS diseases	List of diseases/ disorders	Important aspects
Neurological	Brain tumors (e.g., glioblastoma, meningioma) CNS infections (e.g., Meningitis, Encephalitis) Epilepsy Migraine Multiple sclerosis (MS) Neuronal hyperarousal Spinal cord injury (SCI) Stroke Traumatic brain injury (TBI)	Acute or chronic disorders affecting CNS anatomy (structure) or physiology (function) Major pathologies include trauma, vascular events, neoplasms, infections, autoimmune and demyelinating processes Clinical sequences are usually reversible, relapsing-remitting, or result in stable deficits Common pathological signaling includes inflammation, demyelination, altered neuronal excitability, and mass effects
Neurodegenerative	Alzheimer's disease (AD) Amyotrophic lateral sclerosis (ALS) Dementia with Lewy bodies (DLB) Frontotemporal dementia (FTD) Huntington's disease (HD) Limbic-predominant age-related TDP-43 encephalopathy (LATE) Parkinson's disease (PD) Spinocerebellar ataxias (SCAs)	Irreversible progressive disorders characterized by selective and relentless neuronal loss in specific regions in the CNS The major markers are abnormal protein aggregation, mitochondrial and synaptic dysfunction, neuroinflammation, oxidative stress, apoptosis, excitotoxicity, and neural network disintegration Predominantly age-related, most show insidious onset with gradual cognitive, motor, or behavioral decline

requirements of various neurological or neurodegenerative disorders. Marine bioactives such as bryostatin-1, ziconotide, neuroprotectin D1, fucoidan, and various omega-3 polyunsaturated fatty acids exemplify this prophylactic and therapeutic promise, showing the capacity to modulate synaptic plasticity, protect neurons from oxidative stress, promote axonal regeneration, and influence neuroimmune responses.

Modern Score

The contemporary "score" of marine-derived bioactives in neuroscience reflects a rapidly maturing field defined by the following three conceptual pillars (Table 2).

Table 2 Modern score of marine-derived bioactives in neuroscience

Conceptual pillar	Key components	Representative examples/evidence	Impact on neuroscience
Scientific validation	Strong preclinical evidence in neuroprotection, anti-inflammation, and trauma repair Clinical translation supported by early human trials	Ziconotide (FDA-approved for refractory pain) Bryostatin-1 (in Alzheimer's disease trials) Astaxanthin, Fucoidan (preclinical neuroprotection)	Demonstrates that marine bioactives can alter critical neurodegenerative and repair pathways; bridges bench-to-bedside research
Translational impact	Multimodal "disease-modifying" bioactives challenge single-target therapy Diagnostic use in imaging, biomarker detection, and real-time monitoring	Labeled marine peptides for molecular imaging Conotoxins and marine toxins for CNS diagnostics	Enables precision neurology by combining therapeutic, diagnostic, and monitoring applications in CNS diseases
Technological readiness	Marine bioprospecting and metabolomics for compound discovery Synthetic biology and semisynthetic production for scalability	Marine microbial fermentation for rare metabolite production Structural optimization of bryostatins, manzamines	Converts rare marine compounds into reproducible, scalable resources for pharmaceuticals and nutraceuticals

Scientific Validation

In the last two decades, preclinical studies have robustly demonstrated the ability of marine bioactives to significantly alter critical pathways involved in neurodegeneration, trauma repair, and neuroinflammation. Regulatory success stories, such as ziconotide for refractory pain, and ongoing clinical trials involving bryostatin-1 for Alzheimer's disease, provide strong evidence of translational momentum in neuronal care.

Translational Impact

Marine bioactives are redefining both diagnostics and therapy. Multimodal bioactives challenge the "single-target" paradigm, offering integrated disease-modifying effects. Additionally, labeled marine peptides and specialized toxins emerge as tools for early disease detection, molecular imaging, and real-time therapeutic monitoring, indicating an era of precision health, especially focused on neurology.

Technological Readiness

Advances in marine bioprospecting, metabolomics, and synthetic biology have enhanced novel bioactives discovery, structural optimization, and scalable nutraceutical and pharmaceutical production. This technological leap has transformed once-rare marine bioactives into accessible resources for drug development.

Perspective Necessity
The strategic necessity of pursuing marine-derived natural bioactives in neuroscience is clear. Uniquely, the natural marine world renders unmatched biological diversity, providing molecular scaffolds absent in the terrestrial ecosystems. CNS disorders are inherently multifactorial, demanding therapeutic agents with multimodal mechanisms, a hallmark of many marine bioactives. Additionally, their roles extend beyond the therapeutic efficacy to prophylactic effect and early diagnosis, supporting nutraceutical value (neurovascular health, mitigating oxidative damage), and facilitating early detection of neurodegeneration. In an era of rising global neurological disease burden, these potent and valuable bioactives offer a nutraceutical and clinical route to close critical therapeutic gaps while enabling a shift toward preventive and personalized neurology.

This book, mainly focusing on the role of marine bioactives in the most prevalent CNS disorders, seeks to consolidate this knowledge and provide a comprehensive exploration of the prophylactic, therapeutic, and diagnostic potential of marine bioactives. Each chapter bridges mechanistic pharmacological insights with translational clinical applications, spanning neurodegeneration, traumatic injury, neuropsychiatric syndromes, neuro-oncology, neuroinfections, and the emerging fields of neurodiagnostics and prophylactic neurology. By uniting the disciplines of marine biology, medicinal chemistry, pharmacology, toxicology, and molecular and clinical neuroscience, this book humbly aims to inspire collaborative innovation. The marine world remains a largely untapped reservoir of neurological solutions, and its natural bioactives are poised to transform our healthcare approach to some of the most challenging neurological conditions of our time.

Muscat, Oman — Mullaicharam Bhupathyraaj
Vikravandi, Villupuram, Tamil Nadu, India — K. Reeta Vijayarani
Auburn, AL, USA — Muralikrishnan Dhanasekaran
Bedong-Kedah, Malaysia — Kalander Ameer
Sydney, NSW, Australia — Kamal Dua

Contents

Editors and Contributors

About the Editors

Mullaicharam Bhupathyraaj is Professor of Pharmaceutics at the College of Pharmacy, National University of Science and Technology, Sultanate of Oman. With more than two decades of academic and research excellence, she has made significant contributions to biomedicine, nanopharmaceuticals, herbal formulation development, and artificial intelligence–driven drug delivery systems. She holds ten international patents, numerous publications in high-impact journals, and has received multiple national and international research grants. She serves as external reviewer for the Oman Academic Accreditation Authority and MoHERI-funded research projects and has been a judge for the global Reimagine Education Awards. A registered pharmacist and life member of IPA, APTI, IPGA, and ISTE, she is recognized for her leadership in advancing pharmaceutical innovation and sustainable healthcare.

K. Reeta Vijayarani is a Professor in the Pharmaceutics Department of Surya School of Pharmacy in Vikravandi, Villupuram, Affiliated to the TN Dr. M.G.R. Medical University, Chennai. She has combined experience of 29 years in pharmaceutical research and education, and she has authored and presented papers at both national and international levels. She has built a distinguished career marked by an extensive research portfolio that includes numerous publications, both national and international patents, and the authorship of multiple textbooks and serving as an editorial board member and a reviewer in various peer-reviewed books and journals. Her work has made substantial contributions to pharmaceutical formulations, innovative drug delivery systems, and the integration of Artificial Intelligence in pharmacy. Her research is deeply rooted in the exploration of biodiversity and the development of novel therapeutic techniques. Her scholarly contributions reflect a commitment to understanding ecological and leveraging it for medical and pharmaceutical innovation.

Muralikrishnan Dhanasekaran, MPharm, PGDPM&IR, PhD is an Alumni Professor and Administrative Fellow in the Graduate School at Auburn University, USA. He has mentored numerous undergraduates, Honors College, graduate, and professional (PharmD) students, and has authored nearly 160 peer-reviewed international publications, over 500 scientific abstracts, nearly 80 book chapters, and has served as editor for more than 12 scientific books. He is also an active reviewer for multiple international journals and grant agencies. His research laboratory focuses on preclinical neuropharmacology and drug development, particularly investigating the neuroprotective and neurotoxic effects of natural bioactives and synthetic molecules. His current research aims to discover and develop novel therapeutic agents targeting epilepsy, Alzheimer's disease, and Parkinson's disease. As an Administrative Fellow, he has served as a member and leader on various college, departmental, and university committees, and is actively engaged in national and international graduate student recruitment as well as in advancing the welfare, engagement, and professional development of postdoctoral associates.

Kalandar Ameer is a highly accomplished Senior Associate Professor of Biochemistry with a distinguished career spanning academia and research across international institutions. Currently, he is working in the Faculty of Medicine at AIMST University, Malaysia. Prior to this appointment, he served as Associate Professor of Biochemistry under the UNDP expatriate program at two prominent public universities within the Ethiopian Ministry of Higher Education from 2019 to 2024. His foundational academic and professional experiences were cultivated in various Indian universities from 1996 to 2008.

Kamal Dua has a research experience of over 16 years working in the field of Targeted Drug Delivery for Inflammatory Diseases. He is also a Node Leader of Drug Delivery Research in the Centre for Inflammation at Centenary Institute/UTS, where the targets identified from the research projects are pursued to develop novel formulations as the first step toward translation into clinics. His research is in two complementary areas: drug delivery and immunology, specifically addressing how these disciplines can advance one another helping the community to live longer and healthier. This is evidenced by his extensive publication record in reputed journals.

Contributors

Husni Ahmed Abdullah Al-Goshae International Medical School, Management and Science University, Shah Alam, Malaysia

Manha S. Ameen Sophomore at Thomas S Wootton High School, Rockville, MD, USA

Myiesha S. Ameen Sophomore at University of Maryland, School of Public Health, College Park, MD, USA

Raakin S. Ameen Senior at Thomas S Wootton High School, Rockville, MD, USA

S. Anbazhagan Department of Pharmaceutics, Surya School of Pharmacy, Vikravandi, Villupuram, Affiliated to The TN Dr. M.G.R. Medical University, Chennai, Tamil Nadu, India

Nishi Ann Department of Zoology, Providence Women's College (Autonomous), Calicut, Kerala, India

Vijeyrani Vadipatti Baskarpandian Department of Mathematics, Yadava College, Thiruppalai, Madurai, India

Mullaicharam Bhupathyraaj College of Pharmacy, National University of Science and Technology, Muscat, Oman

Abbi Calhoun Drug Discovery and Development, Harrison College of Pharmacy, Auburn University, Auburn, AL, USA

Jack Carlisle Drug Discovery and Development, Harrison College of Pharmacy, Auburn University, Auburn, AL, USA

Preston Cook Drug Discovery and Development, Harrison College of Pharmacy, Auburn University, Auburn, AL, USA

Muralikrishnan Dhanasekaran Drug Discovery and Development, Harrison College of Pharmacy, Auburn University, Auburn, AL, USA

Graduate School, Auburn University, Auburn, AL, USA

Carson Greer Drug Discovery and Development, Harrison College of Pharmacy, Auburn University, Auburn, AL, USA

Cammi Griffith Drug Discovery and Development, Harrison College of Pharmacy, Auburn University, Auburn, AL, USA

Hannah Griggs Drug Discovery and Development, Harrison College of Pharmacy, Auburn University, Auburn, AL, USA

K. Ilango Crescent School of Pharmacy, B.S. Abdhur Rahman Institute of Science and Technology, Chennai, Tamil Nadu, India

G. Jeyabalan Alwar Pharmacy College, Rajasthan University of Health Sciences, Alwar, Jaipur, India

Yaman Walid Kassab College of Pharmacy, National University of Science and Technology, Muscat, Oman

College of Pharmacy, Al-Bayan University, Baghdad, Iraq

M. Keerthiga Department of Pharmaceutics, Surya School of Pharmacy, Vikravandi, Villupuram, Affiliated to The TN Dr. M.G.R. Medical University, Chennai, Tamil Nadu, India

Tyler King Drug Discovery and Development, Harrison College of Pharmacy, Auburn University, Auburn, AL, USA

S. Daisy Chella Kumari College of Pharmacy, Madras Medical College, Affiliated to The TN Dr. M.G.R. Medical University, Chennai, Tamil Nadu, India

Keyi Liu Drug Discovery and Development, Harrison College of Pharmacy, Auburn University, Auburn, AL, USA

Hailey Lushington Drug Discovery and Development, Harrison College of Pharmacy, Auburn University, Auburn, AL, USA

Kiruba Mohandoss Sri Ramachandra Institute of Higher Education and Research, Chennai, Tamil Nadu, India

Rishi Nadar Drug Discovery and Development, Harrison College of Pharmacy, Auburn University, Auburn, AL, USA

Muath Fahmi Najjar Clinical Pharmacy Department, Al Rayan National College of Health Sciences and Nursing, Al Rayan National Colleges, Al-Madinah al-Munawwarah, Saudi Arabia

Kirubakaran Narayanan Drugs Testing Laboratories, Directorate of Drugs Control, DMS Teynampet, Chennai, Tamil Nadu, India

Jaufar Palekkodan Illham House, Changuvetti kund, Kottakkal, Kerala, India

Suhrud Pathak Drug Discovery and Development, Harrison College of Pharmacy, Auburn University, Auburn, AL, USA

R. Ramadeepa Priya Department of Pharmaceutics, Surya School of Pharmacy, Vikravandi, Villupuram, Affiliated to The TN Dr. M.G.R. Medical University, Chennai, Tamil Nadu, India

Bennett Schrimpf Drug Discovery and Development, Harrison College of Pharmacy, Auburn University, Auburn, AL, USA

Wael Mahmoud Searan Diabetic Foot Center, Center Diabetes & Endocrinology, KFSH, Buraidha Al-Qassim, Saudi Arabia

Hemalatha Selvaraj Karpagam Academy of Higher Education, Coimbatore, Tamil Nadu, India

James Stoddard Drug Discovery and Development, Harrison College of Pharmacy, Auburn University, Auburn, AL, USA

Sophie Mae Talbert Drug Discovery and Development, Harrison College of Pharmacy, Auburn University, Auburn, AL, USA

A. Theresa Thenisaimozhi Central institute of Petrochemical Engineering and Technology, Anna University, Guindy, Chennai, Tamil Nadu, India

K. Reeta Vijayarani Department of Pharmaceutics, Surya School of Pharmacy, Vikravandi, Villupuram, Affiliated to The TN Dr. M.G.R. Medical University, Chennai, Tamil Nadu, India

Peter Weglarz Drug Discovery and Development, Harrison College of Pharmacy, Auburn University, Auburn, AL, USA

Kyle Yeatter Drug Discovery and Development, Harrison College of Pharmacy, Auburn University, Auburn, AL, USA

Marine-Derived Natural Bioactives to Relieve Brain and Spinal Cord Injuries

1

Abbi Calhoun, Preston Cook, Keyi Liu, Suhrud Pathak, Rishi Nadar, Muralikrishnan Dhanasekaran, Husni Ahmed Abdullah Al-Goshae, and Manha S. Ameen

Abstract

Marine-derived natural bioactives have slowly emerged as a potential class of therapeutics in the treatment of complex central nervous system disorders, specifically brain and spinal cord injuries. In this systematic review, key compounds from a multitude of marine organisms will be explored, highlighting their effects as possible therapeutics due to their neuroprotective, anti-inflammatory, regenerative, and oxidative properties. These compounds are shown to modulate crucial pathological mechanisms (i.e., neuroinflammation, blood-brain barrier disruption, etc.), stimulating neurogenesis and axonal regeneration. Many therapeutics, such as ziconotide, bryostatin-1, manzamines, and fucoidan, have exhibited significant efficacy in preclinical and clinical models through targeted modulation of calcium channels, protein kinase C, and inflammatory cytokine pathways. Additionally, preventive agents like astaxanthin, omega-3 fatty acids, neuroprotectin D1, and squalamine promote significant health benefits, combating against neurodegeneration and stroke, through antioxidant and neurovascular

A. Calhoun · P. Cook · K. Liu · S. Pathak · R. Nadar
Drug Discovery and Development, Harrison College of Pharmacy, Auburn University, Auburn, AL, USA

M. Dhanasekaran (✉)
Drug Discovery and Development, Harrison College of Pharmacy, Auburn University, Auburn, AL, USA

Graduate School, Auburn University, Auburn, AL, USA
e-mail: dhanamu@auburn.edu

H. A. A. Al-Goshae
International Medical School, Management and Science University, Shah Alam, Malaysia

M. S. Ameen
Sophomore at Thomas S Wootton High School, Rockville, MD, USA

M. Bhupathyraaj et al. (eds.), *Neuroprotective Marine Bioactives*, Nutritional Neurosciences, https://doi.org/10.1007/978-981-95-5791-2_1

stabilizing effects. Diagnostic applications, including labeled peptides and marine toxins, provide novel avenues for imaging and injury monitoring. Overall, this review highlights ongoing clinical trials and prospects in personalized neurology and integrative CNS care, underscoring the ocean's untapped pharmacological potential. Marine bioactives represent a vital frontier in both therapeutic intervention and preventative strategy for CNS trauma and degeneration.

1.1 Marine-Derived Bioactives

Over the years, marine-derived bioactives have become extremely relevant due to the possibility of treating and relieving brain and spinal cord injuries. There is a rise in neurological and neurodegenerative diseases that have researchers targeting studies on the central nervous system. Your central nervous system (CNS) is a processing center that manages everything that your body does, from your thoughts and feelings to your movements. Your brain and spinal cord are "central" to your CNS because they take in and send out information to your entire body. The central nervous system has a sensory input, an integration center, and a motor output. There are many different reasons for having a brain and spinal cord injury.

Brain and spinal cord injuries are trauma and permanent damage done to your central nervous system. These damages can lead to neurological deficits, disability, or even death. When disrupting the central nervous system, you are also disrupting your motor function, sensory processing, cognition, and autonomic functions.

Your motor functions are your movement controls. When your movement controls are injured, this can cause weakness, paralysis, muscle spasticity, loss of coordination, and even impaired reflexes. Brain injury can affect gross and fine muscle movements. Gross motor movements are large muscle movements as seen in muscle use during exercise or sporting activities. Fine motor movements are movements requiring dexterity and small, very well-controlled movements, e.g., threading a needle.

The brain is "wired" so that damage to the left side of the brain can cause motor deficits on the right side of the body, and damage to the right side of the brain can cause motor deficits on the left side of the body. When there is an injury to the brain, it is going to affect your precision and initiation. This type of injury would take place in your motor cortex and basal ganglia. When there is an injury in your spinal cord, that disrupts your communication with muscles. This takes place on the motor pathways. Loss of feeling and control of movement is known as paralysis. Paralysis from a spinal cord injury can be referred to as tetraplegia, also known as quadriplegia. This means that your arms, hands, trunk, legs, and pelvic organs are all affected by your spinal cord injury. Paraplegia is paralysis that affects all or part of the trunk, legs, and pelvic organs but not the arms.

The sensory processing is also disrupted when the brain or spinal cord is injured. Sensory processing is the brain's ability to receive and interpret sensory input. This refers to your touch, pain, temperature, and position. When injured, this causes numbness or loss of sensation, chronic neuropathic pain, tingling or burning sensations, and loss of proprioception. This can also cause a change in mood swings and tantrums. Your fight or flight response is also affected. Sensory processing disorder (SPD) is a condition that affects how your brain processes sensory information (stimuli). Sensory information includes things you see, hear, smell, taste, or touch. SPD can affect all of your senses, or just one. SPD usually means you're overly sensitive to stimuli that other people are not. But the disorder can cause the opposite effect, too. In these cases, it takes more stimuli to impact you. On the other hand, a lack of sensory input can impact your social and emotional health.

Next is your cognition. This is your memory, attention, language, and emotional regulation. When certain brain injuries occur, this affects your memory loss, difficulty concentrating, impaired decision-making, and personality or behavior changes. This tends to be affected during a traumatic brain injury or a stroke that affects the frontal, parietal, or temporal lobes. Cognitive impairments due to traumatic brain injury (TBI) are substantial sources of morbidity for affected individuals, their family members, and society. Disturbances of attention, memory, and executive functioning are the most common neurocognitive consequences of TBI at all levels of severity. Disturbances of attention and memory are particularly problematic, as disruption of these relatively basic cognitive functions may cause or exacerbate additional disturbances in executive function, communication, and other relatively more complex cognitive functions. Because of the high rate of other physical, neurologic, and psychiatric syndromes following TBI, a thorough neuropsychiatric assessment of the patient is a prerequisite to the prescription of any treatment for impaired cognition.

Lastly, your autonomic function can be altered. These are involuntary functions that are regulated by the autonomic nervous system. The autonomic nervous system has to deal with breathing, heart rate, and digestion. When your autonomic functions are injured, you can have breathing difficulties, abnormal heart rate, digestive problems, and bladder dysfunction. These tend to take place when there is damage to the brainstem or high cervical spinal cord. This can often be life-threatening as well. After a spinal cord injury (SCI), supraspinal influence on the ANS is disrupted, leading to sympathetic blunting and parasympathetic dominance, resulting in cardiac dysrhythmias, systemic hypotension, bronchoconstriction, copious respiratory secretions, and uncontrolled bowel, bladder, and sexual dysfunction. Further, afferent signals to the sympathetic cord elicit unabated reflex sympathetic outflow in response to noxious stimuli below the level of SCI. This article outlines the pathophysiology of SCI on the ANS, clinical ramifications of autonomic dysfunction, and the potential long-term sequelae of these influences following SCI.

Overall, when the brain and spinal cord are injured, this causes many effects in the functions of your body. The motor function can cause paralysis, weakness, spasticity, and loss of coordination. The sensory function can cause numbness, tingling, chronic pain, and loss of proprioception. The cognition function can cause memory loss, poor attention, speech issues, and mood changes. Lastly, when injured, the autonomic function can cause breathing difficulties, heart rate instability, and bladder issues.

Marine-derived bioactives are natural compounds that come from marine organisms that have biological activity. This means they can interact with the living systems to cause beneficial effects. Nature has been the traditional source of new pharmaceuticals. Today, over 50% of the marketed drugs are either extracted from natural sources or produced by synthesis using natural products as templates or starting materials. Since ancient times, early societies used natural medicines, generally as crude extracts from plants, to treat infection, inflammation, pain, and a variety of other maladies. Recently, marine bioactives have become more widely used than in past years. In recent years, a significant number of novel metabolites with pharmacological potential have been discovered from marine organisms, such as polyketides, alkaloids, peptides, proteins, lipids, shikimates, glycosides, isoprenoids, and hybrids, which exhibit biological activity, including anticancer, antitumor, antiproliferative, antimicrotubule, cytotoxic, photo protective, as well as antibiotic and antifouling properties. In the beginning, marine materials were used in a traditional sense for healing. There were not many, or if any, drugs in the early century, so the people of that time just used what they could. The first marine-derived drugs started to develop shortly after. Then came the 1980s and 1990s, where more drugs becam studied, leading to FDA approvals and now still to this day, these drugs being used and tested in clinicals. A large proportion of the current pharmacopoeia can trace its origin back to nature. Biodiversity in the Earth's largest habitat, the ocean, is greater than that on land, suggesting a huge potential for new bioactive chemistry. Compounds isolated from marine organisms have novel structures and modulate human disease targets with novel mechanisms of action. Around 15–20 marine-derived compounds have been approved for clinical use against cancer, pain, viral infection, and heart disease.

Bioactives have novel chemical structures, high potency, and multi-target effects. Those effects include anti-inflammatory, antioxidant, and neuroprotective. Marine-derived bioactives come from many different sources, which differentiate into bioactive classes. Algae, sponges, corals, mollusks, and microbes are a couple of examples of marine sources.

Algae are rich in antioxidants, polysaccharides, and polyphenols that protect neurons, reduce inflammation, and help with recovery. Seaweed and microalgae are types of algae. They appear to have antibacterial, antifungal, antiviral, antioxidative, anticancer, neuroprotective, and chemopreventive activities. These properties confer on microalgae the potential for use in the treatment and/or management of several neurologic and cell dysfunction-related disease conditions, including Alzheimer's disease (AD), AIDS, and COVID-19. Fucoidan, a type of brown algae, is used for anti-inflammatory and antioxidant. Phlorotannins, also brown algae, are used for anti-neuroinflammatory. Astaxanthin, a microalga, is a strong antioxidant that also protects the mitochondria. Lastly, EPA/DHA, another kind of microalgae, helps stabilize the membrane and is also used as an inflammatory. Microalgae are highly diverse and contain significant amounts of various bioactive compounds, including bioactive peptides, PUFA, vitamins, phenolics, carotenoids, sterols, and so on. These bioactive compounds can exhibit considerable health-promoting effects, such as antioxidant, anti-inflammatory, anticancer, and antimicrobial effects.

Sponges are the oldest of the marine sources and are very potent. This bioactive sauce is well known for its therapeutic potential. They are used specifically for neurological, anti-inflammatory, and anticancer applications. The compounds extracted from these sponges are known to exhibit various bioactivities, such as antimicrobial, antitumor, and general cytotoxicity. For example, various compounds isolated from *Theonella swinhoei* have showcased various bioactivities, such as those that are antibacterial, antiviral, and antifungal. Aaptamine, a type of alkaloid, is used as an antioxidant and a neuroprotective and may inhibit neuronal apoptosis. Manzamines, a type of beta-carboline alkaloid, is used as an anti-inflammatory and neuroprotective. Halichondrin B, a type of polyether macrolide, is used to stabilize axon regeneration and neural repair. Discodermolide, a polyketide, is a potent microtubule stabilizer and immunosuppression. At the moment, discodermolide is currently being tested for supporting cytoskeletal integrity in neurons. Discodermolide has been in phase I clinical trials for the treatment of advanced solid malignancies, where patients did not display any severe neuropathy or neutropenia but did show signs of pulmonary toxicity.

Corals produce terpenoids, diterpenes, and steroid-like compounds with anti-inflammatory and neuroactive properties. It has been estimated that the new compounds isolated and identified from soft corals represent more than 20% of the total new marine natural products reported from 2010 to 2011 (Ngandjui et al. 2024; Pope and Deer 2013). Importantly, the natural compounds isolated from soft corals exhibit a spectrum of biological activities, such as antimicrobial, anticancer, anti-inflammatory, antiviral, and antifouling activities (Ngandjui et al. 2024; Pope and Deer 2013; Gao et al. 2021; Ntziora and Falagas 2007; Bălașa et al. 2020). Pseudopterosin, a type of soft coral, is an anti-inflammatory and promotes tissue healing. This is typically used to reduce neuroinflammation after an injury takes place. Sacophytol A, a type of soft coral, is an antitumor and has potential to neuroprotect. It is under clinical now for neurorepair. Sarcophytol A is available only in minute quantities, which is a major problem in carrying out its clinical studies, prompting it to study further on another naturally occurring cembranolide, sarcophine, obtained from red soft coral *Sarcophyton glaucum* (Varijakzhan et al. 2021). Sarcophine is a fish toxin that has inhibitory activity against various vital enzymes, including cholinesterase and phosphofructokinase, and is abundantly available (Varijakzhan et al. 2021).

Marine mollusks are a source of potent neuroactive compounds. There are many different types of mollusks, such as snails, squids, clams, and even mussels. These compounds may contribute to cardiovascular health, tissue repair, immune function, and muscle growth. Albumin extracted from mollusks has anticoagulant and antioxidant properties and is applied as a whipping, suspending, or stabilizing agent. They also produce neurotoxins and peptides for a capture and defense mechanism. There are a few bioactives that come from mollusks. Ziconotide, a type of peptide, blocks N-type voltage-gated calcium channels. It is the only intrathecal, FDA-approved, non-opioid analgesic and is recommended as first-line therapy. Despite these advantages, a small therapeutic window limits ziconotide's clinical utility, with adverse event (AE) challenges that include, but are not limited to, dizziness, nausea, and somnolence. Contulakin-G is a type of neuropeptide mimic, and it binds to neurotensin receptors. Contulakin-G also modulates pain perception and neuronal signaling. Dolastatin-10 is a cytotoxic peptide. Dolastatin-10 disrupts microtubule polymerization. Dol-10, as one of the most medically valuable natural compounds discovered up to now, has brought unprecedented hope for tumor treatment. It is particularly noteworthy that, by modifying the chemical structure of Dol-10 and combining with the application of ADC technology, Dol-10 as a new drug candidate still has great potential for development. Kappa-conotoxin is a type of cone snail and is used to target potassium channels and influence neuronal excitability and synaptic transmission. This is known to be very useful in treating epilepsy, neurotrauma, or hyperexcitability disorders. Last are muscle-derived bioactives. These are a type of blue mussels that are anti-inflammatories and promote nerve cell adhesion and repair.

Marine microbes are bacteria, fungi, and cyanobacteria. These microbes live in extreme environments, which strengthens them, allowing strong pharmacological activity. Currently, antimicrobial drugs for medical treatment derived from natural origin exhibit actinobacteria as the most important secondary metabolite source. Carbohydrates, pigments, polyphenols, peptides, proteins, and essential fatty acids are marine bioactive compounds widely studied for various applications. These compounds have rheological effects and are found helpful in the food industry and diverse biological functions, such as antioxidant, antithrombotic, anticoagulant, anti-inflammatory, antiproliferative, antihypertensive, antidiabetic, and cardioprotective activities (Varidaki et al. 2018). Marine microbes make up a good percentage of bioactive compounds. Salinosporamide A is a proteasome inhibitor. It crosses the blood-brain barrier and inhibits proteasomes. Lynamicins are marinispora species that are antibacterial and neuroactive. Lynamicins are also known for treating central nervous system infections with an inflammatory response. Linezolid may be considered for the treatment of patients with meningitis, brain abscess, ventriculitis, and other CNS infections when previous treatment has failed or where no other treatment options are available. However, linezolid should be avoided, or used with caution, in patients with underlying myelosuppressive disorders, due to the possibility of development of thrombocytopenia and other immunosuppressant adverse effects. Marine natural products are bioactive compounds that come from many different marine organisms, like sponges, algae, corals, and different bacteria. These compounds have been shown to help with neuroprotective, regenerative, and anti-inflammatory roles. This allows them to be used for therapeutic, preventive, and diagnostic purposes in brain and spinal cord injuries.

Therapeutics are a specific type of treatment used to help with different diseases and injuries. The main thing therapeutics treat with brain and spinal cord injuries is neuroregeneration, reducing swelling, and trying to prevent cell death. Natural compounds represent the main treatment strategy for 87% of human diseases (Matrix Neurological n.d.), with 63% of the newly developed drugs being categorized as naturally derived, either modified natural products, unmodified natural products, or synthetic products with a natural compound as the pharmacophore. Moreover, approximately 68% of all anti-infectious drugs, including antibacterial, antiviral, antifungal, and antiparasitic compounds, and 63% of anticancer drugs used between 1981 and 2008 were obtained from natural sources. There are so many specific diseases and injuries that therapeutics are used for treatment. One specifically is anti-inflammatory, which also reduces the chances of getting a brain edema. Numerous chemical compounds, including bioactive peptides, fatty acids, pigments, alkaloids, and polysaccharides, have already been isolated from marine sources that show anti-inflammatory properties, which can be effective in the treatment and prevention of neuroinflammatory disorders.

This is also being used in clinical trials and is still being tested at different clinics. With testing at clinicals, this gives us hope that 1 day, hopefully, marine bioactives could cure all kinds of brain and spinal cord injuries. As was evident in various preclinical studies, algal compounds conferred neuroprotection against a wide range of neurotoxic stressors, such as oxygen/glucose deprivation, hydrogen peroxide, glutamate, amyloid β, or 1-methyl-4-phenylpyridinium (MPP+), and, therefore, hold therapeutic promise for brain disorders. These compounds are being tested and used, hoping to 1 day cure a disorder or an injury.

While we discussed therapies and possible treatment, there are also many marine bioactive products that help and prevent brain and spinal cord injuries. While it is hard to prevent an accidental injury, these preventative applications are more for those prone to stroke, trauma, or neurodegeneration. Astaxanthin, omega-3 fatty acids, and phlorotannins are a few compounds used for preventative measures. Astaxanthin is an antioxidant that goes through the blood-brain barrier and protects neurons from oxidative stress. The main goal of this compound is to prevent brain damage. The effects have been reported as reduced intracellular calcium concentration, reduced T- and B-lymphocyte proliferative capacity, enhanced enzymatic antioxidant capacity, and decreased superoxide anion, hydrogen peroxide, and NO• production. Therefore, the combination of astaxanthin and fish oil may have a slightly beneficial effect in preventing oxidative stress induced by PUFAs. Omega-3 fatty acids are used as an anti-inflammatory that maintains cell membrane integrity in the brain and spinal tissue. Fatty acids (FA) are metabolites universally present in all organisms, where they play a number of biological roles, such as building blocks in biological membranes and signalling molecules. In addition to the known set of FA, marine organisms usually display molecules with a very rich chemistry and have been the source of many novel structures that frequently display marked pharmacological properties. As pharmacological research with marine chemicals continues to be extremely active, this review will focus on the biological role and potential applications of fatty acids from marine organisms, such as sponges, echinoderms, and mollusks, with particular emphasis on their application in cancer, inflammation, tuberculosis, and malaria. Lastly, phlorotannins from brown algae are used to inhibit neuroinflammation and microglial activation. This is used to hopefully prevent chronic neurodegeneration after an injury. These compounds are very abundant in brown seaweeds such as *Fucaceae* and have been demonstrated to possess numerous health-promoting properties, including antioxidant effects through scavenging of reactive oxygen species (ROS) or enhancement of intracellular antioxidant defenses, antidiabetic properties through their acarbose-like activity, stimulation of adipocytes glucose uptake and protection of β-pancreatic cells against high-glucose oxidative stress, anti-inflammatory effects through inhibition of several pro-inflammatory mediators, and antitumor properties by activation of apoptosis on cancerous cells and metastasis inhibition, among others. Overall, these are just a few compounds used for preventative measure.

Diagnostic potential for marine bioactives are molecular probes, biomarkers, or different imaging and monitoring tools for the brain and spinal cord. Marine-derived peptides, marine toxins, and labeled polysaccharide are a few compounds used for diagnostic purposes. Marine-derived peptides are tagged peptides that can cross the blood-brain barrier and bind to injury-related markers for imaging. Indeed, algae are remarkable nanobio factories, and their pragmatic properties reside in their (i) ease of handling; (ii) capacity to absorb/accumulate inorganic metallic ions; (iii) cost-effectiveness; and (iv) capacity of eco-friendly, rapid, and healthier synthesis of NPs. Preclinical and clinical trials shall enable to really define effective algal-based monotherapies. Marine toxins are used in research to selectively block ion channels that help study injury mechanisms and neural activity. However, these are not as sufficient as the other one and are still in testing. The long-lasting interest in bioactive molecules (namely, toxins) produced by (microalga) dinoflagellates has risen in recent years. Exhibiting wide diversity and complexity, said compounds are well-recognized for their biological features, with great potential for use as pharmaceutical therapies and biological research probes. Unfortunately, provision of those compounds is still far from sufficient, especially in view of an increasing demand for preclinical testing. Lastly, it is labeled polysaccharides, which target inflammatory sites and can be congested with imaging agents for detecting lesions or scar tissue. Such health effects are delivered by specific diterpenes, pigments (fucoxanthin, phycocyanin, and carotenoids), bioactive peptides, and polysaccharides. Abundant and recent studies have identified valuable biological activities of native algae polysaccharides, but also of their derivatives, including oligosaccharides and (bio)chemically modified polysaccharides. However, only a few of them can be industrially developed and open up new markets of active molecules, extracts, or ingredients.

Overall, therapeutics, preventive measures, and diagnostics are all aspects of marine-derived natural products. Therapeutics are used to repair and recover from injury, while preventive measures are used to protect tissue and prevent injury and damage. Lastly, diagnostic potential is used to identify and monitor certain injuries in the brain and spinal cord.

There is a lot of damage that can be caused when the central nervous system is damaged. The bioactives, however, have the ability to modulate various pathological processes involved in neuroinflammation, oxidative stress, neuronal degeneration, and tissue repair.

Oxidative stress and free radical damage are a pathophysiology of brain and spinal cord injuries. The severity of spinal cord injury is strictly linked to the balance between the production of reactive oxygen species and the ability of the antioxidant system to detoxify these reactive chemical species (Maugeri et al. 2023; Shao et al. 2020). Reactive oxygen species are unstable molecules, containing oxygen, that easily react with other molecules in a cell. These comprise oxygen-free radicals (e.g., superoxide, hydroxyl radicals, lipid radicals) and non-radicals (e.g., hydrogen peroxide, lipid peroxide, peroxynitrite). Reactive oxygen species are controlled by antioxidant mechanisms that include enzymatic and nonenzymatic molecules. Low-molecular-weight nonenzymatic antioxidant compounds include glutathione, vitamins C and E, β-carotene, and uric acid. Conversely, antioxidant enzymes include superoxide dismutase (SOD), catalase, glutathione reductase, and glutathione peroxidase (Bhatnagar and Kim 2017). Overall, oxidative stress is due to an imbalance that leads to the damage of lipids, proteins, and DNA in the central nervous system. Marine-derived natural compounds have an antioxidant property that helps balance out reactive oxidative species and prevent cellular damage. Brown algae and fish oils are two marine bioactives specific to this. The antioxidative effect of the mixture (P4F6) can be understood to be due to phlorotannin as an antioxidant itself and fucoidan as the activator of endogenous antioxidants against Aβ-induced oxidative stress in brain tissue. Fucoidans and phlorotannins have potent antioxidants that reduce oxidative stress and prevent further damage. Omega-3 fatty acids also help with oxidative damage as they reduce the production of reactive oxidative species and regulate the body's natural antioxidant defenses. There are a few symptoms that come with oxidative stress. Fatigue, memory loss, and sensory deficits are all symptoms. The severity of these symptoms all vary depending on the circumstance.

Neuroinflammation is a significant contributor to brain and spinal cord injuries. Neuroinflammation is defined as an inflammatory response that occurs within the brain or spinal cord. Upon damage to the blood-spinal cord barrier (BSCB) after a physical trauma, neuroinflammation is one of the key components during the primary phase, which persists toward the secondary phase of injury (National Research Council (US) Committee on the Ocean's Role in Human Health 1999; Emerson et al. 2015). The acute period of neuroinflammation is characterized by an infiltration of neutrophils and monocytes to the site of injury (Varijakzhan et al. 2021), whereas in the chronic phase, the progressive tissue degeneration that takes place across a period of months is primarily driven by lymphocytes (Varidaki et al. 2018). Inflammatory responses play a central role in regulating the pathophysiology after SCI, which greatly contributes to the repair of damaged tissues (Sang et al. 2019; Matsunaga and Kubo 2006). However, excessive inflammation may also lead to apoptosis of neurons and oligodendrocytes, resulting in a decline in neuronal functions (Matsunaga and Kubo 2006). Inevitably, changes within the spinal cord microenvironment during neuroinflammation may aggravate and accelerate the course of SCI. There are a few symptoms of neuroinflammation after an injury. These symptoms are loss of movement and sensation, emotional changes, and cognitive impairments. Your brain holds a lot of data and basically is the control center of your body. Depending on where the injury is, there can be a lot of damage caused that affects your body in the long term. There are several marine bioactives that mediate neuroinflammation and the symptoms that come with it. These natural bioactives can also reduce the chance of a secondary injury. Fucoidans, cyanobacteria-derived phycocyanin, and omega-3 fatty acids are a few bioactives that assist when relieving neuroinflammation. Fucoidans, natural algae-derived polysaccharides, have shown an important role in several pharmaceutical applications, including neurodegenerative diseases (AD, PD, Huntington disease) and ischemia-reperfusion injury (IRI), due to their potent immunomodulatory and anti-inflammatory properties. These compounds are involved in three important biochemical processes, such as inhibiting the over-activation of microglia, reducing the pro-inflammatory molecules, and regulating signaling pathways. Cyanobacteria-derived phycocyanin is a type of red algae that is significant for anti-inflammatory activity in the brain and spinal cord. Brain diseases are considered due to the destruction of homeostasis in the brain neural network, which in turn causes abnormal metabolism and apoptosis of neuronal cells. These courses are usually related to inflammatory damage. The antioxidant and anti-inflammatory effects of phycocyanin can lessen the oxidative stress response of neurons, reduce inflammation and regulate immunity, and then alleviate the damage caused by brain diseases to a certain extent. Omega-3 fatty acids help reduce the levels of inflammatory markers and assist in the activation of microglia and macrophages to prevent further inflammation.

Excitotoxicity is a process that releases excitatory neurotransmitters in large amounts. This leads to a calcium overload in neurons, which can cause cell death. This often takes place after a spinal cord or brain injury. The adult spinal cord contains a population of ependymal-derived neural stem/progenitor cells (epNSPCs) that are normally quiescent, but are activated to proliferate, differentiate, and migrate after spinal cord injury. The mechanisms that regulate their response to injury cues, however, remain unknown. Here, we demonstrate that excitotoxic levels of glutamate promote the proliferation and astrocytic fate specification of adult spinal cord epNSPCs. We show that glutamate-mediated calcium influx through calcium-permeable alpha-amino-3-hydroxy-5-methyl-4-isoxazole propionic acid (AMPA) receptors (CP-AMPARs) in concert with Notch signaling increases the proliferation of epNSPCs via pCREB and induces astrocytic differentiation through Hes1 upregulation. Furthermore, the in vivo targeting of this pathway via positive modulation of AMPARs after spinal cord injury enhances epNSPC proliferation, astrogliogenesis, and neurotrophic factor production and increases neuronal survival. Our study uncovers an important mechanism by which CP-AMPARs regulate the growth and phenotype of epNSPCs, which can be targeted therapeutically to harness the regenerative potential of these cells after injury. The symptoms that come with excitotoxicity vary depending on the severity of the injury. However, a couple of them include weakness, loss of coordination, and possibly even death. Omega-3 fatty acids and marine-derived peptides help regulate this pathological mechanism and the symptoms. Omega-3 helps reduce glutamate receptors and calcium influx, protecting the neurons. Marine-derived peptides help in regulating ion channels and neurotransmitter balance and prevent neuronal stimulation.

Neurogenesis and neural repair take place after a brain or spinal cord injury. This is typically when the central nervous system is faced with repairing damaged tissues. NSCs are primitive cells with self-renewal and multidirectional differentiation capabilities in the central nervous system and have the potential to proliferate and differentiate into neurons, astrocytes, and oligodendrocytes (Bălașa et al. 2020).

However, the apoptosis of neurons is one of the main pathological mechanisms of secondary SCI injury (Bai et al. 2023). Therefore, researchers expect that inducing NSC proliferation and differentiation into neurons to compensate for the loss of neurons is an effective approach to reduce pathological damage and promote nerve regeneration (Zhang et al. 2020).

Difficulty learning, memory loss, and possible depression behaviors are all symptoms one can experience with neurogenesis. Fucoidans and phlorotannins, collagen peptides, and chitosan are natural marine bioactive compounds that enhance neurogenesis and axonal growth and help with tissue repair as well as ease the symptoms that come with it. Fucoidans and phlorotannins help stimulate neurogenesis. This mixture of phlorotannin and fucoidan (P4F6) might be a substance for improving cognitive function by effectively regulating cognition-related molecules. Collagen peptides have been proven to improve tissue regeneration, which is essential for neural repair. There are nearly 30 recognized types of collagens, and their most commonly understood role is to provide tissues with structural and mechanical integrity. In the human body as a whole, collagen is the most abundant protein, in particular types I and III (Zhao et al. 2023). Collagen renders biological stiffness, flexibility, and strength to tissues (including the ECM), influencing the degree to which stress deforms tissue and the maximum stress that can be applied before breakdown (Gao et al. 2021; Zhao et al. 2023). Several members of the collagen superfamily, particularly collagens I, IX, and XVIII, are involved in the development of the CNS, playing important roles in neuronal maturation, neural circuit formation, axon guidance, and synaptogenesis. Chitosan is from crustaceans and has the ability to promote axonal growth and stimulate Schwann cell proliferation. Neurotrophin-3 (NT3)-coupled chitosan biomaterial, when inserted into a 5-mm gap of completely transected and excised rat thoracic spinal cord, elicited robust activation of endogenous NSCs in the injured spinal cord. Through slow release of NT3, the biomaterial attracted NSCs to migrate into the lesion area, differentiate into neurons, and form functional neural networks, which interconnected severed ascending and descending axons, resulting in sensory and motor behavioral recovery.

Apoptosis and cell death are pathological mechanism that can occur after brain and spinal cord injuries. This can occur due to the loss of function as well as oxidative stress, excitotoxicity, and inflammation. There is now strong morphological and biochemical evidence from a number of laboratories demonstrating the presence of apoptosis after SCI. Apoptosis occurs in populations of neurons, oligodendrocytes, microglia, and, perhaps, astrocytes. The death of oligodendrocytes in white matter tracts continues for many weeks after injury and may contribute to post-injury demyelination. The mediators of apoptosis after SCI are not well understood, but there is a close relationship between microglia and dying oligodendrocytes, suggesting that microglial activation may be involved. There is also evidence for the activation of important intracellular pathways known to be involved in apoptosis in other cells and systems. There are not really any symptoms that are shown externally. However, most of the symptoms happen on the inside of the body such as cell shrinkage and nuclear fragmentation. Fucoidans and phlorotannins, omega-3 fatty acids, and marine peptides are marine-derived bioactives that help prevent apoptosis. Fucoidans and phlorotannins help protect neurons by inhibiting apoptosis pathways. Fucoidan induces apoptosis in HT-29 and HCT116 human colon cancer cells, and this phenomenon is mediated via both the death receptor-mediated and mitochondria-mediated apoptotic pathways.

Omega-3 fatty acids contribute to neuroprotective effects by regulating cell survival pathways. Omega-3 FAs increased the levels of antiapoptotic proteins like Bcl-2 and Bcl-extra large (Bcl-x(L)), known to be repressed in hypothyroidism. Omega-3 FAs also restored levels of cerebellar phospho (p)-AKT, phospho-extracellular regulated kinase (p-ERK), and phospho-c-Jun N-terminal kinase (p-JNK), which were altered by hypothyroid insults, without interfering with the expression of TH-responsive gene, myelin basic protein (mbp). Taken together, these results supplement an insight into the molecular mechanism of action of omega-3 FAs in the developing brain that involves the regulation of apoptotic signaling pathways under stress. Marine peptides from fish collagen can prevent apoptosis and help with the recovery of damaged tissues.

Blood-brain barrier integrity becomes compromised when an injury takes place with the brain or spinal cord. The blood-spinal cord barrier (BSCB), a physical barrier between the blood and spinal cord parenchyma, prevents toxins, blood cells, and pathogens from entering the spinal cord and maintains a tightly controlled chemical balance in the spinal environment, which is necessary for proper neural function. When injured, this allows immune cells and toxins to infiltrate the central nervous system, which causes inflammation and tissue damage. Fucoidans have been proven to stabilize the blood-brain barrier by reducing the permeability of cells. Fucoidan-based nanocarriers target endothelial P-selectin to induce caveolin-1-dependent transcytosis and thus nanocarrier transport into the brain tumor microenvironment in a selective and active manner, the efficiency of which is increased by radiation treatment. Omega-3 fatty acids improve the integrity of the blood-brain barrier by modulating the expression of tight junctions. A major challenge in treating neurological diseases is getting drugs across the blood-brain barrier, a layer of tightly packed cells that lines the brain's blood vessels and zealously blocks toxins, pathogens, and some nutrients from entering the brain. Unfortunately, the layer also blocks many drugs that are otherwise promising candidates to treat neurological disorders. Essential nutrients like omega-3 s require the assistance of dedicated transporter proteins that specifically recognize them and get them across this barrier. "The transporters are like bouncers at a club, only letting molecules with invites or backstage passes in."

Axonal regeneration and synaptic plasticity is a pathological mechanism in the central nervous system that is often limited due to the loss of synaptic connections and impairing functions after a brain or spinal cord injury. Axon regrowth remains the major prerequisite for plasticity, regeneration, circuit formation, and eventually functional recovery. The attributed functional relevance of axon regrowth, however, will depend on several subsequent conditional neurobiological modifications, including myelination and synapse formation and also pruning of aberrant connectivity. Marine bioactives, fucoidans and phlorotannins, have been used to promote axonal growth and synapse formation by activating signaling pathways. Chitosan is another marine bioactive that is used to support Schwann cell function and myelin repair. Guide microfibres made of chitosan with T cross sections were designed as a topographic guide for axon regeneration, and ZIF-8 nanoparticles were loaded onto their surface in situ. The microfibres were then filled in the lumen of the nerve conduit to act as topographic guides for axon regeneration to construct a novel tissue-engineered neural graft. The biocompatibility of the graft was verified in vitro, and its effect on cell proliferation was investigated. The effect of ZIF-8 nanoparticles on Schwann cell migration has also been studied. Furthermore, the abilities of the prepared grafts to repair 10-mm rat sciatic nerve defects were evaluated. The results showed that the nerve conduit loaded with ZIF-8 microfibres had a significantly better effect on defective nerve repair than the hollow conduit and the conduit filled with untreated microfibres, showing similar performance to that of autologous nerve grafts. The results of this study shed new light on the synthesis and construction of functional tissue-engineered nerve grafts and lay the foundation for optimizing the clinical treatment of peripheral nerve injury. There are many pathophysiologies of brain and spinal cord injuries that can be helped with marine-derived natural bioactives. These bioactives have the ability to regulate oxidative stress, neuroinflammation, excitotoxicity, apoptosis, and neurogenesis. While these are the most common pathophysiologies, there are many more that are modulated with many different marine bioactives. Therapeutic bioactive compounds are used all over the body to heal and treat different diseases and injuries. They can reduce inflammation, help with neuroprotection, and even assist the nerves to relieve damage to the tissues. Ziconotide, bryostatin-1, manzamines, aaptamine, and fucoidans are just a few compounds that we are going to discuss.

Ziconotide comes from cone snail and is a type of peptide. It blocks N-type calcium channels to prevent pain signals from transmission. Ziconotide is an intrathecal analgesic approved by the FDA for chronic severe pain in patients who are intolerant or refractory to systemic analgesics or intrathecal morphine. This medication demonstrates efficacy in managing intractable, severe chronic pain associated with both malignant and nonmalignant conditions. By targeting specific calcium channels, ziconotide provides a unique alternative for patients with refractory pain syndromes. Clinical use requires careful consideration of contraindications, adverse events, toxicity, and drug interactions to ensure safe and effective therapy. A lot of the time this compound is delivered in intrathecal which is an injection into the spinal fluid.

Bryostatin-1 is very interesting due to its potential in neurological and neurodegenerative diseases including brain and spinal cord injuries. This comes from a marine bryozoan and was studied for anticancer effects. Bryostatin-1 is used for neuronal survival, synaptic plasticity, inflammation, and memory formation. Bryostatin is currently in an ongoing clinical trial. In an expanded access trial, three severely symptomatic AD patients received multiple bryostatin infusions for 5 to 9 months. The authors describe behavior improvements that occurred rapidly after the first dose and were maintained for weeks. Two participants were nonverbal, so no cognitive testing was possible. After the long clinical process, these were the results collected. Bryostatin completed a clinical trial for eradication of HIV/AIDS. It has been studied extensively for antitumor activity against multiple types of cancer in more than 20 Phase 2 trials, but none progressed to Phase 3. It has FDA Orphan Drug Designation for treatment of Fragile X syndrome. In terms of brain and spinal cord injuries, Bryostatin is used for different things. In traumatic brain injuries, Bryostatin is used to reduce swelling on the brain and helps improve cognitive recovery. However, in spinal cord injuries it enhanced neuroplasticity and promotes functional recovery. Bryostatin provides a therapeutic benefit in reducing brain injury and improving stroke outcome using a clinically relevant model of cerebral ischemia.

Manzamines are marine sponges and are a type of beta-carboline alkaloid. It has therapeutic potential in neurological diseases and potentially brain and spinal cord injuries. It was first discovered for antimalarial and anticancer activities; however, it is now mostly studied for neurological application. In terms of brain and spinal cord injuries, manzamines suppresses Nf-kB which reduces neuroinflammation. It also supports neural survival, which promotes neural regeneration. The manzamine alkaloids are a unique class of alkaloids possessing a characteristic heterocyclic system and exhibit a diverse range of bioactivities including cytotoxicity, antimicrobial activity, antimalarial activity, antiviral and antiinflammatory activities, anti insecticidal activity, and proteasome inhibitory activity. About 100 manzamine alkaloids have been isolated from more than 16 species of marine sponges belonging to five families.

Aaptamine is a marine sponge that is from the genus Aaptos, thus is a type of alkaloid. These are used as a chemical defense but has potential for central nervous system applications. This bioactive can be used as an antioxidant, neuroprotector, antidepressant, and even a cytoprotector. In terms of spinal cord and brain injury, aaptamine is used in preclinical studies. It prevents glial cell apoptosis an dimproves mitochondrial function. Aaptamine also reduces inflammation and oxidative damage. In the present study, we used a chronic constriction injury (CCI)-induced peripheral neuropathic rat model to explore the analgesic effects of intrathecal aaptamine administration. We also investigated cellular angiogenesis and lactate dehydrogenase A (LDHA) expression in the ipsilateral lumbar spinal cord after aaptamine administration in CCI rats by immunohistofluorescence. The results showed that aaptamine alleviates CCI-induced nociceptive sensitization, allodynia, and hyperalgesia. Moreover, aaptamine significantly downregulated CCI-induced vascular endothelial growth factor (VEGF), cluster of differentiation 31 (CD31), and LDHA expression in the spinal cord.

Fucoidan is a brown seaweed and a type of sulfated polysaccharide. Fucoidan is studied mainly for anti-inflammatory, antioxidant, and anticoagulant. It applies in many ways to brain and spinal cord injuries. In the case of a spinal cord injury, fucoidan reduces inflammation and enhances motor function recovery. Excessive neuroinflammation, apoptosis, glial scar, and demyelination triggered by spinal cord injury (SCI) are major obstacles to SCI repair. Fucoidan, a natural marine plant extract, possesses broad-spectrum anti-inflammatory and immunomodulatory effects and is regarded as a potential therapeutic for various diseases, including neurological disorders. In traumatic brain injuries, fucoidan is helpful in neuronal damage and improves blood-brain barrier integrity. When it comes to neurodegenerative diseases though, fucoidan inhibits amyloid-beta aggregation and reduces oxidative neuronal stress. Alzheimer's disease (AD) is a progressive neurodegenerative disorder that causes memory and cognitive deficits. The present study was carried out to evaluate the protective effects of fucoidan in monocrotophos-induced AD in *Drosophila melanogaster*. In silico studies showed that fucoidan exhibited binding energy of −9.3 kcal with proteins. Consistent with this, fucoidan, in a dose- and time-dependent fashion, had inhibitory activity against cholinergic and monoamine-metabolized enzymes in vitro. Fucoidan inhibited the increase in total mRNA and protein in monocrotophos-fed flies and prevented changes in biochemicals, neurochemicals, and latency time of locomotor, learning, and memory induced by monocrotophos.

Overall, marine-derived bioactive compounds are an extremely important part in therapeutic treatment. Each compound is very unique and different in each way. This makes it so these compounds are able to treat so many different diseases and injuries, specfically brain and spinal cord injuries. Compounds such as ziconotide, fucoidan, bryostatin-1, manzamine A, aaptamine, and salinosporamide-A have shown high potential in the activity of neuroprotective, anti-inflammatory, antioxidant, and regenerative properties. These activities take place in preclinical and even clinical ongoing studies. Each compound is able to target a specific mechanism, whether that is inhibiting neuroinflammation, preventing damage of neurons, and enhancing cognitive and motor function recovery. Marine-derived bioactive compounds could be a decent choice/alternative against SARS-CoV-2. The utilization of marine organisms to prevent or treat various infections has been done for a long time and is still being done in treating various complex diseases such as cancer, various viral diseases, and malaria.

Earlier, preventative bioatives were discussed. Preventative bioactive compounds are used to reduce risk hopefully before injury, strengthen the body's natural defenses, delay neurodegeneration, and improve long-term quality of life. Preventative bioactive compounds exist to protect the brain and spinal cord before the damage happens. The mentioned compounds earlier were astaxanthin, fucoidan, omega-3, and phlorotannins. They are used for a broad-spectrum neuroprotection. However, two more important preventative compounds, neuroprotectin D1 and squalamine.

Neuroprotectin D1 comes from an omega-3 fatty acid, which is found in microalgae and fish oils. This is also produced in the body in response to cellular stress in the brain and spinal cord. Neuroprotectin D1 is used as an anti-inflammatory, anti-apoptotic. This is also an oxidative stress reduction. In terms of the brain and spinal cord, neuroprotectin D1 is involved in Alzheimer's disease, stroke, and ischemia. NPD1 promotes homeostatic regulation of the integrity of these two cells, particularly during oxidative stress, and this protective signaling may be relevant in retinal degenerative diseases. Moreover, neurotrophins are NPD1-synthesis agonists, and NPD1 content is decreased in the CA1 region of the hippocampus of Alzheimer's patients. Overall, NPD1 promotes brain cell survival via the induction of antiapoptotic and neuroprotective gene-expression programs that suppress Abeta42 production and its neurotoxicity. Thus, NPD1 elicits potent cell-protective, anti-inflammatory, prosurvival repair signaling. NDP1 also can be used for spinal cord injuries, and studies are suggesting that this compound can prevent secondary damage in the spinal cord. Mechanistically, NPD1/PD1 treatment blocks nerve injury-induced long-term potentiation, glial reaction, and inflammatory responses, and reverses synaptic plasticity in the spinal cord. Thus, NPD1/PD1 and related mimetics might serve as a new class of analgesics for preventing and treating neuropathic pain.

Squalamine is a preventative compound that comes from dogfish shark. The main things this bioactive compound does are antiangiogenic, anti-inflammatory, antimicrobial, and neurovascular protection. Squalamine prevents chronic inflammation, fights infections, and supports a healthy blood-brain. The antibacterial activity of squalamine was found to correlate with that of other antibiotics, such as colistin and polymyxins. Still, in the field of microbiology, evidence has shown that squalamine and its derivatives have antifungal activity, antiprotozoa effect against a limited list of protozoa, and could exhibit antiviral activity against both RNA- and DNA-enveloped viruses. In terms of brain and spinal cord injuries, squalamine takes part in neurovascular injury prevention, neuroinflammation, and neuroinfection prevention. This compound, however, is still being test in clinical trials today where they are still collecting data. The drug has not yet been tested as an antiviral agent in humans, but it has been given to hundreds of people enrolled in clinical trials designed to test its usefulness for other conditions. Researchers say they hope to begin human trials to test the compound's antiviral activity within the next year.

In conclusion, preventive bioactive compounds are particularly derived from many different marine organisms. These preventive compounds have a promising tactic in assisting with brain and spinal cord injuries beforehand. These compounds, including fucoidan, astaxanthin, omega-3 fatty acids, neuroprotectin D1, squalamine, and phlorotannins, help and assist in various different effects. These effects include anti-inflammatory, antioxidant, neurovascular stabilizing, and anti-apoptic actions. Being able to process oxidative stress, chronic inflammation, and neurovascular dysfunction, all of these preventative compounds can reduce the risk in stroke and traumatic injuries.

Overall, marine ecosystems offer a vast and largely untapped source of bioactive compounds with remarkable chemical diversity and therapeutic potential. Among these, marine-derived molecules have shown increasing promise in the prevention and treatment of central nervous system (CNS) disorders, including traumatic brain and spinal cord injuries (BSCI), neurodegenerative diseases, and neuroinflammatory conditions. These compounds, sourced from marine algae, sponges, mollusks, microbes, and fish, possess unique structures and multifunctional properties that enable them to modulate a wide range of biological processes relevant to neural health.

Therapeutically, compounds such as ziconotide, bryostatin-1, manzamines, aaptamine, and salinosporamide A have demonstrated significant efficacy in preclinical and early clinical models of CNS injury. Their mechanisms often include anti-inflammatory, neuroprotective, antioxidant, and antiapoptotic effects—key factors in limiting secondary damage after neural trauma and promoting recovery. Many act on crucial molecular targets such as protein kinase C (PKC), glycogen synthase kinase-3β (GSK-3β), calcium channels, and inflammatory cytokines.

On the preventive side, marine compounds like fucoidan, astaxanthin, omega-3 fatty acids (DHA and EPA), neuroprotectin D1, and squalamine have shown the ability to enhance brain resilience, reduce baseline oxidative stress, and support vascular and cellular homeostasis. These agents are especially relevant for populations at higher risk of cognitive decline, stroke, or injury due to aging, genetic predisposition, or environmental exposure.

Overall, marine-derived bioactives offer a holistic approach to CNS health—working both preventively and therapeutically. While many compounds remain in the early stages of research, their safety profiles, natural origin, and multi-targeted actions make them attractive candidates for future drug development and integrative neurological care. Continued investment in marine bioprospecting, pharmacological studies, and clinical trials will be essential to fully realize their potential in improving outcomes for patients with brain and spinal cord conditions.

References

Alzheimer's Forum (n.d.) Bryostatin 1. Alzheimer's Forum. Retrieved April 13, 2025, from https://www.alzforum.org/therapeutics/bryostatin-1

Bai X, Li Z, Li J (2023) Anti-inflammatory effects of marine bioactive compounds and their potential as functional food ingredients in the prevention and treatment of neuroinflammatory disorders. Molecules 28(1):2. https://doi.org/10.3390/molecules28010002

Bălașa AF, Chircov C, Grumezescu AM (2020) Marine biocompounds for neuroprotection—a review. Mar Drugs 18(6):290. https://doi.org/10.3390/md18060290

Bhatnagar I, Kim S-K (2017) Fucaceae: a source of bioactive phlorotannins. Int J Mol Sci 18(6):1327. https://doi.org/10.3390/ijms18061327

Emerson NM, Zeidan F, Lobanov OV, Hadsel MS, Martucci KT, Quevedo AS, Starr CJ, Nahman-Averbuch H, Weissman-Fogel I, Granovsky Y, Yarnitsky D, Coghill RC (2015) Pain sensitivity is inversely related to regional grey matter density in the brain. NeuroImage 118:219–227. https://doi.org/10.1016/j.neuroimage.2015.06.039

Gao G, Wang Y, Hua H, Li D, Tang C (2021) Marine antitumor peptide dolastatin 10: biological activity, structural modification and synthetic chemistry. Mar Drugs 19(7):363. https://doi.org/10.3390/md19070363

Matrix Neurological (n.d.) Motor deficits. Matrix Neurological. https://www.matrixneurological.org/information/deficits-of-acquired-brain-injury/physical/motor-deficits/

Matsunaga S, Kubo I (2006) A possible mechanism of action of the chemopreventive effects of sarcotriol on skin tumor development in CD-1 mice. Mar Drugs 4(4):274–281. https://doi.org/10.3390/md200600400

Maugeri G, Amato A, Sortino M, D'Agata V, Musumeci G (2023) The influence of exercise on oxidative stress after spinal cord injury: a narrative review. Antioxidants 12(7):1401. https://doi.org/10.3390/antiox12071401

National Research Council (US) Committee on the Ocean's Role in Human Health (1999) Marine-derived pharmaceuticals and related bioactive agents. In: From monsoons to microbes: understanding the ocean's role in human health. National Academies Press. https://www.ncbi.nlm.nih.gov/books/NBK230700/

Ngandjui YAT, Kereeditse TT, Kamika I, Madikizela LM, Msagati TAM (2024) Nutraceutical and medicinal importance of marine molluscs. Mar Drugs 22(5):201. https://doi.org/10.3390/md22050201

Ntziora F, Falagas ME (2007) Linezolid for the treatment of patients with central nervous system infection. In: Database of abstracts of reviews of effects (DARE): quality-assessed reviews. Centre for Reviews and Dissemination (UK). Available from https://www.ncbi.nlm.nih.gov/books/NBK73759/

Sang VT, Dat TTH, Vinh LB, Cuong LCV, Oanh PTT, Ha H, Kim YH, Anh HLT, Kim YH (2019) Coral and coral-associated microorganisms: a prolific source of potential bioactive natural products. Mar Drugs 17(8):468. https://doi.org/10.3390/md17080468

Shao X, Wang T, Guo Z (2020) Role of oxidative stress in neurological disorders and the effects of bioactive natural products. Antioxidants 9(8):748. https://doi.org/10.3390/antiox9080748

Varidaki A, Hong Y, Coffey ET (2018) Repositioning microtubule stabilizing drugs for brain disorders. Front Cell Neurosci 12:226. https://doi.org/10.3389/fncel.2018.00226

Varijakzhan D, Loh J-Y, Yap W-S, Yusoff K, Seboussi R, Lim S-HE, Lai K-S, Chong C-M (2021) Bioactive compounds from marine sponges: fundamentals and applications. Mar Drugs 19(5):246. https://doi.org/10.3390/md19050246

Zhang Y, Li Y, Zhang Z, Zhang Z (2020) Neuroprotective potentials of marine algae and their bioactive metabolites: pharmacological insights and therapeutic advances. Mar Drugs 18(7):347. https://doi.org/10.3390/md18070347

Zhao C, Li X, Wang X (2023) Marine algae-derived bioactive compounds: a new wave of nanodrugs? Mar Drugs 21(9):484. https://doi.org/10.3390/md21090484

Marine-Derived Natural Bioactives in Neuroscience: An Innovative Constructive Scientific Approach to Alleviate Movement Disorder

2

Kyle Yeatter, Suhrud Pathak, Muralikrishnan Dhanasekaran, Mullaicharam Bhupathyraaj, and Kiruba Mohandoss

Abstract

Natural bioactives derived from marine sources have become a promising area of neuroscience research, providing creative and beneficial scientific approaches to movement disorders. These disorders, which include essential tremor, dystonia, and Parkinson's disease, frequently have intricate pathophysiological underpinnings that make modern treatment strategies difficult to implement. Numerous distinct bioactive substances with neuroprotective, anti-inflammatory, and anti-oxidative qualities are produced by marine species, including algae, sponges, mollusks, and marine microorganisms. This chapter examines how these marine bioactives may alter important molecular pathways, such as neurotransmitter modulation, mitochondrial function, and neuroinflammation, that are implicated in movement disorders. The chapter emphasizes the increasing significance of marine natural products as therapeutic and preventive agents by combining knowledge from marine pharmacology, neurobiology, and therapeutic development. This strategy's interdisciplinary nature provides a fresh foundation for

K. Yeatter · S. Pathak
Drug Discovery and Development, Harrison College of Pharmacy, Auburn University, Auburn, AL, USA

M. Dhanasekaran (✉)
Drug Discovery and Development, Harrison College of Pharmacy, Auburn University, Auburn, AL, USA

Graduate School, Auburn University, Auburn, AL, USA
e-mail: dhanamu@auburn.edu

M. Bhupathyraaj
College of Pharmacy, National University of Science and Technology, Muscat, Oman

K. Mohandoss
Sri Ramachandra Institute of Higher Education and Research, Chennai, Tamil Nadu, India

M. Bhupathyraaj et al. (eds.), *Neuroprotective Marine Bioactives*, Nutritional Neurosciences, https://doi.org/10.1007/978-981-95-5791-2_2

future studies and medication development targeted at reducing the burden of neurological disorders linked to movement.

2.1 Pathology of Movement Disorders

Movement disorders typically stem from neurodegenerative diseases and damage to the brain or nervous system. Disruption of neural pathways in the spinal column and brain stem leads to downstream movement disorders (Galli et al. 2020). These specific disorders are referred to as functional movement disorders (FMD), which can be inherited genetically and developed with or without trauma (Espay et al. 2018; Kamble and Pal 2023). Functional movement disorders manifest in many symptoms, including tremors, dystonia, myoclonus, gait inconsistencies, muscle weakness, and seizures (Ricciardi et al. 2023; Ricciardi and Edwards 2014; Rivest 2003).

Returning upstream of the spinal column and brain stem to the brain, there are several varied neurodegenerative diseases and acute brain injuries that result in movement disorders of varying severity. Physical trauma or extreme mental stress can damage motor control regions of the brain (Schrag and Lang 2005). Affected brain regions include the basal ganglia, cerebellum, motor cortex, and brain stem. General damage can result in cerebral palsy, parkinsonism, ataxia, tremors, dystonia, chorea, akinesia, and post-traumatic epilepsy (Biary et al. 1989; Bologna et al. 2022; Carroll et al. 2006; Conn and Jankovic 2024; Hayes 2019; Holm et al. 2023; Imai 1996; Jang and Lee 2021; Mathewson and Lieber 2015; Mestre 2016; Niedermeyer 2008; Upadhyay et al. 2020; Wimalasundera and Stevenson 2016). More specifically, damage in these areas can result in the dysfunction of dopamine neurotransmitters vital for movement (Broeder et al. 2022). Dopaminergic pathway disruption can result in movement disorders like Parkinson's disease and Huntington's disease or a movement/phonic disorder, Tourette syndrome (Johnson et al. 2023; Kalia and Lang 2015; Klein and Westenberger 2012; Langbehn et al. 2019; Walker 2007).

2.1.1 Functional Movement Disorders

Functional tremors are the most commonly occurring disease of functional movement disorders. Diagnosing functional tremors is gathered at the bedside by observing their inconsistent and incongruent features—particularly distractibility, involuntary contractions, frequency variability, entrainment, and suppression (Schwingenschuh and Espay 2022). Select cases require further neurological investigation using accelerometry and surface electromyography to diagnose. Treatments for functional tremors currently fall under cognitive behavioral therapy and psychotherapies to help control the erratic movement symptoms and address any underlying mental traumas.

As the second most common functional movement disorder, functional dystonia is characterized as an acute onset of fixed limb, truncal or facial posturing, involuntary speech, and facial weakness. Decreased intracortical and spinal inhibition leads to abnormal muscle activation that may be sustained by abnormal sensorimotor processing, abnormal selection of movements, and a hypoactive sense of agency under circumstances of normal movement preparation but abnormal connectivity between the limbic and motor networks (Ricciardi et al. 2023). Treatment for functional dystonia involves rehabilitation of both physical and emotional states. Physical and occupational therapy are recommended for physical tasks, along with psychotherapy and cognitive behavioral therapy, to address stressors and other maladaptive behaviors.

Functional myoclonus falls under functional movement disorders. Functional myoclonus describes brief muscle jerks triggered by neuron discharges (Rivest 2003). These sudden and brief jerks are expressed mainly in the arms and legs. The causes of functional myoclonus are numerous and uncertain, ranging from stress, anxiety, and fatigue to medication side effects or physical trauma. Potential origin regions for functional myoclonus are the cerebral cortex, the brain stem, and the spinal cord (Rivest 2003). Certain medications, such as haloperidol or sulpiride (Both Anti-psychotics), can be used to treat the symptoms of functional myoclonus (Ricciardi and Edwards 2014). Cognitive Behavioral therapy can be used to address other underlying psychological factors, and physical therapy is recommended to address any motor impairments.

2.1.2 Cerebral Palsy

Cerebral palsy is a classification of several permanent movement disorders that develop due to abnormal brain development or brain injury in early life, often preceding or during birth. Classified as a nonprogressive disorder, the brain injury or abnormal development that causes cerebral palsy does not degrade over time (Wimalasundera and Stevenson 2016). However, the symptoms may evolve as a child grows. Cerebral palsy is caused by damage to some brain areas responsible for coordinating muscle tone, strength, and coordination. The regions that contain these characteristics are the motor cortex, basal ganglia, or cerebellum, with the motor cortex being damaged most often. Damage to the brain can occur for a few reasons, including asphyxiation due to inadequate oxygen shared with the brain (as seen in birth asphyxia), infections during pregnancy (such as meningitis), or trauma to the region (Upadhyay et al. 2020). The location of the affected brain area is typically the source of the disorder's symptoms. Symptoms often come as muscle spasticity (contractility), weakness, and difficulty with coordination, which ultimately can bring forth difficulties with movements such as walking, speaking, and controlling the facial muscles, along with abnormal postures and/or involuntary movements (Mathewson and Lieber 2015). Damage to the basal ganglia may result in significant intellectual deficits, so much so that intellectual development, emotional processing, motivation to pursue learning tasks, and learning habits are severely impacted. Cerebellar damage can also cause language and memory impairments. Treatment for cerebral palsy is multidisciplinary and involves improving muscular function and general quality of life. Prescriptions such as muscle relaxants or anti-seizure medication may be used to manage spasticity or seizures. Surgery may be used to release tight muscles, correct bone deformities, or implant devices like a baclofen pump to manage spasticity in some instances. Devices, such as wheelchairs, braces, and communication aids, may be used as support and can help facilitate daily routines. Physical therapy can be used to maintain or regain muscle strength. Occupational therapy can help with day-to-day activities, and speech therapy may help resolve communication problems.

2.1.3 Parkinsonism

Parkinsonism is a group of neurological disorders with symptoms of Parkinson's disease, in which both are explicitly characterized by the degeneration of brain neurons responsible for synthesizing dopamine (Hayes 2019). The most prevalent cause of parkinsonism is Parkinson's disease itself. Parkinsonism also refers to other diseases that produce similar motor symptoms, such as drug-induced parkinsonism, vascular parkinsonism, or progressive supranuclear palsy. The cause of parkinsonism typically involves basal ganglia dysfunction, most notably the loss of dopamine-producing cells in the substantia nigra region of the brain responsible for the control of movement. Dopamine is a vital neurotransmitter that coordinates smooth, controlled muscle movements, and its depletion characterizes the motor symptoms of parkinsonism. In other forms of parkinsonism, such as drug-induced parkinsonism or vascular parkinsonism, the cause is linked to factors like the use of antipsychotic medications or poor blood flow to the brain due to strokes, respectively (Conn and Jankovic 2024; Holm et al. 2023).

Parkinsonism is commonly characterized by tremors (typically in the hands and feet), bradykinesia (slow movement), muscular stiffness, and posture instability (Bologna et al. 2022). Other potential symptoms are cognitive-based and consist of sleep disturbances, sensory changes, difficulty concentrating, autonomic dysfunction, and impaired thinking due to dysfunction in the dopaminergic pathway. Dopaminergic therapy is the treatment for parkinsonism associated with Parkinson's disease that seeks to increase the activity of dopamine levels in the brain. The most prescribed medication for Parkinson's disease is levodopa. Levodopa operates as a dopamine precursor, which is converted to dopamine in the brain. There are other options, such as dopamine agonists and MAO-B inhibitors, that improve symptoms by either mimicking dopamine activity or by being available to support dopamine activity (Tan et al. 2022). Deep brain stimulation (DBS), a surgical intervention that implants cortical electrodes in certain brain regions, can also potentially reduce motor symptoms in patients with severe symptoms. For non-Parkinson's forms of parkinsonism, treatment varies based on the cause and symptoms. For example, in antipsychotic-induced parkinsonism, the offending medications would be addressed, while the treatment of vascular parkinsonism could involve treating cardiovascular risk factors (Conn and Jankovic 2024; Holm et al. 2023). Physical, occupational, and speech therapy can also aid in improving movement, strength, and daily function.

2.1.4 Post-traumatic Ataxia

Post-traumatic ataxia is a type of movement disorder related to an injury to the brain, typically in the cerebellum or pathways of the cerebellum, that helps with one's coordination and balance (Jang and Lee 2021). Post-traumatic ataxia can occur after a traumatic brain injury in the form of a concussion, whiplash, or more severe brain injury through accidents or falls. The injury to the cerebellum affects the ability of one's brain to regulate and coordinate movement after the injury. This trauma can directly impact the cerebellum or the neural pathways connecting to the other parts of the brain (via the cortico-ponto-cerebellar tract) and spinal cord (Jang and Lee 2021). Post-traumatic ataxia is characterized by a lack of coordination, difficulty maintaining balance, and a wide gait. This gait is characterized by the wide stance needed for stability. The lack of coordination can lead to fine motor skill impairments, such as writing. Treatment for post-traumatic ataxia typically focuses on rehabilitation and symptom management. Physical therapy rehabilitates fine motor skills, while occupational therapy will help to have more normalcy in daily activities. Medications may be prescribed as needed to relieve tremors and dizziness.

2.1.5 Post-traumatic Tremors

Post-traumatic tremors are involuntary rhythmic shaking or oscillations of the body that occur after a traumatic brain injury, such as from a concussion, car accident, or severe head trauma. Traumatic damage occurs in the brain regions that control motor function, including the basal ganglia, cerebellum, and brainstem. Direct trauma to the basal ganglia is often the cause of post-traumatic tremors. Alternatively, damage to another part of the brain could cause inflammation in the basal ganglia or a neurotransmitter imbalance. Post-traumatic tremors are characterized as rhythmic shaking, typically in the hands, arms, or legs, but can affect any body part (Biary et al. 1989). The tremors can occur at rest, during movement, or when trying to maintain a specific posture. They may vary in intensity, from minimal tremoring, which varies from slightly perceivable, to more obvious tremors that seriously affect daily functioning. Sometimes, tremors may exacerbate when the person becomes stressed, anxious, or fatigued. For treatment, beta-blockers such as propranolol, anti-seizure drugs like gabapentin, or general muscle relaxants may be prescribed to help reduce the tremors (Biary et al. 1989).

2.1.6 Post-traumatic Dystonia

Post-traumatic dystonia is a movement disorder that develops after a traumatic brain injury, resulting in abnormal muscle contractions, postures, or movements. Specifically, the basal ganglia and other regions that control motor function can cause post-traumatic dystonia when damaged (Carroll et al. 2006). Trauma can disrupt regular coordination, resulting in sustained muscle contractions and uncomfortable postures. Additional damage to the motor cortex or cerebellum can create abnormal processing of movement signals. In combination, involuntary muscle contractions cause twisting, repetitive movements, or abnormal postures in one or more places in the body. Post-traumatic dystonia can be painful, and abnormal movements can affect the individual's ability to perform basic activities such as walking, eating, or writing. For individuals with post-traumatic dystonia, muscle relaxants, anticholinergics, or dopaminergic agents may be prescribed for muscle spasms and to help lessen the severity of dystonic movements. Botulinum toxin injections can also temporarily relieve a specific muscle group involved in the dystonic movements (Carroll et al. 2006). Physical or occupational therapy is necessary for reducing movement complications and adapting to daily activities.

2.1.7 Chorea

Chorea is a movement disorder characterized by involuntary, irregular, and jerky movements that are often unpredictable and appear to flow from one body part to another. Chorea can result from trauma to the basal ganglia, autoimmune disorders, metabolic disorders, and certain infections reaching the brain (Mestre 2016). Chorea is also caused by Huntington's disease, a neurodegenerative disease that causes uncontrolled movements characteristic of Chorea (Huntington's chorea) (Mestre 2016). These factors can lead to neurotransmitter imbalance and loss of dopaminergic receptors due to dopamine receptor autoantibodies. Quick, involuntary movements described as "dancelike" or "flowing" affect any part of the body, including the limbs, face, and trunk, resulting from these dysfunctions. These movements may worsen with stress or emotional upset and often fluctuate in intensity. Chorea can lead to difficulties with fine motor tasks, such as writing, eating, or dressing, and problems with walking and balance due to the erratic, unpredictable nature of the movements. Treatment for chorea depends on the symptoms and where the trauma is located. In cases where chorea is caused by a neurodegenerative disease such as Huntington's disease, individuals can only treat the symptoms as there is no cure for Huntington's disease. Anticonvulsants or benzodiazepines can also be used to reduce the severity of symptoms (Mestre 2016). In cases where chorea results from an underlying infection or autoimmune disorder, addressing the primary condition with antibiotics, corticosteroids, or immunotherapy may alleviate the choric movements. In addition to medication, physical and occupational therapy can be beneficial in helping individuals improve coordination, mobility, and daily functioning.

2.1.8 Akinesia

Akinesia describes the inability to move voluntarily, which primarily affects patients who have Parkinson's disease and other neurological disorders (Niedermeyer 2008). Damage to the basal ganglia region of the brain results in uncontrolled movements while preventing patients from making smooth, controlled movements. Degeneration of dopamine-producing neurons in the substantia nigra leads to a reduction in dopamine levels, which disrupts the normal functioning of the basal ganglia (Niedermeyer 2008). This imbalance impairs the brain's ability to initiate and control movement, resulting in akinesia and other motor symptoms like rigidity and tremors. Akinesia symptoms can include difficulty initiating movement, a marked decline in facial expression, and overall slowness in bodily movement. Someone with akinesia may struggle to begin activities like walking, talking, or even blinking. Akinesia may also present as a "freezing" phenomenon, whereby the person feels momentarily as if their feet are glued to the ground, and movement becomes difficult to initiate (Imai 1996). Treatment for akinesia consists of a few different medications designed to improve dopaminergic activity in the brain. The most common treatment is levodopa (L-DOPA), a precursor to dopamine in the brain, which alleviates motor symptoms (Imai 1996). The second line of therapy for akinesia is dopamine agonists based on a similar mechanism, such as pramipexole or ropinirole. MAO-B inhibitors like selegiline may sometimes be prescribed to avoid the breakdown of dopamine and increase its availability (Tan et al. 2022).

2.1.9 Post-traumatic Epilepsy

Post-traumatic epilepsy is a condition characterized by the induction of seizures or the progression to an epileptic state after a traumatic brain injury. The initial trauma can alter the balance of biochemical processes, causing the production of scar tissue that can become a focal source of atypical electrical activity or induce inflammatory processes that lead to the inherited syndrome of epilepsy (Fordington and Manford 2020). These processes could include disruption of oscillatory activity that affects neurotransmitters like glutamate and GABA, which can lead to an increased state of excitability and result in spontaneous seizure activity.

The symptoms of post-traumatic epilepsy typically are recurrent seizures that occur after some time following the head injury, often months to years later (Fordington and Manford 2020). These types of seizures vary but commonly include focal seizures, generalized tonic-clonic seizures, or absence seizures. These seizures range from mild (brief lapse of consciousness) to severe (significant loss of consciousness and convulsions). All seizures pose a health risk, with the more severe potentially causing cognitive difficulties (Golub and Reddy 2022). Treatment typically involves anti-convulsion medication to stabilize neuronal activity and suppress atypical discharges of electricity in the brain.

2.1.10 Parkinson's Disease

Parkinson's disease is characterized as a chronic, degenerative neurodisorder that mainly targets motor function, with some non-motor neurodegeneration. This disease is typically rooted in genetic predisposition, with mutations occurring on the LRRK2, PRKN, PINK1, and SNCA genes increasing the risk of development (Klein and Westenberger 2012). It's been observed that environmental toxins can increase the risk factor and chance of development of Parkinson's disease. Parkinson's disease is focused on the degeneration of dopaminergic neurons in the substantia nigra pars compacta, which is a part of the basal ganglia involved in the regulation of movement (Kalia and Lang 2015). Degeneration of dopaminergic receptors leads to an imbalance between dopamine and acetylcholine, causing impaired motor control. Separating Parkinson's disease from parkinsonism is the presence of Lewy bodies aggregating in neurons, contributing to dysfunction and the death of neuron cells. Lewy bodies are characterized as aggregates of alpha-synuclein, a protein involved in neurotransmitter function, as it contributes to synaptic vesicle trafficking and neurotransmitter release (Simon et al. 2021).

Symptoms of Parkinson's disease are expressed as motor control dysfunction and specific non-motor symptoms. Motor symptoms include tremors, rigidity, akinesia, bradykinesia, and postural instability (Kalia and Lang 2015). Associated with motor symptoms typically follow depression, anxiety, sleep disturbances, loss of smell, cognitive impairment, or dementia. Other non-motor symptoms in the body involve constipation and autonomic dysfunction. Parkinson's disease focuses on the therapeutic treatment of symptoms since there isn't a cure for the disease. Levodopa is the cornerstone treatment in association with carbidopa to manage dopamine symptoms. Other dopamine agonists like pramipexole and ropinirole directly stimulate dopamine receptors, while MAO-B inhibitors, like selegiline and rasagiline, and COMT inhibitors, like entacapone, act by prolonging the action of dopamine in the brain. Deep brain stimulation may also be used to reduce medication side effects and help with some motor fluctuations. Physical and occupational therapy may be offered to individuals to facilitate normal daily activities.

2.1.11 Tourette's Syndrome

Tourette's syndrome is a developmental disorder characterized by phonic (vocal) and motor tics that start in childhood. The exact cause is unknown, but abnormalities in the basal ganglia are believed to cause motor control dysfunction. Dopamine dysfunction in the basal ganglia causes the aforementioned tics. This dopamine dysfunction likely stems from genetic mutations or another comorbidity (Johnson et al. 2023). There may also be influence from prenatal or perinatal events that increase the risk of development.

Symptoms typically manifest as several motor tics with the presence of a vocal tic. Motor tics can include random blinking, facial gestures, shoulder jerks, or rapid head movements (Johnson et al. 2023). Vocal, or phonic, tics can present as throat clearing, grunting, sniffling, or coprolalia (swearing and inappropriate language). These symptoms can commonly cause Tourette's syndrome to be associated with ADHD, OCD, anxiety, and other learning disabilities. Treatment is based on the severity of the individual's symptoms, particularly how much they interfere with daily life. For more problematic tics, comprehensive behavioral intervention may be used for therapy. Dopamine-depleting agents like tetrabenazine or dopamine blockers, such as risperidone and aripiprazole, may also be used to control tic frequency and intensity.

2.1.12 Huntington's Disease

Huntington's disease is an inherited neurodegenerative disorder characterized by degeneration of neurons in the basal ganglia, which leads to dysfunction in motor control. The cause of Huntington's disease stems from mutation in the HTT gene (Walker 2007). The HTT gene mutation is received in an autosomal dominant pattern, giving the children of an affected parent a 50% chance of inheritance (Langbehn et al. 2019). This genetic mutation causes CAG trinucleotide repeats, producing a mutant protein toxic to neurons called the huntingtin protein (Langbehn et al. 2019).

Symptoms of Huntington's disease express later in life and vary in severity and expression. The most common symptoms are motor dysfunction (chorea), cognitive degeneration, and psychiatric episodes. In the onset of Huntington's disease, motor symptoms reflect chorea: fidgeting, facial twitching, and muscle jerks (Walker 2007). However, as Huntington's disease progresses, voluntary motor control deteriorates, leading to rigidity, dystonia, and difficulty speaking. Treatment for Huntington's follows symptom management, as there is no cure currently. Dopamine-depleting agents, such as tetrabenazine, can manage the symptoms of chorea. Antipsychotics like risperidone or olanzapine can help relieve psychiatric symptoms and movement symptoms. Depending on the psychiatric symptoms, antidepressants may also be prescribed to address emotional disturbances.

2.2 Marine Natural Products: Therapeutic, Preventative, and Diagnostic Purposes

Marine natural products (MNP) have become an essential source of bioactive compounds with great potential for modern medicine. Derived from marine organisms such as algae, sponges, mollusks, and microbes, these compounds exhibit a broad spectrum of biological activities used for therapeutic, preventive, and diagnostic applications. Therapeutically, MNPs have shown promise in the treatment of cancerous cells, infectious diseases, inflammatory disorders, and neurological diseases and disorders (Gogineni and Hamann 2018; Karthikeyan et al. 2022; Santhiravel et al. 2024; Tian et al. 2023). Due to their marine environments, MNPs can offer mechanisms and biological compositions that differ significantly from terrestrial natural products. Many marine-derived products also exhibit preventative activities, most significantly antioxidant, anti-inflammatory, and immune-modulating activity capable of reducing the risk of chronic disease. Moreover, biotechnological advances and molecular imaging have enabled the integration of marine compounds into diagnostic agents, aiding in the early detection and monitoring of various diseases. This can be achieved through specific marine-derived molecules that serve as molecular probes or contrast agents in imaging technologies.

2.2.1 Marine Natural Product Therapeutics

Infectious microbes and cancer are the leading causes of death globally, urging a need for new sources of therapy as existing treatments become less viable. Marine life offers a rich source of incredibly varying structural and unique natural products with potential applications in multiple disciplines (Santhiravel et al. 2024). As microbial infections continue to develop resistance to existing therapeutics, more development is made in the research of MNPs. MNPs can be identified as secondary metabolites with potent antibacterial, antifungal, and antiviral activities (Karthikeyan et al. 2022).

For example, the MNP salinosporamide A has been identified to provide antibacterial activities and provide a potential therapy for autoimmune disorders (Lee et al. 2017; Lee and Jeong 2020). Salinosporamide A from *Salinispora tropica* indirectly interferes with proteasome activity in bacteria with proteasome-like complexes (Lee et al. 2017). Salinosporamide A inhibits protein degradation as a proteasome inhibitor, eventually leading to cellular apoptosis. This process has also been efficacious in regulating T cell activation, particularly suppressing the MAPK pathway (Lee and Jeong 2020). Suppressing activation through inhibiting T cell proliferation offers insight into potential future developments in therapeutics for autoimmune disorders. Along with these functions, salinosporamide A has shown efficacy in preclinical studies treating cancers, such as multiple myeloma, glioblastoma, and solid tumors (Lee et al. 2017). Cancerous cells are much more sensitive to proteasome inhibition, making cancer cells selective targets for salinosporamide A to act upon.

Marine natural products have also shown efficacy in antifungal therapeutics. For example, MNPs scopularide A and B from marine *Scopulariopsis* species have shown strong antifungal properties against *Candida albicans* (Gogineni and Hamann 2018). Scopularide A and B are structurally similar to other lipopeptide antifungals, such as polymyxin B, creating pores and degrading fungal membrane integrity (Hsu et al. 2017). The ability to disrupt fungal membranes derives from the amphipathic structure of scopularide A and B, meaning it consists of both hydrophilic and hydrophobic regions. Once inserted into the fungal membrane, scopularide A and B can induce cell death as fungal homeostasis is disrupted and the contents lysed into the surrounding environment (Gogineni and Hamann 2018).

Neurological disorders, including motor and non-motor disorders, can also be treated with MNPs. Focusing on non-motor neurological disorders, bryostatin-1 from *Bugula neritina* has shown potential efficacy as a therapeutic for Alzheimer's disease (Tian et al. 2023). Bryostatin-1 modulates protein kinase C activity to inhibit amyloid plaque aggregations in regions of the brain. This modulation is achieved by bryostatin-1, which promotes the production of nontoxic beta-amyloid from amyloid precursor protein (Tian et al. 2023). By preventing amyloid plaque formation, bryostatin-1 can improve neuroplasticity and has shown enhanced cognition in preclinical models (Tian et al. 2023).

2.2.2 Marine Natural Product Preventatives

Marine natural products with antioxidant activity have attracted interest because they can neutralize oxidative stress, a significant mechanism associated with aging and most chronic diseases, including neurodegenerative disorders, cardiovascular disease, and cancer. Various marine organisms, including algae, sponges, mollusks, and marine bacteria, biosynthesize a wide variety of antioxidant compounds as a protective measure against the extreme conditions of the marine environment, which include high salinity, ultraviolet radiation, and fluctuating oxygen levels. Phlorotannin, for example, is an MNP derived from brown algae that exhibits strong antioxidant properties. Brown algae species maintain a higher hydroxyl concentration within their structures due to their metabolism of iodine involving halogenation (La Barre et al. 2010; Kumar et al. 2022). The higher hydroxyl concentration allows for the efficient donation of hydrogen atoms to reactive oxygen species (Kumar et al. 2022). The donation of hydrogen atoms neutralizes the reactive oxygen species, thus lowering the oxidative stress of the system. Another example of MNP antioxidant action is astaxanthin, produced by microalgae. Like phlorotannin, astaxanthin neutralizes reactive oxygen species by donating hydrogen atoms from its large structural concentration. However, astaxanthin consists of an amphipathic structure, allowing the MNP to span the cell membrane and support the expression of intracellular enzymes (Nair et al. 2023).

2.2.3 Marine Natural Products in Diagnostic Applications

Marine natural products are increasingly being explored in their applications for disease diagnosis and biomedical imaging. MNPs for diagnostic tools exhibit unique fluorescent, radiolabeling, or molecular recognition properties. Fluorescent structures have revolutionized diagnostics on the molecular level and live-cell imaging. More specifically, green fluorescent proteins have proven to be valuable tools for monitoring gene expression, protein localization, and disease progression at the cellular level (Lin et al. 2019). MNPs have also seen growing efficacy in diagnostic usage as biosensors and detectors for tumor cells or other desired targets. Marine-derived lectins from algae have seen wide usage in constructing both optical and electrochemical biosensors due to lectins' specific binding affinity to carbohydrates (Wang and Anzai 2015). This binding affinity allows the monitoring and detection of cancer cell biomarkers and desired pathogens with high specificity (Wang and

Anzai 2015). MNPs structural and biochemical diversity offers a firm basis for developing novel diagnostic agents capable of facilitating improved early detection and personalized medical interventions.

2.3 Marine Natural Products in Treating Movement Disorders

Marine natural products have grown increasingly relevant to the therapeutic treatment of movement disorders. Generally, movement disorders are characterized by motor dysfunction caused by neurodegeneration or dysfunction in neurotransmitter communication. Several MNPs exhibit neuroprotective, anti-inflammatory, and antioxidant activities that target the underlying pathophysiological mechanisms of the disorders.

2.3.1 Astaxanthin

Astaxanthin (3,3-dihydroxy-β, β-carotene-4,4-dione) is a ketocarotenoid synthesized by *Haematococcus pluvialis/lacustris*, *Chromochloris zofingiensis*, *Chlorococcum*, *Bracteacoccus aggregatus*, *Coelastrella rubescence*, *Phaffia rhodozyma, Paracoccus carotinifaciens*, yeasts, and lobsters (Nair et al. 2023). Harvested primarily from *Haematococcus lacustris* (freshwater green algae), astaxanthin is predominantly present in the 3S and 3’S isomers, with (*trans- and cis-)* geometrical isomers dependent on the configuration of the double bond in the polyene chain (Nair et al. 2023; Šimat et al. 2022) (Fig. 2.1). Astaxanthin is classified under the xanthophylls, relating it to lutein, zeaxanthin, canthaxanthin, and beta-cryptoxanthin (Si and Zhu 2022). Due to the arrangement of two hydroxyl groups, two keto groups, and the long hydrophobic hydrocarbon chains, astaxanthin can pass through the blood-brain barrier as a lipid-soluble compound. Astaxanthin accumulation in specific regions of the brain (cerebral cortex) can be crucial in maintaining and strengthening cognitive abilities. Various studies have shown that astaxanthin administration can promote neuronal regeneration and increase the expression of genes for proteins essential for neural rehabilitation (El-Agamy et al. 2018; Damodara Gowda et al. 2020). The glial fibrillary acidic protein (GFAP) and proteins like it serve a significant role in the central nervous system (CNS). GFAP contributes to general repair and cell communication within the CNS and alleviates blood-brain barrier damage (Cullen et al. 2007) (Fig. 2.1).

Parkinson’s disease is a common neurodegenerative disease caused mainly by oxidative stress and neuroinflammation (Kalia and Lang 2015; Klein and Westenberger 2012). The majority of genes related to Parkinson’s disease are observed to be mediated by miRNAs, specifically alpha-synuclein and LRRK2. Alpha-synuclein (SNCA) has been determined to be the primary gene responsible for the development and initiation of Parkinson’s disease, with the accumulation of SNCA being linked to dopaminergic neuron loss (Bahbah et al. 2021). Dysregulation

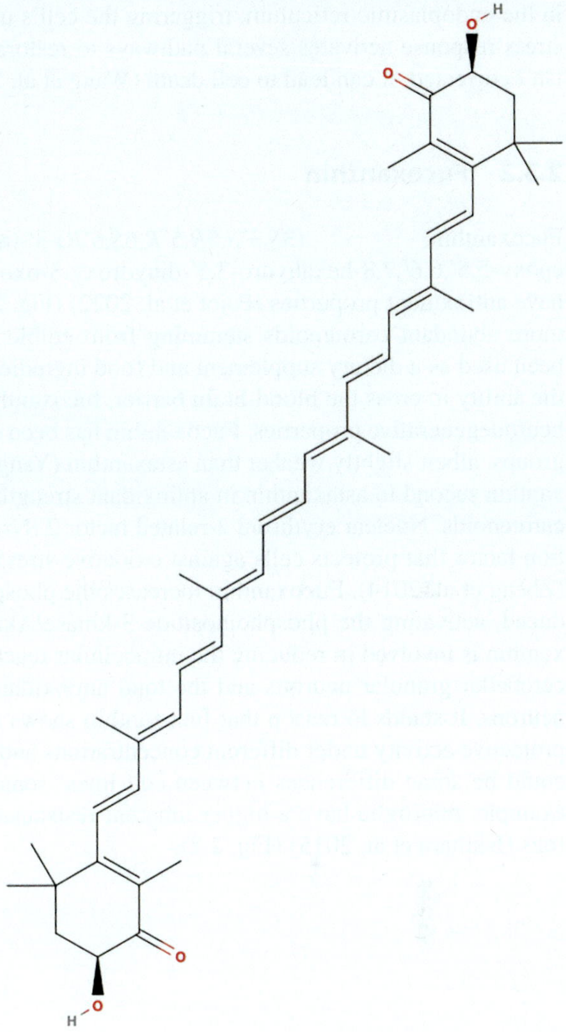

Fig. 2.1 Structure of 3S,3'S astaxanthin

of SNCA and the associated miR-7 in a miR-7/SNCA axis will lead to Lewy body aggregation. Astaxanthin has been observed to protect against Lewy bodies by supporting the expression of miR-7 to regulate SNCA (Shen et al. 2021). In vivo, astaxanthin has been observed to have a protective effect against neuron degradation by MPTP (1-methyl-4-phenyl-1,2,3,6-tetrahydropyridine) through the miR-7/SNCA axis (Shen et al. 2021). MPTP is the precursor to phenylpyridinium cation (MPP+), a neurotoxin that induces dopaminergic neuron death (Bahbah et al. 2021). Astaxanthin further reduces endoplasmic reticulum (ER) stress, which may be observed to help as a preventative measure against Parkinson's disease. MPP+ causes endoplasmic reticulum stress by disrupting mitochondrial function, causing increased oxidative stress, which then creates the aggregation of misfolded proteins in the endoplasmic reticulum, triggering the cell's unfolded protein response. This stress response activates several pathways to restore homeostasis, but if the stress isn't corrected, it can lead to cell death (Wang et al. 2021).

2.3.2 Fucoxanthin

Fucoxanthin, (3*S*,3'*S*,5*R*,5'*R*,6*S*,6'*R*)-3′-(acetyloxy)-6′,7′-didehydro-5,6-epoxy-5,5′,6,6′,7,8-hexahydro-3,5′-dihydroxy-8-oxo-β, β-carotene, is observed to have antioxidant properties (Pajot et al. 2022) (Fig. 2.2). Fucoxanthin is one of the more abundant carotenoids stemming from edible brown seaweeds, which have been used as a dietary supplement and food ingredient (Yang et al. 2022). Through the ability to cross the blood-brain barrier, fucoxanthin is observed to express anti-neurodegenerative properties. Fucoxanthin has been observed to scavenge hydroxyl groups, albeit slightly weaker than astaxanthin (Yang et al. 2022). This leaves fucoxanthin second to astaxanthin in antioxidant strength but is still very potent among carotenoids. Nuclear erythroid 2-related factor 2 (Nrf2) works as a major transcription factor that protects cells against oxidative stress, which is upregulated by Akt (Zheng et al. 2014). Fucoxanthin increases the phosphorylation of Akt when introduced, activating the phosphoinositide 3-kinase/Akt pathway. Furthermore, fucoxanthin is involved in reducing the intracellular reactive oxygen species in primary cerebellar granular neurons and the total antioxidant capacity in primary cortical neurons. It stands to reason that fucoxanthin shows similar antioxidant and neuroprotective activity under different concentrations and study models. However, there could be some differences between cell lines' sensitivity to oxidative stress. For example, microglia have a higher inherent resistance to oxidative stress than neurons (Ishihara et al. 2015) (Fig. 2.2).

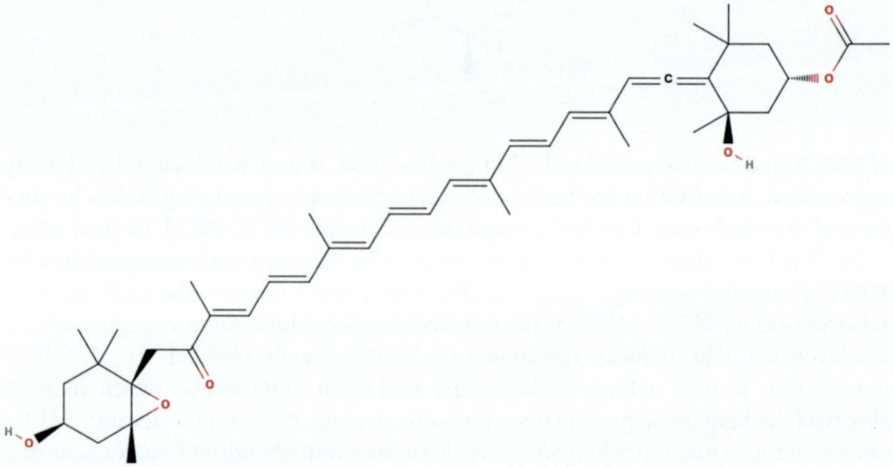

Fig. 2.2 Structure of fucoxanthin

Neuroinflammation is described as the activation of the brain's immune response, with microglia comprising most of the immune response (Subhramanyam et al. 2019). The activation of microglia can protect the brain against pathogens and damaged tissues by secreting neurotrophic factors, but over-activation of microglia can secrete cytotoxic and inflammatory cytokines. The inflammatory response can induce interleukin-1beta, interleukin-6, and tumor necrosis factor-alpha, making it necessary to regulate and balance the neuro-immune response in neurodegenerative disorders (Yang et al. 2022). Fucoxanthin suppresses lipopolysaccharide (LPS) peripheral inflammation by inhibiting the triggering of nuclear factor kappa-B and mitogen-activated protein kinases in macrophages (Yang et al. 2022). Fucoxanthin works with a similar mechanism in the CNS, as it can reduce the secretion of interleukin-1beta, interleukin-6, and tumor necrosis factor-alpha induced by LPS through inhibiting the nuclear factor kappa-B and mitogen-activated protein kinase pathway (Yang et al. 2022).

Fucoxanthin has shown some efficacy in other dopaminergic therapies. It was observed to bind to dopaminergic receptors via hydrogen bond interactions with Ser196 and Thr115 of the D3 receptor and Ser196 and Asp115 of the D4 receptor (Paudel et al. 2019). Fucoxanthin was observed to reverse the MPTP-induced reduction of dopamine levels in C57BL/6 mice, suggesting that fucoxanthin could produce anti-PD neuroprotective effects by enhancing dopaminergic transmission in mice (Pang et al. 2020). Furthermore, fucoxanthin exhibits hydrogen bond affinity with MAO, an enzyme that catalyzes the oxidative deamination of neurotransmitters in the CNS. This bonding affinity indicates that fucoxanthin is a reversible competitive MAO inhibitor, making this MNP unique amongst other carotenoids.

2.3.3 Marine-Derived PUFAs

PUFAs (polyunsaturated fatty acids) contain two or more double bonds within their hydrocarbon chain and are identified by the relation of the first double bond to the omega methyl group at the end. Omega-3 (marine-derived) and omega-6 (terrestrial-derived) PUFAs are characterized by a double bond positioned three or six atoms away from the omega methyl group, respectively (Das 2021; Kousparou et al. 2023). These PUFAs maintain amphipathic regions with hydrophilic phosphate-rich heads and hydrophobic lipid tails (Kousparou et al. 2023). Due to the amphipathic structure, PUFAs incorporate themselves into cell membranes from which they can act upon the cell while maintaining membrane fluidity. PUFAs are made of linoleic acid (LA) and a-linoleic acid (ALA), making them essential fatty acids, as the human body can only produce limited amounts of LA and ALA. Higher consumption of omega-3 PUFAs, vegetables, and low meat intake has been observed to lower the risk of development of Parkinson's disease (Kerdiles et al. 2025). The most prominent version of omega-3 in the CNS is docosahexaenoic acid, 22:6 (DHA) (Fig. 2.3).

Diets with high DHA concentrations retain higher concentrations of this essential fatty acid in the CNS and, conversely, lower levels of omega-6 fatty acids. In Kaindlstorfer et al., 2019, the relationship between DHA and voluntary exercise to

Fig. 2.3 Structure of docosahexaenoic acid (DHA)

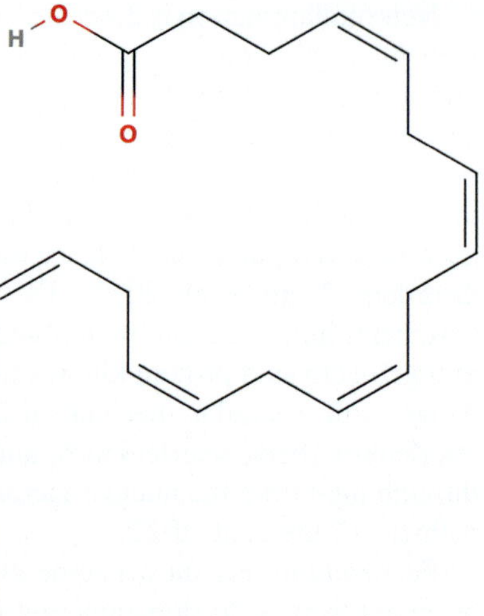

improve dexterity in forelimbs in the mice was observed (Kaindlstorfer et al. 2019). This could show potential fine motor therapy with DHA intake for diseases such as Parkinson's and Huntington's, as voluntary movement could help to increase the effects of DHA. High intake of DHA was also noted to upregulate dopamine levels. In this same experiment, mice with depleted dopaminergic nigral cells were shown to partially recover dopaminergic terminals (Kaindlstorfer et al. 2019). Along with that, DHA, in coordination with exercise, reduced the dopaminergic and 5HT receptor turnover rates by downregulating MAO activity. This diet also leads to a general reduction of MAO-B activity in the frontal cortex (Chalon et al. 1998).

Omega-3 PUFAs have shown some potential efficacy in slowing the rate of loss or even increasing the rate of muscle mass gain (Smith et al. 2015). Given omega-3's effect on muscle mass, there may be observed results in supplementation to individuals with Huntington's disease and potentially Parkinson's disease if they're suffering from muscle atrophy. While this may not serve as a cure, it can provide some therapeutic relief for maintaining muscle movement.

2.3.4 Sargachromenol

Sargachromenol (SC) is an MNP isolated from a marine brown seaweed species, *Sargassum horneri* (Fig. 2.4). SC stands out for its antioxidant, anti-inflammatory, and neurotrophic properties, with its potential therapies for movement disorders. SC has an affinity for modulating apoptosis-mediated proteins, with an observed efficacy for mediating glutamate neurotoxicity (Han et al. 2022). Glutamate toxicity can stem from various neurological disorders or trauma to the brain, as the ability to reuptake the neurotransmitter is reduced (Iovino et al. 2020).

High concentrations of compound SC significantly increase cell viability and inhibit glutamate-induced damage in HT22 neural cells. There are potential

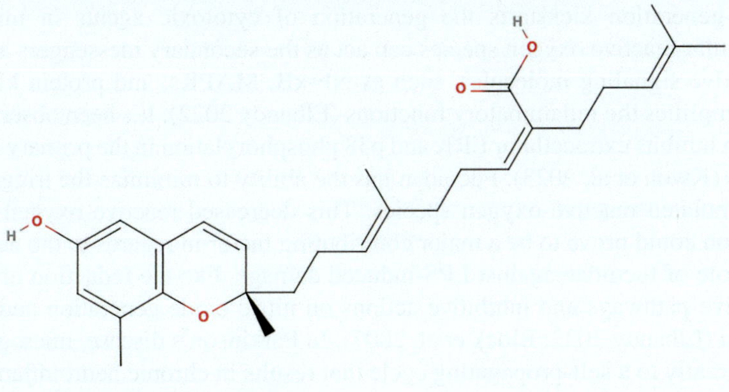

Fig. 2.4 Structure of (S) sargachromenol

benefits associated with the effects of SC on the preservation of protective characteristics s provided by HT22 against cellular apoptosis (Han et al. 2022). DNA fragmentation and nuclear condensation elicited by glutamate can be decreased with the treatment of SC. The ability of SC to attenuate apoptosis caused by the reduction of sub-G1 populations in the cell cycle indicates SC's protective effects against glutamate-induced apoptosis in HT22 cells. It's observed that SC likely prevents glutamate-induced normocellular apoptosis due to the favorable regulation of apoptosis-mediated protein expression (Han et al. 2022). Furthermore, SC's ability to suppress the MAPK/NF-κB signaling pathway and activate the Nrf2/HO-1 signaling pathway also shows efficacy in protecting against oxidative stress. With SC's ability to regulate glutamate-induced apoptosis, there could be efficacy in the treatment of movement disorders where glutamate-induced apoptosis is interfering with motor control. Specifically in Huntington's disease or Parkinson's disease, with neurodegeneration causing increased activity of glutamate and the prevention of glutamate's reuptake (Fig. 2.4).

2.3.5 Marine Polysaccharides

Polysaccharides extracted from brown seaweed possess multiple monosaccharide units linked to each other via numerous glycoside linkages, forming an intricate atomic arrangement. Polysaccharides are a well-studied component in pharmaceutical applications due to their immunomodulatory and anti-inflammatory effects (Elbandy 2022). Seaweeds are a reservoir of polysaccharides that can possess MNP characteristics, with fucoidans being the prominent sulfated polysaccharide. With some structural variation across species, fucoidans all exhibit antioxidation, neuroprotective interactions, and inflammatory inhibitive characteristics.

Triggering microglia can force the generation of intracellular and extracellular oxidative stress, with intracellular reactive oxygen species needed for microglial operations and inflammatory interactions (Block et al. 2007). Reactive oxygen species generation kickstarts the generation of cytotoxic agents in microglia. Intracellular reactive oxygen species can act as the secondary messengers affecting consecutive signaling molecules, such as NF-κB, MAPKs, and protein kinase C, which amplifies the inflammatory functions (Elbandy 2022). It's been observed that fucoidan inhibits extracellular ERK and p38 phosphorylation in the primary microglia Field (Kwon et al. 2023). Fucoidan has the ability to minimize the triggering of LPS-stimulated reactive oxygen species. This decreased reactive oxygen species generation could prove to be a major contributing factor in regards to the neuroprotective role of fucoidan against LPS-induced damage, like the reduction of MAPK conductive pathways and inhibitive actions on nitric oxide generation and TNF-α secretion (Elbandy 2022; Block et al. 2007). In Parkinson's disease, microglia contribute greatly to a self-propagating cycle that results in chronic neuroinflammation and potential neurodegeneration. Fucoidan has been reported to suppress LPS-induced microglial activation, which can protect dopaminergic neurons and regulate inflammation (Elbandy 2022).

Carrageenans are another set of polysaccharide MNPs descending from specific types of red seaweeds, helping to show an increased absorption efficiency compared to other polysaccharides. Kappa-carrageenan oligosaccharides (KOS) exhibited immunomodulating properties through the modification of lipopolysaccharide LPS-triggered microglial cells, which can mitigate neurodegenerative conditions with the involvement of inflammatory processes. KOS-specific binding proteins may exist on the surface of microglia membranes that can regulate KOS in the cell, suggesting that there may be a competitive relationship between KOS and LPS in the microglia (Elbandy 2022). Pre-treatment with KOS suggests that they can outcompete LPS for binding with microglia. However, KOS has a comparatively weaker bond due to its lower level of charge compared to LPS (Elbandy 2022). This observation suggests that KOS may not be very effective as a protective measure against neurodegeneration, but KOS could be effective in preventing the processes of CNS diseases.

2.3.6 Phlorotannin

Phlorotannins are marine natural products produced by brown marine algae and are currently found as dietary supplements. While known to cross the blood-brain barrier, these compounds' specific neuropharmacological action remains unclear. When phlorotannins are taken as supplements, cognitive benefits include resistance to stress and improved memory performance (Kwon et al. 2023).

Two phlorotannin derivatives, eckol and dieckol, were shown to be able to inhibit the enzymatic activity of MAO isoforms, MAO-A and MAO-B (Jung et al. 2017). Both phlorotannins were observed to noncompetitively inhibit MAO-B (Jung et al. 2017). Dieckol would also noncompetitively inhibit MAO-A, while eckol reversibly inhibited MAO-A. These results show that phlorotannins can operate as orthosteric modulators of the MAO enzymes and can likely be used to help improve motor skills and non-cognitive functions in neurodegenerative diseases (Table 2.1).

Table 2.1 In-silico evaluation of ADME parameters of phlorotannin derivatives, eckol and dieckol

		Eckol	Dieckol
Physiochemical properties	Formula	C18H12O9	C36H22O18
	Molecular weight	372.28 g/mol	742.55 g/mol
	Num. heavy atoms	27	54
	Num. arom. heavy atoms	18	36
	Fraction Csp3	0.0	0.0
	Num. rotatable bonds	2	6
	Num. H-bond acceptors	9	18
	Num. H-bond donors	6	11
Lipophilicity	Molar refractivity	91.69	181.42
	TPSA	149.07	287.14
	Log Po/w (iLOGP)	1.78	2.65
	Log Po/w (XLOGP3)	2.46	4.87
	Log Po/w (WLOGP)	3.61	7.62
	Log Po/w (MLOGP)	0.25	0.04
	Log Po/w (SILICOS-IT)	1.07	1.75
	Consensus Log Po/w	1.83	3.39
Water solubility	Log S (ESOL)	−4.06	−7.61
	Solubility	3.25e-02 mg/mL; 8.72e-05 mol/L	1.83e-05 mg/mL; 2.46e-08 mol/L
	Class	Moderately soluble	Poorly soluble
	Log S (Ali)	−5.23	−10.63
	Solubility	2.17e-03 mg/mL; 5.83e-06 mol/L	1.72e-08 mg/mL; 2.32e-11 mol/L
	Class	Moderately soluble	Insoluble
	Log S (SILICOS-IT)	−3.14	−6.32
	Solubility	2.69e-01 mg/mL; 7.23e-04 mol/L	3.55e-04 mg/mL; 4.79e-07 mol/L
	Class	Soluble	Poorly soluble

(continued)

Table 2.1 (continued)

		Eckol	Dieckol
Pharmacokinetics	GI absorption	Low	Low
	BBB permeant	No	No
	P-gp substrate	No	No
	CYP1A2 inhibitor	Yes	No
	CYP2C19 inhibitor	No	No
	CYP2C9 inhibitor	Yes	Yes
	CYP2D6 inhibitor	No	No
	CYP3A4 inhibitor	Yes	No
	Log Kp	−6.82 cm/s	−7.37 cm/s
Druglikeness	Lipinski	Yes: violation: NHorOH>5	No; 3 violations: MW > 500, NorO>10, NHorOH>5
	Ghose	Yes	No; 4 violations: MW > 480, WLOGP>5.6, MR > 130, #atoms>70
	Veber	No; 1 violation: TPSA>140	No; 1 violation: TPSA>140
	Egan	No; 1 violation: TPSA>131.6	No; 2 violations: WLOGP>5.88, TPSA>131.6
	Muegge	No; 1 violation: H-don>5	No; 5 violations: MW > 600, TPSA>150, #rings>7, H-acc > 10, H-don>5
Medicinal chemistry	Bioavailability score	0.55	0.17
	PAINS	0 alert	0 alert
	Brenk	0 alert	0 alert
	Leadlikeness	No; 1 violation: MW > 350	No; 2 violations: MW > 350, XLOGP3 > 3.5
	Synthetic accessibility	3.26	4.68

2.4 Conclusion

Marine Natural products have seen an exponential rise in research in recent years for new applications to existing problems. Particularly, new developments are needed in the field of marine bioactives to counter the ever-growing problems of anti-bacterial resistance and cancer development. As experiments and pre-clinical trials continue, there may be new applications of MNPs that work as more powerful therapeutics for movement disorders, hoping to bring more normalcy to individuals' lives. Potentially, developments in MNPs like astaxanthin and fucoxanthin can bring new techniques to treat chronic inflammation. Currently, there's promise in

MNPs like astaxanthin and PUFAs to provide better therapeutics for neurodegenerative disorders. Further developments in interactions with DHA could eventually lead to full regeneration of dopaminergic pathways. Other MNPs like bryostatin-1 have the potential to strengthen neuroplasticity and maintain motor control pathways.

References

Bahbah EI et al (2021) Molecular mechanisms of astaxanthin as a potential neurotherapeutic agent. Mar Drugs 19(4):201

Biary N et al (1989) Post-traumatic tremor. Neurology 39(1):103–103

Block ML, Zecca L, Hong J-S (2007) Microglia-mediated neurotoxicity: uncovering the molecular mechanisms. Nat Rev Neurosci 8(1):57–69

Bologna M, Truong D, Jankovic J (2022) The etiopathogenetic and pathophysiological spectrum of parkinsonism. J Neurol Sci 433:120012

Broeder S et al (2022) Novel insights into the effects of levodopa on the up- and downstrokes of writing sequences. J Neural Transm 129(4):379–386

Carroll CG, Hawley JS, Ney JP (2006) Post-traumatic shoulder dystonia in an active duty soldier. Mil Med 171(6):494–496

Chalon S et al (1998) Dietary fish oil affects monoaminergic neurotransmission and behavior in rats. J Nutr 128(12):2512–2519

Conn H, Jankovic J (2024) Drug-induced parkinsonism: diagnosis and treatment. Expert Opin Drug Saf 23(12):1503–1513

Cullen DK, Simon CM, Laplaca MC (2007) Strain rate-dependent induction of reactive astrogliosis and cell death in three-dimensional neuronal–astrocytic co-cultures. Brain Res 1158:103–115

Damodara Gowda KM, Suchetha Kumari N, Ullal H (2020) Role of astaxanthin in the modulation of brain-derived neurotrophic factor and spatial learning behavior in perinatally undernourished Wistar rats. Nutr Neurosci 23(6):422–431

Das UN (2021) "Cell membrane theory of senescence" and the role of bioactive lipids in aging, and aging associated diseases and their therapeutic implications. Biomolecules 11(2):241

El-Agamy SE et al (2018) Astaxanthin ameliorates doxorubicin-induced cognitive impairment (Chemobrain) in experimental rat model: impact on oxidative, inflammatory, and apoptotic machineries. Mol Neurobiol 55(7):5727–5740

Elbandy M (2022) Anti-inflammatory effects of marine bioactive compounds and their potential as functional food ingredients in the prevention and treatment of neuroinflammatory disorders. Molecules 28(1):2

Espay AJ et al (2018) Current concepts in diagnosis and treatment of functional neurological disorders. JAMA Neurol 75(9):1132

Fordington S, Manford M (2020) A review of seizures and epilepsy following traumatic brain injury. J Neurol 267(10):3105–3111

Galli S et al (2020) Functional movement disorders. Rev Neurol 176(4):244–251

Gogineni V, Hamann MT (2018) Marine natural product peptides with therapeutic potential: chemistry, biosynthesis, and pharmacology. Biochim Biophys Acta Gen Subj 1862(1):81–196

Golub VM, Reddy DS (2022) Post-traumatic epilepsy and comorbidities: advanced models, molecular mechanisms, biomarkers, and novel therapeutic interventions. Pharmacol Rev 74(2):387–438

Han E-J et al (2022) Sargachromenol isolated from Sargassum horneri attenuates glutamate-induced neuronal cell death and oxidative stress through inhibition of MAPK/NF-κB and activation of Nrf2/HO-1 signaling pathway. Mar Drugs 20(11):710

Hayes MT (2019) Parkinson's disease and parkinsonism. Am J Med 132(7):802–807

Holm H, Gundersen V, Dietrichs E (2023) Vascular parkinsonism, Tidsskr Nor Legeforen 143(7)

Hsu L-H et al (2017) The antibiotic polymyxin B exhibits novel antifungal activity against Fusarium species. Int J Antimicrob Agents 49(6):740–748
Imai H (1996) Clinicophysiological features of akinesia. Eur Neurol 36(1):9–12
Iovino L, Tremblay ME, Civiero L (2020) Glutamate-induced excitotoxicity in Parkinson's disease: the role of glial cells. J Pharmacol Sci 144(3):151–164
Ishihara Y et al (2015) Dual role of superoxide dismutase 2 induced in activated microglia. J Biol Chem 290(37):22805–22817
Jang SH, Lee HD (2021) Ataxia due to injury of the cortico-ponto-cerebellar tract in patients with mild traumatic brain injury. Medicine 100(48):e28024
Johnson KA et al (2023) Tourette syndrome: clinical features, pathophysiology, and treatment. Lancet Neurol 22(2):147–158
Jung HA et al (2017) Evaluation of the inhibitory effects of eckol and dieckol isolated from edible brown alga Eisenia bicyclis on human monoamine oxidases A and B. Arch Pharm Res 40(4):480–491
Kaindlstorfer C et al (2019) L-dopa response pattern in a rat model of mild striatonigral degeneration. PLoS One 14(6):e0218130
Kalia LV, Lang AE (2015) Parkinson's disease. Lancet 386(9996):896–912
Kamble N, Pal PK (2023) Electrophysiology in functional movement disorders: an update. Tremor Other Hyperkinet Mov 13(1)
Karthikeyan A, Joseph A, Nair BG (2022) Promising bioactive compounds from the marine environment and their potential effects on various diseases. J Genet Eng Biotechnol 20(1):14
Kerdiles O et al (2025) Additive neurorestorative effects of exercise and docosahexaenoic acid intake in a mouse model of Parkinson's disease. Neural Regen Res 20(2):574–586
Klein C, Westenberger A (2012) Genetics of Parkinson's disease. Cold Spring Harb Perspect Med 2(1):a008888–a008888
Kousparou C et al (2023) DHA/EPA (Omega-3) and LA/GLA (Omega-6) as bioactive molecules in neurodegenerative diseases. Int J Mol Sci 24(13):10717
Kumar LRG et al (2022) Phlorotannins–bioactivity and extraction perspectives. J Appl Phycol 34(4):2173–2185
Kwon YJ et al (2023) Therapeutic effects of phlorotannins in the treatment of neurodegenerative disorders. Front Mol Neurosci 16
La Barre S et al (2010) The halogenated metabolism of brown algae (Phaeophyta), its biological importance and its environmental significance. Mar Drugs 8(4):988–1010
Langbehn DR et al (2019) Association of CAG repeats with long-term progression in Huntington disease. JAMA Neurol 76(11):1375
Lee H-S, Jeong G-S (2020) Salinosporamide a, a marine-derived proteasome inhibitor, inhibits T cell activation through regulating proliferation and the cell cycle. Molecules 25(21):5031
Lee Y, Phat C, Hong S-C (2017) Structural diversity of marine cyclic peptides and their molecular mechanisms for anticancer, antibacterial, antifungal, and other clinical applications. Peptides 95:94–105
Lin C-Y et al (2019) Unified model for photophysical and electro-optical properties of green fluorescent proteins. J Am Chem Soc 141(38):15250–15265
Mathewson MA, Lieber RL (2015) Pathophysiology of muscle contractures in cerebral palsy. Phys Med Rehabil Clin N Am 26(1):57–67
Mestre TA (2016) Chorea. Continuum 22(4):1186–1207
Nair A et al (2023) Astaxanthin as a king of ketocarotenoids: structure, synthesis, accumulation, bioavailability and antioxidant properties. Mar Drugs 21(3):176
Niedermeyer E (2008) Akinesia and the frontal lobe. Clin EEG Neurosci 39(1):39–42
Pajot A et al (2022) Fucoxanthin from algae to human, an extraordinary bioresource: insights and advances in up and downstream processes. Mar Drugs 20(4):222
Pang Q et al (2020) Fucoxanthin attenuates behavior deficits and neuroinflammatory response in 1-methyl-4-phenyl-1,2,3,6 -tetrahydropyridine-induced parkinson's disease in mice. Pharmacogn Mag 16(67):51

Paudel P et al (2019) Characterizing fucoxanthin as a selective dopamine D3/D4 receptor agonist: relevance to Parkinson's disease. Chem Biol Interact 310:108757

Ricciardi L, Edwards MJ (2014) Treatment of functional (psychogenic) movement disorders. Neurotherapeutics 11(1):201–207

Ricciardi L et al (2023) Dysfunctional networks in functional dystonia. Springer, pp 157–176

Rivest J (2003) Myoclonus. Can J Neurolog Sci 30(S1):S53–S58

Santhiravel S, Dave D, Shahidi F (2024) Bioactives from marine resources as natural health products: a review. Pharmacol Rev:PHARMREV-AR-202

Schrag A, Lang AE (2005) Psychogenic movement disorders. Curr Opin Neurol 18(4):399–404

Schwingenschuh P, Espay AJ (2022) Functional tremor. J Neurol Sci 435:120208

Shen D-F et al (2021) Astaxanthin suppresses endoplasmic reticulum stress and protects against neuron damage in Parkinson's disease by regulating miR-7/SNCA axis. Neurosci Res 165:51–60

Si P, Zhu C (2022) Biological and neurological activities of astaxanthin (Review). Mol Med Rep 26(4)

Šimat V et al (2022) Astaxanthin from crustaceans and their byproducts: a bioactive metabolite candidate for therapeutic application. Mar Drugs 20(3):206

Simon C et al (2021) α-Synuclein-mediated neurodegeneration in Dementia with Lewy bodies: the pathobiology of a paradox. Cell Biosci 11(1)

Smith GI et al (2015) Fish oil–derived n−3 PUFA therapy increases muscle mass and function in healthy older adults. Am J Clin Nutr 102(1):115–122

Subhramanyam CS et al (2019) Microglia-mediated neuroinflammation in neurodegenerative diseases. Semin Cell Dev Biol 94:112–120

Tan Y-Y, Jenner P, Chen S-D (2022) Monoamine oxidase-B inhibitors for the treatment of Parkinson’s disease: past, present, and future. J Parkinsons Dis 12(2):477–493

Tian Z et al (2023) Bryostatin-1: a promising compound for neurological disorders. Front Pharmacol 14

Upadhyay J, Tiwari N, Ansari MN (2020) Cerebral palsy: aetiology, pathophysiology and therapeutic interventions. Clin Exp Pharmacol Physiol 47(12):1891–1901

Walker FO (2007) Huntington's disease. Lancet 369(9557):218–228

Wang B, Anzai J-I (2015) Recent progress in lectin-based biosensors. Materials 8(12):8590–8607

Wang X et al (2021) PHB blocks endoplasmic reticulum stress and apoptosis induced by MPTP/MPP+ in PD models. J Chem Neuroanat 113:101922

Wimalasundera N, Stevenson VL (2016) Cerebral palsy. Pract Neurol 16(3):184–194

Yang M et al (2022) Fucoxanthin has potential for therapeutic efficacy in neurodegenerative disorders by acting on multiple targets. Nutr Neurosci 25(10):2167–2180

Zheng J et al (2014) Fucoxanthin enhances the level of reduced glutathione via the Nrf2-mediated pathway in human keratinocytes. Mar Drugs 12(7):4214–4230

Exploring the Promising Potential of Marine-Derived Bioactives and Their Effects on Neuropsychiatric Disorders

3

Hannah Griggs, Preston Cook, Keyi Liu, Suhrud Pathak, Rishi Nadar, Jaufar Palekkodan, Myiesha S. Ameen, and Muralikrishnan Dhanasekaran

Abstract

Marine-derived bioactives represent an emerging frontier as potential therapeutics in the treatment of neuropsychiatric disorders because of their unique chemical structures, diverse mechanisms, and various physiological benefits. This systematic review explores the potential utilization of many marine-derived bioactives as preventives, diagnostics, and therapeutics in the treatment and management of neuropsychiatric disorders (i.e, depression, anxiety, schizophrenia, etc.). All of the compounds discussed exemplify antioxidant, anti-inflammatory, neuroprotective, neurotransmission modulating, and neurogenic effects, aiding in halting pathological processes, such as oxidative stress, inflammation, disrupted neurotransmitter signaling, and impaired neuroplasticity. Omega-3 fatty acids (DHA/EPA), fucoidan, astaxanthin, phlorotannins, and marine peptides are agents accentuated due to their ability to infiltrate the blood-brain barrier, influencing many cognition processes and aiding in synaptic resilience. This review also explores the historical division between neurology and psychiatry,

H. Griggs · P. Cook · K. Liu · S. Pathak · R. Nadar
Drug Discovery and Development, Harrison College of Pharmacy, Auburn University, Auburn, AL, USA

J. Palekkodan
Illham House, Changuvetti kund, Kottakkal, Kerala, India

M. S. Ameen
Sophomore at University of Maryland, School of Public Health, College Park, MD, USA

M. Dhanasekaran (✉)
Drug Discovery and Development, Harrison College of Pharmacy, Auburn University, Auburn, AL, USA

Graduate School, Auburn University, Auburn, AL, USA
e-mail: dhanamu@auburn.edu

M. Bhupathyraaj et al. (eds.), *Neuroprotective Marine Bioactives*, Nutritional Neurosciences, https://doi.org/10.1007/978-981-95-5791-2_3

highlighting the increase in neuropsychiatry, resulting in a need for more integrative therapeutic strategies. Additionally, advances in marine biotechnology, such as single-cell genomic and high-throughput screening, are rapidly advancing compound discoveries and clinical translation. Though challenges persist in bioavailability, scalability, and standardization, marine-derived bioactives offer innovative therapeutic strategies for healthcare providers and patients, potentially addressing the subpar clinical care in the mental health field. In efforts to fully implement marine-derived bioactives into clinical care, continued research surrounding marine organisms is imperative to unlock the full potential of the marine as a pharmacological epicenter.

3.1 Marine-Derived Bioactives

Marine-derived bioactives have shown promising potential in treating neuropsychiatric disorders. The oceans contain habitats and habitants that possess great biodiversity and chemical qualities that have proven to be related to overall health and modern medicine.

Marine-derived bioactives are naturally occurring compounds with biological activity that are extracted from marine organisms like fish, algae, bacteria, marine invertebrates, or plants in the ocean. They include fatty acids, peptides, polysaccharides, amino acids, alkaloids, flavonoids, terpenoids, and steroids. These compounds often have unique chemical structures that are not found in terrestrial organisms. Because of this, marine-derived bioactives are studied for their potential use in developing drugs for cancer, inflammation, infection, diseases, and their potential to positively affect medicine today.

Among the many positive effects marine-derived bioactives have on health, their effect on neuropsychiatric disorders is still being deeply studied. "The complex interaction between the mind and the brain has been a topic of deep reflection throughout medical history. The division between neurology and psychiatry, which each encompass separate domains of medical specialization, has been a fundamental aspect of the medical field for a significant period. This research explores the historical context of the division between neurology and psychiatry. It clarifies the moment neuropsychiatric illnesses emerged as a separate field of study. The division between neurology and psychiatry can be traced back to the nineteenth century, a period characterized by significant changes in medical disciplines. The distinction between the two fields was initially conceived to offer a systematic method for comprehending and addressing problems of the neurological system. Neurology, which concentrates on the biological and structural elements of the brain, seeks to decipher the enigmas of physical disorders that impact the neurological system. Psychiatry, on the other hand, developed as a field focused on investigating mental and emotional illnesses, frequently linked to intricate behavioral expressions." "At the beginning of the twentieth century, there was a significant change in how medical professionals thought about the relationship between neurology and psychiatry. They realized that the borders between these two fields were more relaxed than they had previously believed." "The critical point in the acknowledgment of neuropsychiatric diseases as a separate discipline arose in reaction to the increasing occurrence of problems that resisted straightforward classification. Conditions such as epilepsy, neurodevelopmental disorders, and traumatic brain injuries were recognized as requiring a comprehensive understanding that connects the detailed structure and function of the brain with the intricate aspects of behavior and cognition. This insight sparked a fundamental change in thinking, resulting in the development of neuropsychiatry as a connection between the seemingly separate fields of neurology and psychiatry. As we explore the historical development of the division between neurology and psychiatry and the subsequent rise of neuropsychiatric disorders as a separate field, it becomes clear that integrating these disciplines is crucial for a thorough comprehension of brain-related disorders." While studies relating marine bioactives to neuropsychiatric disorders are showing promising results, further research is needed to better understand the full potential and long-term effects of these findings. The complexity of neuropsychiatric conditions emphasizes the need for intense clinical testing in order to establish evidence-based treatments. The deep interconnection between neurology and psychiatry continues to unfold in research studies today.

"A neuropsychiatric disorder is a complex and incompletely understood neurological condition that affects the brain's structure, synaptic circuits, and neurotransmission. It is characterized by abnormalities in the cerebral cortex and limbic system, reduced neurogenesis, altered neural plasticity, and increased expression of susceptibility genes. Examples of neuropsychiatric disorders include depression, anxiety, autism, bipolar disorder, and schizophrenia." Neuropsychiatric disorders are complex and multifaceted conditions that take into account both biological and psychological factors. This complicates diagnostic and therapeutic approaches towards treatments. The biological mechanisms are better understood than the psychological mechanisms, like emotional regulation and cognitive processing.

The cerebral cortex is the part of the brain that is responsible for complex functions like decision-making, memory, and processing sensory information. This part of the brain is impacted in conditions like schizophrenia and bipolar disorder. The limbic system is the part of the brain involved in emotional processing and regulation. It is often affected or becomes unbalanced in conditions like depression, anxiety, and bipolar disorder. Neurotransmission is a process that plays a critical role in regulating mood, cognition, and behavior. Serotonin, dopamine, and glutamate are key examples of neurotransmitters. A deficiency in serotonin is a well-known cause of depression, insufficient dopamine transmission is known to give characteristics related to schizophrenia, and disruption of glutamate pathways is involved in mood changes in bipolar disorder.

Marine-derived bioactives can interact within the brain and nervous system, which makes them a possible method of treatment for neuropsychiatric disorders. These interactions can lead to neuroprotection, neurogenesis, and modulation of neurotransmitter systems.

The discovery of marine-derived bioactive compounds has experienced significant growth in recent decades. This growth has been driven mainly by technological advancements, increased research funding, and a growing recognition of the ocean's potential as a source of therapeutics. The factors driving this increasing discovery include technological advancements, collaborative research, and symbiotic microbes.

The increase in attention and resources focused on marine-derived bioactive compounds is leading to more frequent and effective discoveries. "Every year, hundreds of new compounds are discovered from the metabolites of marine organisms. Finding new and useful compounds is one of the crucial drivers for this field of research." A study done on the increasing rate of marine bioactive discovery in 2015 focused on the years from 1985 to 2012. They "analyzed temporal trends of the number of annualized novel bioactive products obtained from marine organisms from 1985 to 2012 and compared the total number of compounds found from marine organisms." Data before 1985 was not included because the number of novel products discovered was less than 100 in those years. The number of novel marine compounds greatly increased after 1985. Since then, the number stabilized at about 500 products per year in the late 1990s, and the second sharp rise in yield was found after 2005. In 2006, the total number of novel compounds discovered annually saw a huge increase; however, the number of bioactive compounds did not increase significantly. Also, they noticed that the annual proportion was not constant. "The data observed in the last decade suggests that the average proportion of bioactive compounds among the new compounds is declining, though there is a large number of marine natural products yet to be explored. This may indicate that the research level of bioactivity is not keeping up with the discovery of new compounds." This suggests the need for new technologies and methods to reach further success in discovering marine-derived bioactives (Fig. 3.1).

This same study provided specific methods and advancements that need to be explored using technology in order to increase the discovery of marine natural products. "The first improvement is about methods of bioactivity-guided separation and purification of marine secondary metabolites, which combine the discovery of new compounds with bioactivity assays. These improvements in technology are dependent upon the automation in NMR and mass spectroscopy (MS), which also allow the study of the functions of new compounds extracted from marine organisms' secondary metabolites. Second, for the discovery of new lead compounds for drug development, empirical cell-based screening for cytotoxicity has evolved to a more mechanistic approach that targets specific molecular lesions, though there are still shortcomings. Combined with phenotypic and high content screening through a large number of drug targets, bioactivity research will be very effective in revealing the potentially useful biological properties of natural products. In addition, a high-throughput strategy was established to sequence and assemble single-cell genomes of novel uncultivated microorganisms from marine samples. Using single-cell genomic approaches allows us to study uncultured microbes with partial to near complete genomes. The highly accurate sequence data will generate new knowledge in several areas relevant to medicinal chemistry, including the identification and validation of new drug targets in human pathogens. Furthermore, the discovery of new bioactive compounds from marine metabolites will form the basis for new

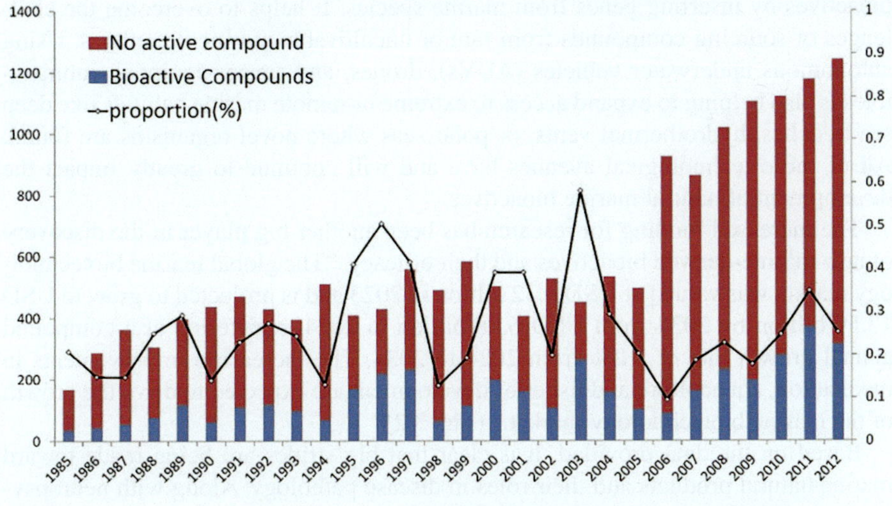

Fig. 3.1 Variation in the number of new marine natural products for 1985–2012

drug leads. As mentioned above, these technological advances are crucial to remove the bottlenecks slowing the discovery of potent secondary metabolites. More interdisciplinary approaches and innovative manipulations are urgently needed to improve bioactivity studies."

The study went on to further clarify the context and viewpoint of their data. They stated, "it is worth mentioning that, in this article, the number of bioactive compounds is somewhat imprecise because the bioactivity record updates and the discoveries of new bioactivity on these compounds cannot be reflected in our data. Of the 16617 new compounds, the number of bioactive compounds is 4196, or 25.25%. However, this does not mean that the other 74.75% are inactive. Indeed, their bioactivity may be discovered in future studies."

Overall, technology has supercharged marine bioactive discovery through genomics, next-generation sequencing, mass spectrometry, nuclear magnetic resonance, synthetic biology, and robotics. Genomics involves sequencing the DNA of marine organisms to understand their genetic potential. Metagenomics goes a step further by analyzing genetic material directly from environmental samples which has revealed unculturable microbes and their biosynthetic pathways. "Genome-based research of marine microorganisms will maximize the biodiscovery process and solve the problems of supply and sustainability of drug molecules for medical treatments." Next-generation sequencing allows rapid, cost-effective DNA and RNA analysis, and enables massive parallel sequencing of marine organisms and their microbiomes. This helps scientists pinpoint genes involved in natural product biosynthesis. Mass spectrometry and nuclear magnetic resonance are used to determine the structure and molecular composition of compounds. Modern systems can identify and characterize complex marine compounds with minute sample sizes. Synthetic biology is the idea of engineering microorganisms to produce marine bioactives by inserting genes from marine species. It helps to overcome the challenges of sourcing compounds from rare or uncultivable marine organisms. Using autonomous underwater vehicles (AUVs), drones, and sensor-equipped submersibles is also helping to expand access to extreme or remote marine habitats like deep sea trenches, hydrothermal vents, or polar seas where novel organisms are found. All of these technological avenues have and will continue to greatly impact the development of natural marine bioactives.

The increased funding for research has been another big player in the discovery of new marine-derived bioactives and their efficacy. "The global marine biotechnology market was valued at USD 6.32 billion in 2023 and is projected to grow to USD 13.59 billion by 2034 from USD 6.78 billion in 2024, registering at a compound annual growth rate of 7.2% from 2024 to 2034. The increasing improvements in medication, aquaculture and fisheries development are expected to drive the growth of the marine biotechnology market." (Fig. 3.2)

Based on the data provided, it is clear that big strides are being made toward marine natural products and their roles in disease pathology. Along with neuropsychiatric potential, marine natural products have yielded promising treatments regarding anticancer agents, antimicrobial activity, antiviral activity, neuroprotective and neurotoxic studies, anti-inflammatory effects, and immunomodulatory effects.

Trabectedin and halichondrin B are marine natural products that have yielded some promising cancer therapies. Trabectedin is derived from tunicate *Ecteinascidia turbinata* and is used to treat soft tissue carcinomas and ovarian cancer. It interacts with the minor groove of DNA and disrupts the cell cycle as well as DNA repair. Halichondrin B inhibits microtubule dynamics, which can lead to mitotic blockage in cancer cells. This can lead to several outcomes like cell death, cell cycle arrest,

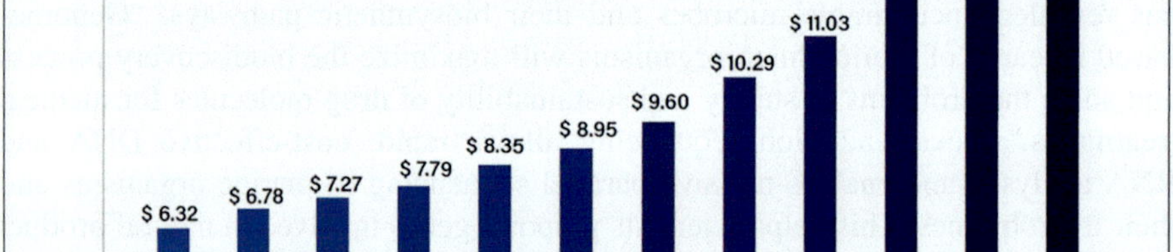

Fig. 3.2 Marine biotechnology market size 2023 to 2034 (USD billion)

and chromosomal instability. "The dynamic pharmaceutical pipeline consisting of approved and developmental marine-derived anticancer agents offers new hopes and new tools in the treatment of patients afflicted with previously intractable types of cancer." "While the current review focuses on marine-derived anticancer agents already approved or in clinical development, it is expected that the increasing number of early candidates in preclinical investigations will not only be translated in new chemotherapeutic tools, but also in the discovery of additional molecular targets."

"Infectious diseases have always threatened human health, survival, and development. Antibiotics represent the first solution to combat bacterial infections. Generally speaking, antibiotics take effect through disrupting cell wall synthesis, cell membrane integrity, protein synthesis (e.g., 30S ribosome and 50S ribosome inhibitors), nucleic acid synthesis, or folate synthesis. However, accompanying their use, drug-resistant bacteria have emerged." Cyanobacterial metabolites are marine natural products that offer promising research for antimicrobials. Marine cyanobacterial metabolites show promise against MRSA, TB, and even viruses like HSV and HIV. There is also recent interest in marine peptides for potential activity against SARS-CoV-2. These compounds show patent activity against drug-resistant bacteria and fungi.

Tetrodotoxin (TTX) and ziconotide (Prialt) are marine natural products that have an effect on the nervous system. This makes them useful for neuropathological research. Tetrodotoxin (TTX) is from pufferfish and some marine bacteria. It is a potent neurotoxin, but used in small doses for pain management research and studying ion channel pathology. The most promising therapeutic use of TTX is perhaps in the treatment of certain pains. The use of TTX has been investigated for medical purposes other than pain mitigation in animal models." "This toxin is a promising tool as therapeutic drug, especially for pain treatment." "Because TTX blocks voltage-gated sodium channel and causes paralysis, it can also be applied as a potential pain relief and some researchers are trying to make use of the analgesic activity of TTX to treat various types of pains such as severe cancer pain, or to help in reducing cue-induced increases in heroin craving and associated anxiety." Ziconotide (Prialt) is from the venom of *Conus magus* (cone snail). "The cone snail venom peptide that ultimately became Prialt®, an approved non-opioid therapeutic for intractable pain was discovered as part of a basic research investigation of two fish-hunting cone snails, *Conus geographus* (the Geographer Cone), a species that can cause human fatality, and *Conus magus*, a smaller species with a narrower biogeographic distribution in the Western Pacific. It treats severe chronic pain by blocking N-type calcium channels in spinal cord neurons." "The identification of a non-opioid drug lead, particularly when the molecular target is defined, can quickly lead to a very considerable accompanying structure/function database that accelerates preclinical development."

Fucoidan, from brown algae, has anti-inflammatory, anticoagulant, and immune-boosting properties. It is studied for therapeutic effects in colitis, arthritis, and even COVID-19-related inflammation. Fucoidan brings great impetus to the immune system by enhancing phagocytosis. In clinical and animal studies, fucoidan has been demonstrated to improve innate immunity and reduce inflammation. "In vitro and in vivo data have shown that the polysaccharide is effective against influenza viral infection, and specific inhibitory activity against novel coronavirus infection in vitro was demonstrated for fucoidan. Fucoidan can also protect lipopolysaccharide-induced acute lung injury by regulating signal pathways to inhibit inflammation and oxidation.

In recent years, the blood-lipid reduction, anti-atherosclerosis, and anti-thrombus effects of fucoidan have been widely investigated by researchers. However, there are much fewer systematic reports on the properties of fucoidan in relation to CVD. In this article, we review the progress of research into the use of fucoidan for the treatment of CVD in recent years, including its hypolipidemic, anti-atherosclerotic, and antithrombotic effects and mechanisms. The benefits of treating CVD patients with fucoidan are emphasized, and we provide some theoretical bases and inspiration for the development of cardiovascular drugs." "Fucoidan is a complex marine acidic polysaccharide with diverse biological activities that is found in a wide range of organisms, and it can play an important therapeutic role at all stages of cardiovascular disease: initial hyperlipidemia, the formation of fat streaks in the arteries, lipid accumulation, fibrous plaques, atherosclerotic plaques, plaque hemorrhage and rupture, thrombosis, and cerebral infarction. Hyperlipidemia is the basis of many cardiovascular and cerebrovascular diseases, and fucoidan has shown particularly diverse and potent effects on lipid levels. The anti-inflammatory, lipid-lowering, and antioxidant effects of fucoidan can have a positive impact over the entire course of atherosclerosis, and its prospects as a treatment for cardiovascular and cerebrovascular diseases are good." "Although fucoidan has a wide range of biological activities and promising preclinical effects, there are still some problems that need to be urgently solved before fucoidan can be developed as a drug. As a marine polysaccharide, its specific structure is still unclear, and its sulfate content, monosaccharide content, and molecular weight vary greatly among the different sources and extraction methods. Furthermore, the relationship between its composition and activity remains uncertain. Research on polysaccharides brings certain difficulties, and clarifying the molecular structure of fucoidan is still a major challenge." Marine-derived polysaccharides, peptides, and lipids are increasingly explored for new health treatments and discoveries.

Marine natural products have gained increasing attention in recent decades due to their unique structures and diverse therapeutic potential. Their application extends through various therapeutic, preventative, and diagnostic purposes in medicine, including the treatment of neurodegenerative diseases, cancer, infections, inflammatory disorders, and more. Marine natural products are secondary metabolites produced by marine organisms as a defense mechanism against predators, competition, or environmental stress. These compounds are often chemically distinct from terrestrial bioactive substances, and many of them have evolved to interact with specific biological targets in both marine and terrestrial organisms. As a result, marine natural products have been the source of numerous novel therapeutic agents and are being actively investigated for their pharmacological properties. Many of these organisms produce bioactive compounds with antibacterial, antiviral, anti-inflammatory, antioxidant, neuroprotective, and antitumor activities. The complexity and novelty of marine natural products make them attractive candidates for drug discovery and development.

Marine natural products have shown promising therapeutic potential in the treatment of a variety of diseases, including cancer, infectious diseases, neurodegenerative diseases, cardiovascular disorders, and immune system diseases. "The marine ecosystems are a rich source of bioactive products and have been explored for lead drug molecules that have proven to be novel therapeutic targets. Over the last 70 years, many structurally diverse drug products and their secondary metabolites have been isolated from marine sources. The drugs obtained from marine sources have displayed an exceptional potential in the management of a wide array of diseases, ranging from acute to chronic conditions. A beneficial role of marine drugs in human health has been recently proposed. The current review highlights various marine drugs and their compounds and role in the management of chronic diseases such as cancer, diabetes, neurodegenerative diseases, and cardiovascular disorders, which has led to the development of new drug treatment approaches."

The therapeutic routes for delivering marine-derived bioactives are an essential part of ensuring their efficacy, bioavailability, and safety when used in treating neuropsychiatric disorders or other conditions. These routes vary based on the chemical properties of the bioactive compound, the condition being treated, and the desired therapeutic effect. The oral route is the most common and noninvasive way to administer therapeutic agents. Marine bioactives, such as omega-3 fatty acids, polyphenols, and peptides, are often formulated into capsules, tablets, liquids, or powders for oral consumption. The intravenous (IV) route is used when rapid delivery of marine bioactives is necessary. This method delivers compounds directly into the bloodstream, bypassing the digestive system, which can be beneficial for compounds that have low bioavailability when taken orally. The topical route involves the application of bioactive compounds directly onto the skin or mucous membranes. This method is particularly useful for treating local conditions or for transdermal absorption when the bioactive can pass through the skin barrier. The inhalation route involves the delivery of therapeutic agents via the respiratory tract into the lungs. This method is used primarily for the treatment of lung conditions or to ensure direct access to the bloodstream via the lungs for systemic effects. The sublingual or buccal route involves the placement of a bioactive compound under the tongue (sublingual) or between the gums and cheek (buccal) for absorption through the mucous membranes of the mouth. The therapeutic routes for delivering marine-derived bioactives are varied and depend on several factors, such as the bioactive compound's chemical structure, target disease, and the desired therapeutic effects. Each route has its own set of advantages and challenges, and the bioavailability, formulation, and pharmacokinetics of the bioactive need to be carefully considered to optimize their clinical effectiveness. The development of advanced drug delivery systems is crucial for improving the efficacy of marine bioactives. As research continues, the delivery routes and formulations will be refined to maximize the potential of marine-derived compounds for treating a wide range of neuropsychiatric disorders and chronic diseases.

The preventative route for delivering marine-derived bioactives focuses on using these bioactive compounds to prevent disease or maintain health before any symptoms of illness or disorder arise. This approach is particularly relevant in chronic conditions, age-related diseases, and neuropsychiatric disorders, where early intervention can help slow disease progression, enhance mental health, and maintain overall well-being. The goal is to enhance physiological processes and strengthen the body's natural defenses against potential health threats, using marine bioactives. The oral route is the most common and convenient method for preventative health with marine-derived bioactives. Many marine bioactives are incorporated into dietary supplements, functional foods, and beverages to promote health maintenance and disease prevention. The topical route involves the application of bioactive compounds directly onto the skin. Marine-derived bioactives, especially collagen, astaxanthin, and marine oils, are often used in skincare products, such as creams, gels, and serums. This route is ideal for preventing skin aging, wound infections, and UV damage, and can also offer neuroprotective effects for the skin. The sublingual or buccal route involves placing a bioactive compound under the tongue or in the cheek pouch for absorption through the mucous membranes. This route allows for rapid absorption into the bloodstream, bypassing the digestive system and liver metabolism. It is typically used for fast-acting supplements, such as neuroprotective compounds and anti-inflammatory agents. The inhalation route is less commonly used for the prevention of diseases, but there is growing interest in pulmonary delivery of bioactives, especially for lung health, immunomodulation, and neurological protection. For example, marine-derived omega-3s and marine collagen may be delivered via aerosols or nebulizers for lung regeneration or to support the blood-brain barrier. The preventative routes for delivering marine bioactives offer a variety of noninvasive methods to maintain health, enhance well-being, and prevent diseases. Oral supplementation is the most common method, but topical, sublingual, and inhalation routes also provide targeted, fast-acting options for maintaining cardiovascular, neuroprotective, and skin health. While the effectiveness of these routes depends on the bioactive's chemical properties, proper formulation, and delivery systems, marine bioactives have significant potential for preventative health strategies, particularly in combating chronic diseases, neurodegenerative disorders, and skin aging.

Diagnostic management refers to the use of marine-derived bioactives not only as therapeutic agents but also as tools for early detection, monitoring, and diagnosis of diseases, especially neuropsychiatric disorders and neurodegenerative diseases. Marine bioactives hold promise in diagnostic management by aiding in the identification of biomarkers, improving the understanding of disease mechanisms, and providing insights into disease progression. Here's a detailed discussion on how marine-derived bioactives contribute to diagnostic management. Early detection of neurodegenerative and neuropsychiatric disorders is crucial for effective intervention and slowing disease progression. Marine bioactives, particularly those with neuroprotective, anti-inflammatory, and antioxidant properties, can aid in early diagnosis and screening of these diseases. Marine bioactives can be integrated into molecular diagnostic tools and imaging techniques to help in disease diagnosis and monitoring. These tools are particularly useful for detecting brain diseases, neuropsychiatric disorders, and cognitive decline. Once a diagnosis is made, the ability to monitor disease progression and evaluate treatment efficacy is vital for adjusting therapeutic strategies. Marine bioactives can play an important role in this phase of diagnostic management by influencing biomarkers that indicate the response to treatment. Marine bioactives can also be used in preventative diagnostics to help identify individuals at risk of developing neurodegenerative or neuropsychiatric diseases. Early markers of cognitive decline or increased inflammation can be detected through the measurement of specific bioactive levels or their metabolites. Marine-derived bioactives offer significant potential for diagnostic management in neuropsychiatric and neurodegenerative diseases. These bioactives contribute to the development of biomarkers, improve diagnostic imaging, and assist in monitoring disease progression and treatment responses. Additionally, marine bioactives are useful in preventative diagnostics, enabling early detection and risk stratification for conditions such as Alzheimer's disease, Parkinson's disease, and depression. Their molecular action, ability to influence inflammatory pathways, and capacity to enhance brain health offer opportunities for integrating marine bioactives into routine diagnostic protocols, both for identifying and monitoring neuropsychiatric and neurodegenerative diseases.

Marine natural products offer a wide range of applications in therapeutic, preventative, and diagnostic purposes. From their role in treating neurodegenerative diseases to their potential in preventing cardiovascular disorders and offering antioxidant protection, marine-derived compounds have a unique and valuable position in modern medicine. However, challenges such as bioavailability, sustainability, and standardization need to be addressed to maximize their clinical potential. Continued research into the therapeutic properties of marine bioactives will likely lead to the development of novel drugs, preventive measures, and diagnostic tools, further expanding the utility of the marine environment in human health.

Neuroprotection is the process of protecting the health and function of neurons in the brain and spinal cord. Marine-derived bioactives such as omega-3 fatty acids are known to protect neurons from damage from stress, inflammation, and cytotoxicity. These are all factors that are commonly involved in neuropsychiatric disorders. Docosahexaenoic acid (DHA) and eicosapentaenoic acid (EPA) have proven to protect neurons from oxidative damage and enhance cellular membrane integrity. This protection helps reduce inflammation and oxidative stress, which are both related to mood disorders. Some studies show low levels of DHA and EPA and may be associated with an increased risk of depression. An article discussing the studies on omega-3 fatty acids relating to neuropsychiatric disorders states, "Omega-3 polyunsaturated fatty acids have received considerable attention in the field of mental health, in particular regarding the treatment of depression. The existing body of evidence demonstrates that omega-3 fatty acids, in particular eicosapentaenoic acid (EPA) and docosahexaenoic acid (DHA), have antidepressant effects that can be attributed to their modulation of neuroinflammation, neurotransmitter function, and neuroplasticity. Nevertheless, clinical trials of omega-3 supplementation have yielded inconsistent results. Some studies have demonstrated significant reductions in depressive symptoms following omega-3 treatment, whereas others have shown minimal to no beneficial impact. A range of factors, encompassing dosage, the ratio of EPA to DHA, and baseline nutritional status, have been identified as having a potential impact on the noted results. Furthermore, it has been suggested that omega-3 fatty acids may act as an adjunctive treatment for those undergoing antidepressant treatment. Notwithstanding these encouraging findings, discrepancies in study designs and variability in individual responses underscore the necessity of further research in order to establish uniform, standardized guidelines for the use of omega-3 fatty acids in the management of depressive disorders." While some findings look promising, the variability in study results emphasizes the need for further research. More consistent clinical trials are needed in order to better discover how marine-derived omega-3 fatty acids can be used in treating neuropsychiatric disorders.

Antioxidants are also known to play a role in the treatment of neuropsychiatric disorders due to their ability to combat stress. Oxidative stress is an imbalance between reactive oxygen species production and the body's ability to neutralize them. "Low levels of reactive oxygen species are considered essential for neuronal development and function, but excessive levels are hazardous and can lead to cellular damage." "The brain, with its high oxygen consumption and lipid-rich content, is highly susceptible to oxidative stress. Therefore, oxidative stress–induced damage to the brain has a strong potential to negatively impact normal CNS functions. Although oxidative stress has historically been considered to be involved mainly in neurodegenerative disorders such as Alzheimer disease, Huntington disease, and Parkinson disease, its involvement in neuropsychiatric disorders, including anxiety disorders and depression, is beginning to be recognized. This review is a discussion of the relevance of cerebral oxidative stress to impairment of emotional and mental well-being." Antioxidants help protect neurons from free radicals, which reduces inflammation and prevents oxidative damage to cellular components. Many neuropsychiatric disorders are associated with chronic low-grade inflammation in the brain. Antioxidants may also help maintain neuroplasticity, which promotes better cognitive function and emotional regulation. Since antioxidants can reduce inflammation and oxidative stress, this may be able to help the progression or severity of certain psychiatric conditions. Astaxanthin is a specific antioxidant found in marine algae that is believed to help neutralize free radicals and prevent cellular damage. "The beneficial effects of astaxanthin are linked to its oxidative, anti-inflammatory, and anti-apoptotic characteristics. Studies have also demonstrated that astaxanthin can easily cross the blood-brain barrier to protect the brain from acute injury and chronic neurodegeneration." Because of these characteristics, astaxanthin is a great option for treating neurological diseases even though there is still a lack of research in some areas. Some other promising antioxidants are vitamins E and C, glutathione, coenzyme Q10, and polyphenols. Glutathione imbalances may be a primary cause of brain disorders. Increasing the concentration of glutathione in the brain is a promising approach for the treatment of brain disorders. Coenzyme Q10 is an antioxidant and a vital factor in the mitochondrial electron transport chain. Coenzyme Q10 depletion may play a role in neuropsychiatric disorders due to its ability to protect the mitochondria and relieve oxidative stress. Lastly, polyphenols are natural compounds that have gained attention due to their anti-inflammatory and antioxidant benefits. Because of these benefits, they may potentially be able to treat various neuropsychiatric disorders like depression. They have been demonstrated to influence several mechanisms involved in the formation of depression, like neurotransmitter systems, intracellular signaling, neuroplasticity, and cell survival. "Polyphenols are present in fruits, vegetables, tea, and cocoa have also been reported to influence the function and expression of key signaling molecules implicated in inflammatory processes in depression, indicating their potential significance in the development of innovative antidepressant medicines for the treatment of inflammation-associated depression. Recent evidence suggests that plant sources and their bioactive phytochemicals exert therapeutic effects on mental illnesses with few side effects as well as with a wide range of efficacy. Consumption of polyphenol-rich diets may hold the dietary manipulation as a non-invasive, natural, and cost-effective approach of supporting a healthy brain. It also acts as neuroprotection that is attributed to their ability to increase cerebral blood flow and inhibit platelet aggregation." The neuroprotective benefits and their ability to enhance cerebral blood flow open doors for polyphenols not only as a potential treatment for depression but also as a preventive measure for other brain-related disorders. Their potential as a noninvasive, side-effect-minimal, and cost-effective alternative to traditional antidepressant treatments could significantly influence future strategies for treating depression. Their benefits could be explored further, perhaps by comparing the bioavailability and effectiveness of polyphenols from various sources and studies that support these claims. This could better provide a well-rounded argument for using polyphenols in the treatment and management of mental health.

Polyphenols are found in marine algae, seaweed, and marine plants. Brown algae contain polyphenolic compounds like phlorotannins. Phlorotannins are polyphenolic compounds that are unique to brown algae and are known for their strong antioxidant properties. They play a significant role in reducing oxidative stress, which is linked to various chronic diseases, including neurodegenerative conditions. Phlorotannins have been shown to exhibit neuroprotective effects by scavenging free radicals and reducing inflammation in the brain. This can be particularly useful in conditions like Alzheimer's disease and other cognitive impairments associated with oxidative damage. Red algae also contain polyphenolic compounds, such as flavonoids and phenolic acids. These compounds have antioxidant, anti-inflammatory, and anticancer properties. Green algae, like chlorella and spirulina, are also rich in polyphenols, particularly flavonoids and phenolic acids. These polyphenols contribute to the algae's ability to combat oxidative stress and inflammation. Red, brown, and green seaweeds are considered one of the richest marine sources of polyphenolic compounds. Polyphenols in seaweeds are mostly in the form of flavonoids, phenolic acids, and tannins. They exhibit antioxidant and anti-inflammatory effects and are thought to contribute to human health benefits. Marine plants also contain polyphenolic compounds that have antioxidant and anti-inflammatory properties. Studies have shown that the polyphenols in marine algae and seaweeds may help stimulate adult neurogenesis, particularly in the hippocampus, which is crucial for memory and learning. Marine-derived polyphenols, by improving neurogenesis, enhancing cerebral blood flow, and reducing oxidative stress, could provide potential treatments for depression, anxiety, and other mood disorders. Marine polyphenols could also work synergistically with other nutrients found in marine sources (like omega-3 fatty acids) to boost brain health and function, offering a natural and noninvasive way to combat mental health issues without the side effects associated with many pharmaceutical treatments. While the benefits of polyphenols from marine sources are well-documented, much of the research is still in the early stages. More clinical trials and human studies are needed to fully understand the bioavailability, efficacy, and mechanisms by which these compounds work in the brain. Future research should also focus on identifying the specific marine polyphenols that are most effective in treating or preventing cognitive decline and neuropsychiatric disorders, as well as developing targeted therapies based on marine-derived bioactives. Polyphenols in marine sources, such as algae, seaweeds, and marine plants, offer a promising area of research for promoting brain health and treating neurodegenerative diseases and mental health disorders like depression. Their antioxidant, anti-inflammatory, and neuroprotective properties make them potential candidates for future therapeutic interventions aimed at enhancing neurogenesis, improving cognitive function, and supporting emotional health. The continued exploration of marine polyphenols may uncover new, natural, and effective treatments with fewer side effects compared to conventional pharmaceuticals.

Neurogenesis is the process of developing new neurons in the brain. "The mammalian brain is known to generate its full capacity of neurons during their embryonic development stage and perinatal stage." However, there are a few regions in the brain that continue to generate newborn cells throughout life. This process is termed adult neurogenesis, which mainly occurs within the hippocampus and the lateral ventricles." "Newborn neurons have been implicated to have a crucial role in brain functions such as learning and memory." They are thought to play a role in synaptic plasticity and cognitive flexibility. Synaptic plasticity is the process by which connections between neurons are strengthened or weakened based on experiences. This is critical for memory formation and learning. Cognitive flexibility is the ability to adapt behavior based on new information or changes in the environment. Studies have found that certain areas of neurogenesis are impaired in several neurodegenerative and neuropsychiatric diseases. The incorporation of new neurons also plays a role in emotional regulation, helping the brain to adapt to stress and negative emotions. This can make neurogenesis a critical factor in maintaining mental health and cognitive function throughout life. In various neurodegenerative diseases, particularly Alzheimer's Disease (AD), Parkinson's Disease (PD), and Huntington's Disease (HD), there is a marked decline in neurogenesis, particularly in the hippocampus. This impairment in neurogenesis is thought to contribute to the progressive cognitive decline, memory loss, and emotional disturbances observed in these conditions. In Alzheimer's and depression, hippocampal neurogenesis is affected. Marine-derived bioactives have been shown to stimulate neurogenesis. The link between cognitive decline and changes in hippocampal neurogenesis is poorly understood, but enhancing neurogenesis could potentially reverse some of the cognitive and emotional damage. While there is growing evidence that reduced neurogenesis in the hippocampus is linked to cognitive decline and emotional disturbances, the exact mechanisms underlying this relationship are still not fully understood. Several factors contribute to the decline of hippocampal neurogenesis, including oxidative stress, inflammation, neurotrophic factor deficits (such as brain-derived neurotrophic factor (BDNF)), and genetic factors. It is hypothesized that a loss of neurogenesis may create a vicious cycle in neurodegenerative and psychiatric diseases, where the lack of new neurons contributes to further cognitive decline, which then leads to an even greater impairment of neurogenesis. This makes it crucial to understand and intervene in the pathways that regulate neurogenesis to slow or reverse cognitive deterioration. Studies suggest that enhancing neurogenesis could potentially reverse some cognitive deficits and improve emotional regulation. For example, marine-derived compounds that stimulate hippocampal neurogenesis may improve memory, learning, and emotional resilience, thus offering a novel approach to treating both cognitive and emotional symptoms of depression. Despite the promising findings on marine-derived bioactives, there are still challenges in translating these findings into effective treatments for human patients. One of the main challenges is understanding the bioavailability of these compounds—whether they can effectively reach the brain and have sufficient potency to stimulate neurogenesis. In conclusion, neurogenesis, particularly in the hippocampus, plays a pivotal role in maintaining cognitive function, learning, memory, and emotional regulation throughout life. Impaired neurogenesis is a common feature of many neurodegenerative diseases, including Alzheimer's disease, and mood disorders like depression. However, marine-derived bioactives offer a promising avenue for stimulating neurogenesis and potentially reversing cognitive and emotional damage. By targeting neurogenesis, these bioactive compounds may provide new, natural therapeutic options for treating cognitive decline and mental health disorders, offering a more holistic and accessible approach to brain health. Future research is needed to further investigate the mechanisms behind neurogenesis and how marine bioactives can be harnessed for therapeutic purposes in neurodegenerative and neuropsychiatric diseases.

Marine-derived peptides are also a potential method of treatment for neuropsychiatric disorders. Bioactive peptides are biomolecules derived from proteins. These peptides are anywhere from 2 to 20 amino acids in length with distinct biological activities that can influence various physiological processes. Marine-derived bioactive peptides have been shown to possess a wide range of bioactivities, including antioxidant, anti-inflammatory, antimicrobial, and neuroprotective properties. "They are considered therapeutic agents for treating and preventing certain diseases, with high specificity, a broad spectrum of action, low toxicity, high structural diversity, and small size." These peptides are often derived from the protein breakdown products of marine species, including collagen, muscle proteins, and enzymes. These peptides can be extracted from marine sources through enzymatic hydrolysis or fermentation processes, which break down the protein into smaller, biologically active fragments. There are a variety of sources of marine bioactive peptides. They can come from fish, crustaceans, or mollusks. Fish, especially species like salmon, cod, tuna, and mackerel, are rich in proteins that can be broken down into bioactive peptides. Fish collagen is particularly rich in peptides that can exert various health benefits, including neuroprotection. Mollusks, such as squid, octopus, and clams, contain peptides that have demonstrated beneficial effects on metabolic health and neuroprotection. These peptides are often derived from the breakdown of muscle proteins. Mollusks, such as squid, octopus, and clams, contain peptides that have demonstrated beneficial effects on metabolic health and neuroprotection. These peptides are often derived from the breakdown of muscle proteins. Crustaceans like shrimp, crab, and lobster contain proteins that can be converted into bioactive peptides. Chitosan, derived from the exoskeletons of crustaceans, has been studied for its potential neuroprotective effects as well. Marine bioactive peptides represent a promising class of compounds with significant potential for neuroprotective applications. Derived from fish, mollusks, and crustaceans, these peptides offer a variety of beneficial effects, including antioxidant, anti-inflammatory, and neurotransmission-enhancing properties. Their ability to modulate neurogenesis, protect neurons from oxidative damage, and influence neurotransmitter systems could make them valuable tools in the treatment and management of neurodegenerative diseases like Alzheimer's and Parkinson's disease, as well as mood disorders like depression. However, further research is needed to fully understand the mechanisms of action of these peptides, their bioavailability, and their efficacy in human clinical trials. With continued advancements in marine biotechnology and neurobiology, marine-derived bioactive peptides may offer novel, natural, and effective therapeutic options for improving brain health and managing cognitive and emotional disorders.

While marine-derived bioactives are a promising and intriguing method of treatment for neuropsychiatric disorders, translating the positive findings from laboratory studies to clinical applications involves a range of challenges. Some of these challenges include safety concerns, variability of bioactive concentrations in marine sources, conducting large-scale clinical trials, and ethical and environmental issues. Overcoming these barriers will require concerted efforts in research, regulation, and technology development, including the establishment of consistent extraction methods, formulation protocols, and bioavailability optimization strategies. To make marine-derived bioactives a viable therapeutic option, ongoing research is crucial in addressing these challenges. In addition, collaborations between researchers, industry stakeholders, and regulatory bodies will be essential for developing safe, effective, and standardized marine-based treatments that can make a real impact in managing neuropsychiatric disorders.

To conclude, neuropsychiatric disorders like depression, anxiety, schizophrenia, bipolar disorder, and neurodegenerative diseases like Alzheimer's and Parkinson's present a lot of challenges to global public health. With a significant proportion of patients not responding to existing therapies and many treatments carrying undesirable side effects, the search for more effective therapeutics and treatments is urgent. Marine natural products have emerged as a highly promising, yet relatively untapped, source of bioactive compounds that may hold the key to addressing these complex disorders and medical diagnoses. Over the past few decades, marine biodiversity has been increasingly recognized as a goldmine of pharmacologically relevant molecules. Marine organisms, which range from sponges and mollusks to cyanobacteria and deep-sea fungi, produce a wide range of secondary metabolites with unique chemical scaffolds and mechanisms of action. Unlike many terrestrial bioactives, marine-derived compounds often feature unusual halogenation patterns, cyclic structures, or hybrid biosynthetic origins, which contribute to their bioactivity and therapeutic potential. In the context of neuropsychiatric research, these features offer particularly valuable opportunities for the development of central nervous system-targeted therapeutics, which must cross the blood–brain barrier and interact with highly specific neuronal targets. Among the most promising marine-derived compounds is ziconotide, a synthetic peptide derived from the venom of the cone snail *Conus magus*. Approved for severe chronic pain, ziconotide is a potent blocker of N-type calcium channels—an essential mechanism in neurotransmission, which has broader implications for mood and cognitive regulation. Its development exemplifies the potential for marine toxins to be repurposed into safe and targeted central nervous system treatments. Similarly, tetrodotoxin, though a neurotoxin in nature, has informed the design of sodium channel blockers for neuropathic pain and epilepsy. Other marine natural products, which are derived from marine dinoflagellates or cyanobacteria, have become essential tools in neuroscience research, helping scientists unravel mechanisms of ion transport, synaptic transmission, and neurodegeneration. Furthermore, marine-derived alkaloids, terpenes, and peptides are showing promise in preclinical models of psychiatric disorders. Certain marine sponge metabolites have demonstrated antidepressant-like and anxiolytic effects, potentially through modulation of dopamine, serotonin, and norepinephrine systems. Marine polyphenols and polysaccharides, particularly fucoidans from brown seaweeds, exhibit neuroprotective, antioxidant, and anti-inflammatory effects, which may be useful in mitigating neuroinflammation, which is a common pathological feature of depression and neurodegenerative diseases. Despite these exciting developments and findings, several challenges remain that currently limit the broader application of marine natural products in neuropsychiatric therapy. First, the complexity of marine ecosystems and the often minute quantities of bioactive compounds make isolation and sustainable harvesting difficult. Many promising organisms are found in extreme or remote environments, such as deep-sea hydrothermal vents, polar regions, or coral reefs, which complicates both access and cultivation. Additionally, many marine compounds are produced by symbiotic microorganisms within their host organisms, and replicating these conditions remains a major hurdle. Technological advancements, however, are rapidly overcoming these limitations. Innovations in metagenomic sequencing and synthetic biology are enabling scientists to identify and clone biosynthetic gene clusters from unculturable organisms, allowing bioactive production in more tractable microbial hosts such as *E. coli* or yeast. Furthermore, machine learning and AI-driven drug discovery platforms are revolutionizing natural product screening by predicting pharmacological activity, toxicity profiles, and blood-brain barrier permeability based on molecular structure. Such tools could greatly accelerate the process from marine compound identification to clinical development, especially in the context of central nervous system drug discovery, which has traditionally suffered from high attrition rates. Another important consideration is the lack of in-depth mechanistic studies on marine compounds with neuropsychiatric potential. While some marine natural products have been well-characterized, many others are still in the early phases of discovery. To advance these compounds through the translational process, more rigorous and invasive studies are required. For example, determining how a marine compound influences neural circuits, neurotransmitter dynamics, or inflammatory cascades in relevant animal models of psychiatric disorders is essential for validating therapeutic potential. Moreover, safety profiles must be carefully assessed, particularly for compounds with neurotoxic origins. In addition to therapeutic applications, marine bioactives offer significant value as neuroscientific tools. Several marine toxins have been instrumental in advancing our understanding of neuronal signaling pathways, synaptic plasticity, and memory formation. Their high specificity for ion channels or intracellular signaling molecules makes them ideal for dissecting the complex pathophysiology of neuropsychiatric disorders. In this way, even if certain marine compounds never reach the clinic as therapeutics, their role in illuminating the biology of mental illness remains profound. Looking forward, the integration of multidisciplinary approaches, which combine marine biology, neuroscience, pharmacology, computational modeling, and systems biology, will be crucial to unlocking the full potential of marine-derived bioactives. Collaborative initiatives between marine ecologists, synthetic chemists, and clinical researchers can help bridge the gap between oceanic biodiversity and therapeutic innovation and discovery. Importantly, these efforts must be grounded in sustainability and ethical bioprospecting, ensuring that marine resources are explored responsibly and preserved for future generations. In conclusion, the marine environment represents one of the most promising and underexplored reservoirs for the discovery of novel therapeutics targeting neuropsychiatric disorders. The structural and functional diversity of marine natural products offers a unique opportunity to develop drugs with mechanisms of action that differ from conventional psychopharmaceuticals, potentially leading to more effective and better-tolerated treatments. While considerable challenges remain, which range from compound isolation and scalability to translational relevance and regulatory approval, the growing body of evidence suggests that marine-derived bioactives will play an increasingly important role in future mental health therapies. As the global burden of neuropsychiatric illness continues to rise, it is crucial that we look beyond the traditional terrestrial array of remedies and toward the vast, bioactive potential of our oceans and marine life.

Use of Marine-Derived Natural Bioactives in the Field of Diagnosis of Neurological and Neurodegenerative Disorders

4

Peter Weglarz, Suhrud Pathak, Muralikrishnan Dhanasekaran, K. Reeta Vijayarani, Raakin S. Ameen, and Mullaicharam Bhupathyraaj

Abstract

Marine ecosystems are a relatively unexplored and abundant source of natural bioactive chemicals with enormous promise for biological uses. Marine-derived compounds have recently received a lot of interest in their potential involvement in the diagnosis and treatment of neurological and neurodegenerative diseases. This chapter investigates the prospective value of marine bioactives as innovative diagnostic tools, emphasizing their specificity, sensitivity, and flexibility to evolving diagnostic demands. As the global prevalence of neurological conditions develops, there is a greater desire for more accurate, early-stage, and even tailored diagnostic treatments. Marine natural materials have unique structural and functional features that can help overcome this diagnostic gap. The chapter

P. Weglarz · S. Pathak
Drug Discovery and Development, Harrison College of Pharmacy, Auburn University, Auburn, AL, USA

M. Dhanasekaran (✉)
Drug Discovery and Development, Harrison College of Pharmacy, Auburn University, Auburn, AL, USA

Graduate School, Auburn University, Auburn, AL, USA
e-mail: dhanamu@auburn.edu

K. R. Vijayarani
Department of Pharmaceutics, Surya School of Pharmacy, Vikravandi, Villupuram – 605 652. Affiliated To The TN Dr. M.G.R. Medical University, Chennai, Tamil Nadu, India

R. S. Ameen
Senior at Thomas S Wootton High School, Rockville, MD, USA

M. Bhupathyraaj
College of Pharmacy, National University of Science and Technology, Muscat, Oman

M. Bhupathyraaj et al. (eds.), *Neuroprotective Marine Bioactives*, Nutritional Neurosciences, https://doi.org/10.1007/978-981-95-5791-2_4

underlines the necessity of further exploration and multidisciplinary study in unlocking the therapeutic and diagnostic potential of marine-derived chemicals. These studies may eventually lead to advances in diagnosing and controlling hitherto untreatable diseases, ushering in a new era in neuromedicine.

4.1 Introduction

For thousands of years, marine natural bioactives have been utilized by populations across the globe as therapeutic applications for a variety of different disorders. Particularly in traditional Mediterranean and Chinese medicine, invertebrate organisms were consumed to alleviate stress and symptoms of disorder. Broths, beverages, juices, and raw flesh from various forms of marine life were cultivated by natives and consumed for medicinal purposes (Catanesi et al. 2021). From ancient times, a multitude of marine species were acknowledged to play a role in one's health. Such included several invertebrates (mollusks, sponges, sea cucumbers, cone snails, etc) as well as forms of algae, specific types of Penicillium fungi, seaweed, and cyanobacteria. Even tuna eyeballs and shark/fish cartilage were considered useful remedies to the illnesses of men and women living on coastal lands to seas, and oceans.

While there is much still unknown about the medicinal value of much of marine life, naturally-derived bioactive products have been found to be an application of "pharmaceuticals, nutraceuticals, cosmetics, antibiotics, anti-fouling products, [as well as] anti-cancer research" (Romano et al. 2022), They have great pharmaceutical potential, serving as a promising source for discovery and developments in the advancement of health care at large. Many derivatives contain medicinal value and therapeutic potential for the prevention of a wide variety of disease states, exhibiting antifungal, antibacterial, antiviral, and antitumor properties depending on the specific drug and disease state (Santhiravel et al. 2025). Such properties are exhibited on a molecular level, where marine organisms produce chemical compound byproducts as a result of a necessary function or response to their environment. For example, sodium oligomannate is a marine-derived oligosaccharide produced from brown algae (*Ecklonia kurome*). It has been studied in Chinese laboratories to "reduce cerebral amyloidosis and reactive microglia in a sex-specific manner" and has also proven to be an effective supplement when consumed orally to improve one's gut microbiome (Bosch et al. 2024). According to another study, the effects of the oligosaccharide were investigated where a group of individuals with Alzheimer's disease would be orally administered the supplement. The control group that wasn't administered sodium oligomannate demonstrated poorer levels of cognitive decline than those who ingested the supplement. Thus, this compound suggests neuroprotective effects against cognitive decline (Syed 2020).

The diversity and adaptability of marine organisms suggest that their natural products may demonstrate great efficacy as preventative, diagnostic, and treatment and adjuvants to neurological diseases. Marine life has been forced to adapt to extreme environmental conditions, some residing at great depths that are highly pressurized and void of light. Other environmental challenges include extreme temperatures (organisms that live near hydrothermal vents at the bottom of oceans, while others adapt to glacial environments). All of these adverse conditions suggest a need for survival and a willingness of such organisms to evolve where they can live in previously uninhabitable conditions.

Since much of marine life has been forced to adapt in unfavorable environments, there are several bioactive compounds and secondary metabolites that are produced through a variety of organismal functions, species to species. *Conus magus,* commonly known as a marine gastropod (cone snail), evolved to produce venom (conotoxin) as a defense mechanism in response to predators. The class of *Tetraodontidae* (pufferfish) also developed a similar mechanism, producing tetrodotoxin while also retaining the ability to ingest water from its environment to swell its shape. These protective adaptations saved both species from becoming easy prey to their respective voracious adversaries.

Neurodegeneration can be defined as the "neuronal loss of function, death and aggregation of misfolded proteins." This results in "the excessive production of extracellular and intracellular deposits" (Catanesi et al. 2021). An increase in reactive oxygen species (ROS), inflammation, cerebrospinal fluid levels, oxidative stress, and nucleic acid oxidation (causes DNA damage and development of diseases like Alzheimer's) are all associated problems with a disruption of balance in the central nervous system, as well as the progression of neurodegenerative disorders. The benefit of marine-derived natural products, however, is that they display great efficacy in combating oxidative stress and are effective in eliminating free radicals and reactive species. In doing so, they may become extremely useful in the prevention of neurological disorders upon further research.

Throughout this chapter, the neuroprotective and therapeutic effects of the marine natural bioactive Bryostatin will be discussed in conjunction with a variety of neurological and neurodegenerative diseases. The biochemical mechanisms and interactions of this drug will be reviewed and analyzed as an effective adjuvant or treatment in the prevention and diagnosis of Huntington's disease, prion diseases (transmissible spongiform encephalopathies), and cerebral amyloid angiopathy. These disease states will be compared to other neurological disorders where bryostatin displayed a significant improvement in cognitive ability corresponding to its associated disorder.

4.2 What Classifies as a Neurological/Neurodegenerative Disorder?

While neurological and neurodegenerative disorders at a glance may appear to be in the same classification of disease state, this isn't necessarily the truth. Neurological disorders are broad and arise from an injury to the brain, spinal cord, or nerves throughout the body (epilepsy, stroke, migraines, traumatic brain injury, etc.). Such dysfunction in the peripheral or central nervous system is indicative of a neurological (or neurodegenerative) disorder. In fact, neurodegenerative disorders are a subset of neurological disorders, characterized by a progression of impaired neuronal function (Lamptey et al. 2022). This can result in more limited nerve impulse conduction and brain activity. Examples of common neurodegenerative conditions include Alzheimer's disease, Parkinson's disease, multiple sclerosis (MS), and Lewy body dementia. Cause for neurodegeneration often coincides with elevated age, mutations in genes, as well as environmental conditions (voluntary or involuntary exposure to chemicals, toxins, pollution, or even lower levels of sunlight).

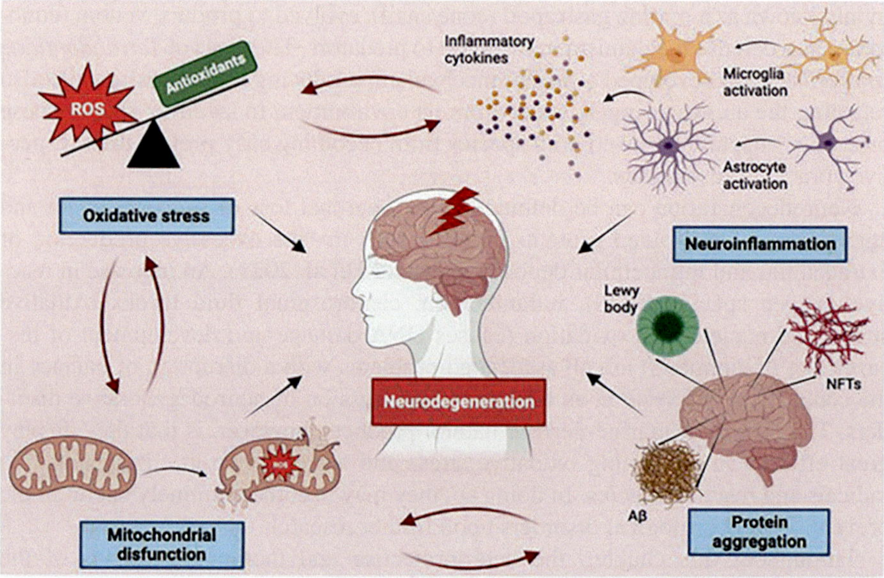

Alzheimer's disease and Lewy body dementia primarily affect one's memory and behavior. Cognitive decline in Lewy body dementia is a result of the accumulation of abnormal protein deposits of alpha-synuclein in the brain (such protein deposits are called Lewy bodies). While there is no concrete evidence as to what specifically predisposes someone to develop Alzheimer's disease, those with accumulations of abnormal proteins, genetic mutations, or elevated age are at higher risk. Parkinson's disease inhibits proper motor function, with patients experiencing stiffness, tremors, and slower movement, while multiple sclerosis is an autoimmune disease where demyelination in neurons slowly impairs movement, vision, balance, and cognitive ability over time. All of these conditions share a few things in common. They are all chronic and progressive (symptoms worsen over time), typically arising from some protein misfunction that affects the action potential of synapses from occurring.

Thus, many therapeutic and diagnostic options must be considered in the treatment of such aggressive disorders. Although there is currently no cure for the neurodegenerative conditions listed above, there have been several advancements in medicine to limit the neuronal damage and symptoms experienced by patients. One source of treatment comes from the "theranostic applications of nanoparticles in neurodegenerative disorders," which combine therapeutic and diagnostic applications of drugs (Ramanathan et al. 2018). A "targeted drug delivery system" is developed to penetrate the blood-brain barrier, manipulating nanoparticles to attach themselves to suitable drugs for treatment. While there is evidence suggesting the efficacy of nanoparticles in conjunction with neurodegenerative drug therapies, the technology is not yet universal or commonly used.

Current modern treatments for neurodegenerative disorders vary depending on the disease state. In Alzheimer's disease, for example, cholinesterase inhibitors (donepezil, rivastigmine, and galantamine) are used to slow the progression of the illness by boosting acetylcholine levels in the brain. These pharmacodynamic actions are effective in slowing cognitive decline in patients. Similarly, treatments for Parkinson's disease have the function of increasing hormone concentration, where levodopa restores the level of dopamine in the substantia nigra while preventing neuron loss. Such therapeutic options have demonstrated success in reducing symptoms but have differing effects patient to patients. Adverse effects caused by drugs in an effort to combat neurodegenerative diseases have been observed, some resulting in more harm/discomfort than in individuals only suffering from the disease.

For example, levodopa is a pharmaceutical administered orally that demonstrates great therapeutic potential in the treatment of Parkinson's disease. In penetrating the blood-brain barrier and alleviating motor symptoms, however, long-term complications may arise with the administration of the drug. The toxicity of levodopa can arise over time, resulting in damage to neuronal cells. Common adverse effects include "nausea, dizziness, headache, somnolence..., low serum vitamin B12, elevated methylmalonic acid levels..., neuropathy..., [and] cardiac arrythmias" (Gandhi and Saadabadi 2023). It is also known that "Abrupt withdrawal or dose reduction of levodopa is associated with an increased risk of neuroleptic malignant syndrome (NMS)" (Gandhi and Saadabadi 2023). Carbidopa and domperidone are inhibitors used to prevent the effects of nausea and dizziness. They can be administered orally alongside levodopa to reduce symptoms. Additional adverse effects (hallucinations and psychosis) may be present in older patients.

The nature of these undesirable and harmful effects suggests a need for better alternative therapeutic agents and diagnostics. Current pharmaceuticals, while proving to have significant benefits in the treatment of neurodegenerative diseases, also maintain a variety of shortcomings that can negatively impact the health of an individual, possibly advancing their impairments. Thus, the effects of natural bioactives must be explored in relation to neurodegeneration. Marine natural bioactives specifically exhibit pharmacological potential, and their mechanisms of action must be investigated.

4.3 Ancient Medicinal and Diagnostic Use of Marine Life: What Can We Expect in the Future?

Marine life has been utilized for its medicinal values in several cultures and societies throughout human history. "Tax records generated in China in the year 2953 BCE" suggest fish-derived use of medicine for treating infertility and baldness, Roman Emperor Claudius "recommended the discharges of electric fish (*Torpedo nobiliana*) to cure migraine and headaches," while ancient Greek societies recognized "the antibiotic effects of certain sponges" and "the benefits of apply brown algae for treating inflammation" (International Society of Ethnobiology 2013). A variety of seaweed-based remedies, derivatives from mollusks, algae, crustaceans, sponges, echinoderms, and fish have all been experimented with for the treatment of several disorders.

There are also countless traditions since humanity's origin that have encouraged the consumption of marine life based on the sustenance and health they provided for several communities. This has spanned across the globe since the early formations of coastal tribes and civilizations, which utilized marine organisms as sources of energy and a focal point of some spiritual practices. Many of these traditions still stand the test of time. For instance, in Hawaiian culture, sharks are deeply revered as protectors of the sea while turtles symbolize peace, wisdom, and longevity. The ocean itself is seen as a place of cleansing, offering spiritual connection. Forms of worship developed in response to the ocean's vastness and powerful nature, serving as a provider to man who lived in its vicinity. Such spiritual beliefs were likely intertwined with a primal desire to hunt and travel, where tribes would develop fishing tactics, build boats to serve as methods of transportation, or for trade and commerce.

The customs and traditions responsible for the discovery of organisms with great medicinal value and potential can be attributed to humans' innate abilities of observation. In doing so, the interactions of humans and marine life have contributed greatly to many advancements in medicine. In fact, the study of ethnobiology (an interdisciplinary field that observes the relationships between humans, marine organisms, and their environments) continues to grow as marine life is further explored. Its vibrancy is enhanced by recent developments in society, where "common glassworts" were used as diuretics (increase urea production of kidneys) in the Netherlands (2007), shrimp species have been discovered to contain derivatives combating asthma (1999 in Brazil), and red mangroves were found to display efficacy against a myriad of ailments (International Society of Ethnobiology 2013). The pharmacopeia of marine natural bioactives is constantly expanding as research in ocean life becomes more advanced. The vast majority of Earth's oceans are unexplored, with new organisms and life forms being discovered every day. It is likely that several marine organisms not yet identified or commonly studied could exhibit great potential as pharmaceutical agents in an environment where the significance of studying marine life is better understood.

There is still little known about the survivability of deep-sea organisms, where evolution has encouraged the development of several abstract features that are rarely found in terrestrial life or organisms that inhabit shallow marine environments

(National Research Council (US) Committee on the Ocean's Role in Human Health 1999). Such organisms exhibit traits of bioluminescence, elongated appendages, buoyancy control, pressure tolerance, and adaptations in temperature that allow them to reside in hypothermal vents and freezing conditions at the bottom of the ocean. The evolutionary methods are widely unknown for most organisms since the research on these forms of marine life is extremely limited.

A common protein currently used as a selective marker/reporter when studying genes and protein localization is the GFP (green fluorescent protein). It originates from a deep-sea jellyfish species, *Aequorea Victoria,* demonstrating green fluorescence as it lives in the depths of the ocean (Chalfie 1995). The GFP protein has become a source of research throughout several organisms, where a protein may be introduced into a particular genome to study gene expression and cell behavior. Yeast is a common organism used in multiple studies where a plasmid may be used to insert a gene of interest paired with a GFP protein. A GFP-fused yeast would then be compared to a wild-type form, where stress conditions could be tested to determine whether the particular gene of interest is upregulated or downregulated.

The existence of the GFP protein emphasizes the necessity for more exploratory methods of research on marine life. Being a somewhat recently discovered species, the known existence of organisms like *Aequorea Victoria* (which is now a diagnostic when measuring levels of gene expression, sometimes in correspondence to the diagnosis of certain diseases) highlights a great importance in furthering research and exploration. There are so many unknown species waiting to be discovered, some of which could demonstrate great pharmacological potential as therapeutic and diagnostic agents. Such discoveries could be especially useful in combating the effects of neurodegenerative diseases, where medicine derived from marine life is limited. While there are some current options involved in these processes, further implementations to study marine life should be developed internationally for discovery in medicine.

4.4 Bryostatin: Where Does It Come From? Why Is It a Potential Pharmacological Agent?

A current potential therapeutic option in the treatment of neurodegenerative disorders originates from a small group of colonial organisms found in aquatic environments across the world. Bryostatin is a marine natural bioactive found in many small invertebrates in freshwater and saltwater environments. It is effective in penetrating the blood-brain barrier (BBB) despite its bulky chemical structure. They are a family of macrolide lactones, most notably produced by the bryozoan species *Bugula neritina,* an invasive sessile marine animal found in tropical waters across the globe. Such organisms are colonial filter feeders, surviving off of phytoplankton, detritus, and other microscopic bacteria and protozoa. "Bryostatins are a family of protein kinase C modulators that have potential applications in biomedicine" (Trindade-Silva et al. 2010). Small amounts of bryostatin analogs have been found present in many different bryozoan species. It is suspected that the presence of bryostatin is sourced from microbial symbionts from which the chemical compounds were isolated into the invertebrates.

"The first bryostatins were discovered using an antineoplastic bioassay-guided fractionation approach," as shown in the image below (Trindade-Silva et al. 2010). Novel bryostatins were detected by chromatography and mass spectroscopy. "Bryostatin 0 is the hypothetical precursor to all the known bryostatins, where compounds which contain octa-2,4-dienoate moieties are found only in the deep *B. neritina* sibling species, whereas shallow *B. neritina* contains bryostatins with other kinds of esterifications. The oxygen groups highlighted in red represent the pharmacophoric elements involved in binding to PKC" (Trindade-Silva et al. 2010).

Bryostatins have been utilized throughout cancer research and most especially in the treatment of HIV. They have proven to have a significant effect on the proliferation and activation of CD8+ T cells, "eliminating reactivated virus and achieving a cure" in patients with acute infections, with cases of chronic HIV (Li et al. 2025). The cytotoxic T lymphocytes typically become exhausted in cases of cancer, where bryostatin has been shown to improve immune function in response to disease. The compound bryostatin-1 is most responsible for these results, restoring CD8+ T cell ability while also contributing to several other important mechanisms in disease management and progression. Bryostatin-1 is the most promising analog as a potential pharmacological/therapeutic agent because of its hydrophobic alkyl chain on C-20, where bryostatins attach to the "diacylglycerol binding site of the C-1 regulatory domain of protein kinase C (PKC)" (Trindade-Silva et al. 2010). In doing so, brystatin activates the PKC enzyme to undergo a variety of regulatory cellular functions. Specifically in the Central Nervous System, PKC plays an important role in the formation and modification of synapses (synaptic plasticity). It modulates neurotransmitters, which are also involved in stimulating neuronal survival and the prevention of memory loss. "Activation of PKC requires the increase in [Ca2+]i and DAG and/or cis-unsaturated fatty acids" (Huang and Huang 1993), where PKC regulates ion channels important in neuronal signaling.

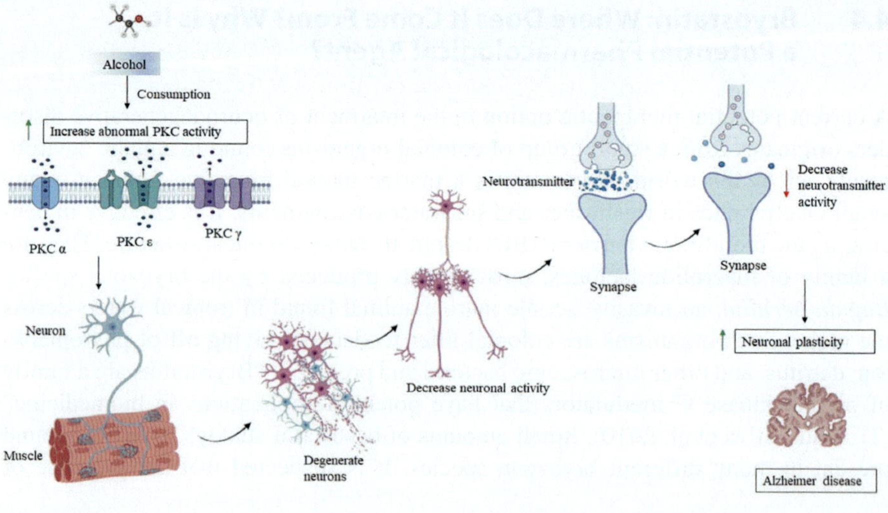

Since the role of PKC is so intertwined with the correct functionality of the central nervous system, it is of utmost importance that this particular enzyme is not dysregulated in the management of neuronal function and synaptic action potential. Thus, the ability of bryostatin to activate and induce proper PKC enzymatic activity demonstrates great pharmacological potential as a treatment option for neurodegenerative diseases, such as Alzheimer's disease (a disorder caused by dysfunction in the central nervous system). While poor availability of bryostatins has hindered its clinical development, new techniques designed for the isolation of the natural compound could be implemented. The complexity of the chemical compound has made it challenging and expensive to synthesize. Its potential as a viable pharmaceutical for a host of antiviral, anticancer, and neurodegenerative disorders, however, suggests a need for more effective extractions of the compound to perform further research.

4.5 Huntington's Disease: Diagnosis, Prevention, Symptoms, and Current Treatment

Huntington's disease is the first of three disorders to be investigated. Like Alzheimer's and Parkinson's disease, it is a chronic neurodegenerative disease that is characterized by the progression of cognitive decline, behavioral changes, and the induction of involuntary movements. It is a genetic condition (an autosomal dominant disorder) that "occurs as a result of cytosine, adenine, and guanine (CAG) trinucleotide repeats on the short arm of chromosome 4p16.3 in the Huntingtin (HTT) gene. This mutation leads to an abnormally long expansion of the polyglutamine in the HTT protein, which leads to neurodegeneration" (Ajitkumar and De Jesus 2023). The expansion of polyglutamine also makes the HTT protein more prone to aggregation, thus accelerating the progression of the disease. The proliferation of this genetic mutation determines the extent of the disorder.

To diagnose the disorder, a blood sample must be taken from a patient, which is processed and analyzed through genetic testing for the presence of CAG repeats (the genetic mutation caused by Huntington's disease). If the mutation is present, the patient will be diagnosed with the disorder. Presymptomatic testing can be utilized if Huntington's disease is suspected to have familial ties. Clinical evaluations may also be performed to test the cognitive fragility and movement of the patient. MRIs and CT scans of the brain display further degradation and prove to be another effective diagnostic for Huntington's disease. The most confirmatory method of diagnosis, however, is genetic testing.

Huntington's disease is most prevalent in individuals aged 30–50 years old (Roos 2010). There are severe cases where early onset of the disease can result in behavioral and learning disabilities in juveniles (known as juvenile Huntington's disease) (Roos 2010). After initial diagnosis, the typical lifespan can range from approximately 15–30 years, depending on the age at which symptoms first arose. The younger the patient is diagnosed, the more aggressive the disease and the shorter their lifespan tends to be.

The main associated symptoms include chorea (hallmark involuntary movements), dystonia (severe muscle spasms and abnormal postures), and loss of coordination/motor function, along with the development of mental disorders such as depression and anxiety. These also tend to progress as the patient ages. There is currently no cure for Huntington's disorder. While there are currently no preventative methods to stop disease onset or progression, effective treatments exist in the management of the disease. The primary purpose of these therapies is to reduce associated symptoms of the disorder and improve quality of life.

Pharmacological agents are utilized to target motor, behavioral, and cognitive dysfunction, respectively. "The most commonly used chorea drugs are antipsychotics and tetrabenazine (TBZ)" (Videnovic 2013). The mechanism of tetrabenazine is used to inhibit the vesicular monoamine transporter 2 (VMAT2). This reduces levels of dopamine and serotonin (as well as several other monoamines) reuptake into the brain, ultimately helping to limit involuntary movements caused by neural degradation. This is performed through the utilization of the "CYP 2D6 isoenzyme," which is involved in the metabolism of TBZ (Yero and Rey 2008). The FDA-approved pharmaceutical has promising but limited effects, significantly decreasing chorea scores by "42% with a standard deviation of 17.8%" in individuals with Huntington's disease from a study group run by the "Unified Huntington's Disease Rating Scale (UHDRS) and the Beck Depression Inventory (BDI)" (Yero and Rey 2008). Associated results also displayed a significant increase "in reports of drowsiness and insomnia" (Roos 2010). Along with this came adverse effects of mood swings. While proving to be an effective pharmacological agent, patients with depression or other psychological disorders are generally advised not to use the drug as treatment since it could potentially worsen symptoms of mental illness.

Antidepressants such as Serotonin reuptake inhibitors (SSRIs) are utilized to regulate behavioral function (mood swings, depression, anxiety, etc.). While no official treatments are addressing the cognitive decline of patients with Huntington's disease, "positive allosteric modulators" of "N-methyl-d-aspartate receptors" such as SAGE-718 have been under investigation for potential pharmacological action in slowing the progression of cognitive decline (Hill et al. 2022). All pharmaceuticals are administered orally in the management of Huntington's disease.

The most common causes of disease are pneumonia, followed by heart conditions, and suicide. Dysphagia (swallowing impairment) may be induced by low motor function, leading to aspiration of food or liquids entering the lungs instead of the esophagus. In doing so, food particles can get trapped, creating an infection from an accumulation of bacteria in the bronchial cavity. Ultimately, individuals with Huntington's disease will eventually succumb to their illness as the disorder will severely impair neuronal and motor function until they can no longer perform vital functions to survive.

4.6 Bryostatin as a Therapeutic and Diagnostic Agent for Huntington's Disease

While Huntington's disease is a chronic and progressive disorder, new innovative medications and technologies can be explored to manage symptoms and improve quality of life. Such can be sought out in marine natural bioactives, where bryostatin can be used as a therapeutic drug to reduce neurodegeneration in individuals with the disorder.

Since Huntington's disease is a genetically inherited condition, bryostatin may not be used for diagnostic or preventative purposes since it can't manipulate or detect one's gene expression. It does, however, possess biochemical properties that can be applied to mechanisms designed to target positive responses in correspondence to disease management. Specifically, in the instance of cognitive decline, bryostatin could be used as a pharmaceutical agent in reducing symptoms of dementia induced by Huntington's disease. Since it is used to target the PKC enzyme, it activates proper functionality, which has been degraded in patients with neurodegenerative disorders (such as Alzheimer's and Parkinson's disease). In a small clinical trial testing the cognitive and physical ability of Alzheimer's disease patients who were administered bryostatin, all experienced a "return of some speech, increased attentional focus, and the ability to swallow" (Denvir et al. 2015). While these improvements were not long-lasting, they represented a temporary improvement in the functionality of the PKC enzyme and some respective processes of the central nervous system. The dose of bryostatin was also meager, matching most other studies involving bryostatin-1, where a concentration of 10 nM was applied via intravenous infusion. Regardless, the improvements in cognitive and motor function are notable and significant, suggesting advancements in research involving bryostatin-1 as a treatment for Huntington's disease as well as other neurodegenerative disorders.

Bryostatin has also demonstrated positive results in promoting the production of BDNFs (brain-derived neurotrophic factors), which are essential in supporting growth, learning, and the formation of long-term memory. They act as "paracrine and autocrine factors," present on "both pre-synaptic and post-synaptic target sites," mediating structural plasticity in the central nervous system while reducing oxidative stress (Colucci-D'Amato et al. 2020). Properly functioning protein kinase C epsilon enzymes play a major role in the expression of BDNFs and their corresponding levels of synaptic density. When these enzymes are downregulated in cases of Huntington's disease, the functionality of BDNF is severely compromised, negatively influencing the learning, memory, and mental status of a patient.

Bryostatin-1, a specific PKC epsilon activator, however, has demonstrated great potential in maintaining neuronal plasticity in patients beginning to display these disabilities. Significant positive changes in this enzymatic function have been studied to exhibit strong correlations in reducing neurodegeneration (Sen et al. 2018). While never explicitly tested in relation to Huntington's disease, there is evidence to suggest an improvement in the symptoms of a patient carrying the disorder.

Although bryostatin-1 may not present itself as a great pharmaceutical in curing Huntington's disease, it demonstrates neuroprotective potential in limiting progression and thus an improvement in patients' quality of life and duration of lifespan. While further research must be conducted to prove these assumptions concretely, the encouraging therapeutic activities displayed by the bioactive are remarkable when used as treatment in other disease states involving neurodegeneration. Use of the marine-derived agent can help improve synaptic plasticity, increase response to neurotransmitters in signal transduction, elevate neurotrophic factors, aid in the activation of transcription factors important for neuronal survival, and correct abnormal PKC signaling in Huntington's disease. There has also been little to no report of adverse effects in relation to bryostatin-1 treatment, further suggesting a need for future studies and trials to be conducted.

4.7 Prion Diseases: Diagnosis, Prevention, Symptoms, and Current Treatment

Prion diseases (or transmissible spongiform encephalopathies) are a type of rare, fatal brain disorder caused by the accumulation of abnormal, misfolded prion proteins. This aggregation of misfolded proteins causes other normal proteins to misfold, thus proliferating abnormal prions that function irregularly. More specifically, "prions reproduce by associating with normal PrP cellular isoforms, converting α-helices into indigestible β-pleated sheets" (Sitammagari and Masood 2025). These particles accumulate and are the ultimate cause of disease. The disorder is progressive, spreading rapidly once an abnormal prion begins to proliferate, creating sponge-like holes in the brain of a patient and almost always resulting in death. The onset of disease, however, typically isn't noticed until a patient starts feeling symptoms, and the abnormal prion can incubate in one's body for decades before causing neurodegeneration. Thus, prion diseases are considered neurodegenerative disorders, "caused by transmissible particles that contain a pathogenic isoform of the prion protein, a normal constituent of cell membranes" (Johnson 2005). They can be found and transmitted by animals and humans alike, but for the purposes of this discussion, only prion diseases present in humans will be addressed.

Several types of prion diseases are acquired and transmitted in their way. Creutzfeldt-Jakob disease (CJD) is the most common human prion disorder, typically originating with no traceable cause for neurodegeneration, where infectious prion proteins (while not considered viral or bacterial) sporadically spread throughout the brain. It primarily affects the central nervous system, damaging neurons and swelling astrocyte cells to degrade their functionality. This results in irreversible damage and severe cognitive decline in patients with the disease. There is, however, a variant of CJD (vCJD) that has been linked to the consumption of meat from "cows with mad cow disease" (mostly in the UK) (de Villemeur 2013). This form of CJD primarily affects younger populations, where the "median age at death is 28 years," and an individual has fallen ill between 13 and 14 months prior (Sitammagari and Masood 2025). CJD primarily onsets in the older population; however, patients first experience symptoms of the disease in their mid-50s to mid-70s on average. This form of prion disease has a very poor prognosis, with death occurring in nearly 70% of patients within a year" (Sitammagari and Masood 2025).

Gerstmann-Sträussler-Scheinker syndrome (GSS) is an inherited form of prion disease that is caused by a mutation in the PRNP gene (which encodes for prion proteins). It is an autosomal dominant disorder, where ataxia (lack of balance and coordination), dementia, and hearing loss can result from neurodegeneration. Dysarthria (speaking difficulties) and nystagmus (involuntary eye movements) may also be present among individuals experiencing further progression of the disease (Shin et al. 2023). The average lifespan is similar to CJD, where an individual typically lives for about a year after their initial onset of symptoms.

Another form of prion disease, called kuru, has been transmitted through the practice of cannibalistic rituals in Papua New Guinea. In fact, "the word 'Kuru' means to tremble due to cold or fever" (Mahat and Asuncion 2023). Like CJD and GSS, it is a non-inflammatory neurodegenerative condition with a similar prognosis, symptoms, and rapid onset of disease. The general symptoms and lifespan of individuals with each disorder are generally the same.

There are several different ways to diagnose prion diseases, including clinical assessments (cognitive and motor function is observed), MRI and EEG scans (show sporadic changes in the brain), cerebrospinal fluid analysis (where elevated levels of prion protein in CSF can indicate disease), as well as lumbar punctures (a spinal tap). The main method of diagnosing a prion disease, however, is by performing a biopsy of brain tissue. Genetic testing can also be utilized when determining a patient with GSS. While there are no current effective preventative methods that can halt the onset of disease progression, therapeutic options are available to lessen symptoms and improve quality of life.

There are no current treatments that can cure Prion diseases. There are, however, pharmaceutical drugs that display neuroprotective effects in slowing the rate of degeneration. Pentosan polysulfate (PPS), for example, is "a large polyglycoside molecule" that can "act in competition with endogenous heparin sulphate proteoglycans as co-receptors for PrP on the cell surface" (Panegyres and Armari 2013). In a study where the chemical compound was "directly administered into the cerebral vesicles" of prion-infected mice, PPS significantly prolonged the onset of disease and inhibited the formation of abnormal prion proteins in cell culture (Panegyres and Armari 2013).

In another study, "a 68-year-old patient with sporadic CJD" was administered a "continuous intraventricular PSS infusion" at a concentration of "1–120 μ/kg/day" (Panegyres and Armari 2013). While the early symptoms didn't improve, there was a "stabilization period" where neurodegeneration was not observed for ~4 months. The patient also survived 2 years longer than the typical prognosis of a year. A younger individual with vCJD was administered the same treatment, surviving over 4 years with the condition and experiencing less progressive rates of degeneration.

These therapeutic effects are incredibly promising, where other polycyclic and polyanionic drugs (quinacrine, thioflavine, amphotericin B, and tetracyclines) have shown efficacy in preventing the proliferation of diseased prion proteins (bryostatin demonstrates therapeutic potential by the nature of its polycyclic form) (Panegyres and Armari 2013). Such compounds may be effective in reducing prion accumulation in the brain and thus slowing overall progress. Other methods of treatment include immunotherapies and the introduction of transcription factors to help alter gene expression in response to prion production. These techniques are not well researched, however, with little to no experimental data showing any significance of a therapeutic effect. The most common cause of death in patients with prion disease involves respiratory failure, complications due to an infection, or pneumonia. These are all a result of severe neurodegeneration.

4.8 Bryostatin as a Therapeutic and Diagnostic Agent for Prion Diseases

As mentioned in the previous section, bryostatin-1 is a polycyclic chemical analog that demonstrates potential neuroprotective effects against the proliferation of abnormal prion proteins. While never explicitly tested in patients with prion diseases, the pharmacological actions it displays may be effective in slowing the aggregation of abnormal proteins, and thus the spread of the disorder. For instance, bryostatin-1 has been identified as a functional antagonist of phorbol ester compounds, where phorbol esters are used to regulate the cleavage of normal prion proteins. In cases of disease, such compounds are utilized in the proliferation of abnormal prions, resulting in rapid neurodegeneration in patients (Vincent et al. 2000). Bryostatin-1 may be effective in reducing this spread; however, it has demonstrated a mechanism of preventing phorbol ester formation. This may result in the potential to disrupt the process of abnormal prion production.

Bryostatin-1 may also be a useful therapeutic option in the progression of prion diseases due to the effects it has on activating PKC enzymes. Similar to Huntington's disease, prion diseases mainly impact the central nervous system, causing a host of symptoms to arise due to neurodegeneration. In doing so, the PKC enzyme is dysregulated, causing neuronal death, vascular damage, and impaired function of the blood-brain barrier. PKC enzymes are primarily responsible for inducing and regulating synaptic plasticity. Since the aggregation of prion proteins lessens synaptic density within the brain of a diseased patient, cognitive and motor impairments can arise.

In fact, the cleavage of prion proteins is regulated by proper PKC function (Alfa Cissé et al. 2008). Bryostatin-1 has the neuroprotective effect of activating and preserving this enzymatic function, helping to promote a healthy regulation of neurons in the brain. While PKCs may be largely dysfunctional in patients with prion diseases, bryostatin-1 may be a useful therapy in combating such dysfunction, restoring neuron viability in the brain (or at the very least, preventing rapid onset of disease and slowing overall progression).

Although prion diseases are incurable neurodegenerative disorders, natural marine derivatives such as bryostatin-1 offer a unique therapeutic option in slowing the progression of the disease. They may also be used as a diagnostic tool, detecting the progress of the disorder based on their functionality in reducing abnormal prion proteins. Its abilities as a polycyclic compound and PKC activator display great pharmacological potential as a treatment option for patients with prion diseases. While the research in such disorders is limited in correspondence to the treatment potential of bryostatin-1 compounds, there is considerable evidence to suggest that bryostatin-1 may exert neuroprotective effects by modulating PKC pathways. In activating specific PKC isoforms, bryostatin-1 can promote synaptogenesis, reduce neuroinflammation, and enhance cognitive function, mechanisms that could be beneficial in the context of prion-induced neurodegeneration. Although clinical applications are still in the early stages, preliminary findings underscore the potential of bryostatin-1 as a therapeutic candidate for slowing or mitigating the progression of prion diseases.

4.9 Cerebral Amyloid Angiopathy: Diagnosis, Prevention, Symptoms, and Current Treatment

Cerebral amyloid angiopathy (CAA) is a neurological and neurodegenerative condition caused by an aggregation of amyloid-beta proteins "within leptomeninges in small and medium-sized cerebral blood vessels" (Kuhn and Sharman 2025). This chronic cerebrovascular disorder causes fragile vessels to form as a result of amyloid deposition, causing lobar intracerebral hemorrhages, cognitive decline, "incidental microbleeds, hemosiderosis (excessive accumulation of iron), [or] inflammatory leukoencephalopathy" (Kuhn and Sharman 2025). The condition can also coexist with Alzheimer's disease, where patients may display signs of dementia but at a much more gradual rate. It is typically sporadic or a genetically related disorder (more commonly sporadic). The beta amyloid deposits weaken blood vessel walls, making microbleeds commonplace in patients with CAA.

Older patients and those with larger hematomas are at greater risk of developing CAA and typically have the worst outcomes. In fact, familial CAA, also known as "presenile CAA" (most common in older patients), is caused by a mutation in the amyloid precursor protein (APP) gene (Kuhn and Sharman 2025). This genetic malfunction is the primary cause of neurodegeneration in patients with CAA, promoting the production of pathogenic "amyloid-β (Aβ) peptides" (known as senile plaques) (Yuksel and Tacal 2019). Sporadic CAA has been associated with the "dysregulation of lipid metabolism," where malfunctions of apolipoprotein E (APOE) catalyze neurodegeneration (Yang et al. 2023). When functioning properly, they are variants of lipid transport proteins. When abnormal APOE epsilon 2 or APOE epsilon 4 alleles are present in patients, they can express a greater risk for cranial hemorrhages in those suffering from CAA.

The fragility of blood vessels caused by the onset of amyloid plaque accumulation can lead to a host of symptoms and discomfort in diseased patients. The size and location of hemorrhages in the brain often determine the severity of neurodegeneration, as asymptomatic forms of the disorder also exist. The most common and advanced symptoms include hypertension, transient focal neurological episodes (brief seizures), slow development of cognitive decline, headaches, nausea/vomiting, and focal neurological deficits. These greatly impair the life of an individual inhabiting the disorder, where the progression of the disease corresponds to worse motor and cognitive function.

If CAA is not promptly treated, it can result in rapid morbidity and mortality. A diagnostic method called the "Boston Criteria" is implemented to evaluate patients suspected of having CAA. It utilizes clinical, pathological, and radiographic findings to determine a "probable diagnosis" of CAA (Charidimou et al. 2022). Concrete diagnoses are not possible without a postmortem examination of the brain. Gradient-echo magnetic resonance imaging techniques are often used to identify CAA. This specific type of MRI is utilized because it can detect iron deposits that may have been left by old hemorrhages in the brain that were caused by microbleeds (which is the case in the images to the left). Findings of this may strongly correlate with symptoms of CAA, and thus, this form of resonance imaging serves a very useful diagnostic purpose.

While there are no current preventative methods/cures for CAA, several therapeutic options exist in slowing the neurodegeneration of individuals with the disease. These pharmaceuticals intend to reduce brain bleeds, lowering hypertension to lessen the risks of recurring intracranial hemorrhages. Since high blood pressure increases the likelihood of developing a hemorrhage, it is important to use antihypertensive medications (ACE inhibitors and ARBs) to maintain a safe range (Piepho 2000). Anticoagulant drugs and antiplatelet agents, such as warfarin and aspirin, respectively, must be avoided to prevent the risk of bleeding. Immunosuppressants can also demonstrate neuroprotective effects, targeting amyloid deposits and limiting levels of inflammation in patients suffering from CAA. In cases where Alzheimer's disease is concurrent with CAA, cholinesterase inhibitors may also be utilized in treatment.

While there are many current useful pharmaceuticals designed to manage the effects of CAA, their adverse effects cannot be overshadowed. ACE (angiotensin-converting-enzyme) inhibitors, for example, are used to treat hypertension in patients with CAA. Although these drugs are effective in the management of disease, adverse effects, causing hypotension and hyperkalemia (increased levels of potassium in the blood), can lead to further complications such as "cardiac arrhythmias, muscle weakness, and paralysis" (Simon et al. 2025). Structural properties of ACE inhibitors also display toxicity, where adverse effects may be related to the binding group of the compound (Piepho 2000). Such poor pharmacological actions demonstrated by current neuroprotective agents encourage a need for more natural therapeutic and diagnostic mechanisms that can be efficacious in treating and managing symptoms of CAA.

4.10 Bryostatin as a Therapeutic and Diagnostic Agent for Cerebral Amyloid Angiopathy

As discussed with previous disease states, bryostatin-1 is a successful modulator and activator of the PKC enzyme used in the regulation of neuronal activity and the rate of synapse formation. It has demonstrated great neuroprotective potential in the treatment of Alzheimer's disease, enhancing synaptogenesis, reducing amyloid production, and regulating neuronal survival. In doing so, this therapeutic agent has been effective in regulating symptoms associated with poor motor and cognitive function.

CAA is known to coexist with Alzheimer's disease, with patients often experiencing similar levels of neurodegeneration (this is especially evident in their cognitive abilities). In fact, CAA and Alzheimer's disease are both caused by amyloid protein accumulations, developing deposits in the walls of blood vessels (making them thin/fragile) and aggregating extracellularly as plaques, respectively (Greenberg et al. 2020). This makes bryostatin-1 an intriguing pharmaceutical in the treatment of CAA since it could potentially be used adjunctively in the management of Alzheimer's disease. In fact, bryostatin-1 has been known to inhibit amyloid-beta protein production throughout a variety of different neurodegenerative disorders. Although the specific amyloid proteins targeted would be slightly different in the case of CAA (Aβ40 subtype), the general mechanistic actions of bryostatin-1 would likely remain the same, with the bioactive displaying phagocytic actions to amyloid depositions.

Bryostatin-1 may also be effective in regulating the effects of the APP gene responsible for producing amyloid precursor proteins, shifting the cleavage of APP precursors and preventing amyloidogenic pathways from prevailing. This mechanism could limit excessive formations of the protein, reducing the progression of the disease. Since CAA is associated with elevated levels of neuroinflammation, bryostatin's pharmacodynamic action and ability to penetrate the blood-brain barrier to administer therapeutic effects may be incredibly useful in preventing the occurrence of microbleeds and recurrent intracranial hemorrhages in patients.

In acting as a PKC activator, bryostatin also promotes levels of vascular competency and integrity, helping to maintain proper function in its respective vascular networks. This is especially relevant in the degeneration of blood vessel walls, which undergo a thinning process in the CAA disease state. Brystatin may help prevent this degeneration of cerebral vessels, signaling for proper blood flow and counteracting vessel damage in instances of disorder (Kizub et al. 2014). Emulations of this PKC activity from the natural compound show great promise in the treatment of neurodegeneration in CAA.

While definitive diagnoses of CAA are impossible to determine without the scope of pathological processes postmortem, bryostatin-1 could potentially contain diagnostic value in measuring the extent of a hemorrhage/microbleed in diseased patients. This could provide significant advances in the probable identification of neurodegeneration, where bryostatin could be utilized alongside MRI technologies to suggest a particular prognosis or disease state based on a particular scan.

PKC modulation is becoming an increasingly researched topic of study in the prevention and management of neurodegenerative disorders. While not being explicitly researched in patients exhibiting cases of CAA, the pharmacological effects that bryostatin-1 provides as PKC modulators and APP gene regulators cannot be understated. Since the adverse effects of the natural bioactive compound are also extremely limited, it should be considered as a potential treatment option in the management of CAA.

4.11 Future of Marine Natural Products: How May They Be Used as Therapeutics and Diagnostics

The oceans cover more than 70% of the surface of our planet, which contains an impressive reservoir of largely untapped biologically active compounds. Although marine natural products have been studied extensively for several decades, current research has shown enormous potential for marine bioactives in both therapeutic and diagnostic approaches towards a wide range of neurodegenerative disorders.

Marine organisms, like sponges, bryozoans, mollusks, and marine microbes, yield compounds with unique chemical diversity and great potency. Compared to most terrestrial bioactive compounds, marine natural compounds contain novel scaffolds, neuroprotective properties, and new modes of action, making them interesting agents in an era of high levels of antibiotic resistance and emerging diseases. Research in the search for treatments for these complex chronic diseases has been ever-increasing as a result.

Marine compounds such as bryostatin-1 have promising applications for the regulation of critical signaling pathways that control synaptic plasticity, neuronal survival, and memory formation. These mechanisms bear strong relevance for memory/acquired cognitive disorders such as Alzheimer's disease. Furthermore, they have applications for Huntington's disease and perhaps other prion diseases and cerebral amyloid angiopathy. The findings show bryostatin's influence on various key signaling pathways, particularly the protein kinase C signaling pathway, beta-amyloid processing, as well as neuroinflammation responses, providing a new way of treating a disease that currently has few potential options.

In addition to potential therapeutics, marine natural products also hold enormous value as diagnostics. Perhaps the most prominent use of marine natural products is green fluorescent protein (GFP), originally a compound found in *Aequorea victoria* (a deep-sea jellyfish species). This product fundamentally changed molecular biology and imaging. This is only one example, however, of how marine-sourced biomolecules may be used as valuable diagnostics. Current emerging marine biosensors and imaging contrast agents present viable ways to incorporate marine natural products into diagnostic devices of future generations. As the field of personalized medicine and testing continues to evolve, marine compounds hold the distinctiveness of specificity, sensitivity, and biocompatibility.

Future advancements in technology and drug discovery have the potential to contribute to further explorations in the treatment of neurodegenerative disorders.

Therefore, sustainable sourcing approaches alongside the cultivation of marine microbes are necessary in maintaining healthy ecosystems and marine biodiversity.

4.12 Conclusion

Marine natural products are a massive, untapped reservoir of bioactive compounds that have the potential to be transformed into life-changing therapeutics/diagnostic agents for neurodegenerative conditions while also providing more opportunities for penetration beyond the closed discipline of research. While advancements in medicine continue to develop, the necessities for therapies and diagnostics will continue to grow, as new diseases may arise. In delving further into the mysteries of the sea, important findings of new marine natural products could be used to treat illnesses that were once considered incurable or irreversible. Moreover, improved, sensitive, and even personalized diagnostic tools may become available as a result in the response to neurodegenerative conditions. Man must venture out to make such discoveries.

References

Ajitkumar A, De Jesus O (2023) Huntington disease. In: StatPearls. StatPearls Publishing

Alfa Cissé M, Louis K, Braun U, Mari B, Leitges M, Slack BE, Fisher A, Auberger P, Checler F, Vincent B (2008) Isoform-specific contribution of protein kinase C to prion processing. Mol Cell Neurosci 39(3):400–410. https://doi.org/10.1016/j.mcn.2008.07.013

Bosch ME et al (2024) Sodium oligomannate alters gut microbiota, reduces cerebral amyloidosis and reactive microglia in a sex-specific manner. Mol Neurodegener 19(1):18. https://doi.org/10.1186/s13024-023-00700-w

Catanesi M, Caioni G, Castelli V, Benedetti E, d'Angelo M, Cimini A (2021) Benefits under the sea: the role of marine compounds in neurodegenerative disorders. Mar Drugs 19(1):24. https://doi.org/10.3390/md19010024. PMID: 33430021; PMCID: PMC7827849

Chalfie M (1995) Green fluorescent protein. Photochem Photobiol 62(4):651–656. https://doi.org/10.1111/j.1751-1097.1995.tb08712.x

Charidimou A, Boulouis G, Frosch MP, Baron JC, Pasi M, Albucher JF, Banerjee G, Barbato C, Bonneville F, Brandner S, Calviere L, Caparros F, Casolla B, Cordonnier C, Delisle MB, Deramecourt V, Dichgans M, Gokcal E, Herms J, Hernandez-Guillamon M et al (2022) The Boston criteria version 2.0 for cerebral amyloid angiopathy: a multicentre, retrospective, MRI-neuropathology diagnostic accuracy study. Lancet Neurol 21(8):714–725. https://doi.org/10.1016/S1474-4422(22)00208-3

Colucci-D'Amato L, Speranza L, Volpicelli F (2020) Neurotrophic factor BDNF, physiological functions and therapeutic potential in depression, neurodegeneration and brain cancer. Int J Mol Sci 21(20):7777. https://doi.org/10.3390/ijms21207777

de Villemeur TB (2013) Creutzfeldt-Jakob disease. Handb Clin Neurol 112:1191–1193. https://doi.org/10.1016/B978-0-444-52910-7.00040

Denvir J, Neitch S, Fan J, Niles RM, Boskovic G, Schreurs BG, Primerano DA, Alkon DL (2015) Identification of the PS1 Thr147Ile variant in a family with very early onset dementia and expressive aphasia. J Alzheimer's Dis 46(2):483–490. https://doi.org/10.3233/JAD-150051

Gandhi KR, Saadabadi A (2023) Levodopa (L-Dopa). In: StatPearls. StatPearls Publishing. https://www.ncbi.nlm.nih.gov/books/NBK482140/

Greenberg SM, Bacskai BJ, Hernandez-Guillamon M, Pruzin J, Sperling R, van Veluw SJ (2020) Cerebral amyloid angiopathy and Alzheimer disease – one peptide, two pathways. Nat Rev Neurol 16(1):30–42. https://doi.org/10.1038/s41582-019-0281-2

Hill MD, Blanco MJ, Salituro FG, Bai Z, Beckley JT, Ackley MA, Dai J, Doherty JJ, Harrison BL, Hoffmann EC, Kazdoba TM, Lanzetta D, Lewis M, Quirk MC, Robichaud AJ (2022) SAGE-718: a first-in-class N-methyl-d-aspartate receptor positive allosteric modulator for the potential treatment of cognitive impairment. J Med Chem 65(13):9063–9075. https://doi.org/10.1021/acs.jmedchem.2c00313

Huang KP, Huang FL (1993) How is protein kinase C activated in CNS. Neurochem Int 22(5):417–433. https://doi.org/10.1016/0197-0186(93)90037-6

International Society of Ethnobiology (2013) A brief history of the human use of marine medicines. International Society of Ethnobiology. https://www.ethnobiology.net/brief-history-human-use-marine-medicines/

Johnson RT (2005) Prion diseases. Lancet Neurol 4(10):635–642. https://doi.org/10.1016/S1474-4422(05)70192-7

Kizub IV, Klymenko KI, Soloviev AI (2014) Protein kinase C in enhanced vascular tone in diabetes mellitus. Int J Cardiol 174(2):230–242. https://doi.org/10.1016/j.ijcard.2014.04.117

Kuhn J, Sharman T (2025) Cerebral Amyloid Angiopathy. [Updated 2023 Jun 5]. In: StatPearls [internet]. StatPearls Publishing, Treasure Island (FL). Available from: https://www.ncbi.nlm.nih.gov/books/NBK556105/

Lamptey RNL, Chaulagain B, Trivedi R, Gothwal A, Layek B, Singh J (2022) A review of the common neurodegenerative disorders: current therapeutic approaches and the potential role of nanotherapeutics. Int J Mol Sci 23(3):1851. https://doi.org/10.3390/ijms23031851

Li L, Zhao M, van Meurs M, Brouwers-Haspels I, den Dekker RJH, Wilmsen MEP, Grashof DGB, van de Werken HJG, Rao S, Rokx C, Mueller YM, Katsikis PD (2025) Bryostatin-1 enhances the proliferation and functionality of exhausted CD8+ T cells by upregulating MAP kinase 11. Front Immunol 15:1509874. https://doi.org/10.3389/fimmu.2024.1509874

Mahat S, Asuncion RMD (2023) Kuru. In: StatPearls. StatPearls Publishing

National Research Council (US) Committee on the Ocean's Role in Human Health (1999) From monsoons to microbes: understanding the Ocean's role in human Health. National Academies Press (US), Washington, DC. Marine-Derived Pharmaceuticals and Related Bioactive Agents. Available from: https://www.ncbi.nlm.nih.gov/books/NBK230700/

Panegyres PK, Armari E (2013) Therapies for human prion diseases. Am J Neurodegener Dis 2(3):176–186

Piepho RW (2000) Overview of the angiotensin-converting-enzyme inhibitors. Am J Health-Syst Pharm 57(Suppl 1):S3–S7. https://doi.org/10.1093/ajhp/57.suppl_1.S3

Ramanathan S, Archunan G, Sivakumar M, Tamil Selvan S, Fred AL, Kumar S, Gulyás B, Padmanabhan P (2018) Theranostic applications of nanoparticles in neurodegenerative disorders. Int J Nanomedicine 13:5561–5576. https://doi.org/10.2147/IJN.S149022

Romano G et al (2022) Biomaterials and bioactive natural products from marine invertebrates: from basic research to innovative applications. Mar Drugs 20(4):219. https://doi.org/10.3390/md20040219

Roos RA (2010) Huntington's disease: a clinical review. Orphanet J Rare Dis 5:40. https://doi.org/10.1186/1750-1172-5-40

Santhiravel S et al (2025) Bioactives from marine resources as natural health products: a review. Pharmacol Rev 77(1):100006. https://doi.org/10.1124/pharmrev.123.001227

Sen A, Nelson TJ, Alkon DL, Hongpaisan J (2018) Loss in PKC epsilon causes downregulation of MnSOD and BDNF expression in neurons of Alzheimer's disease hippocampus. J Alzheimer's Dis 63(3):1173–1189. https://doi.org/10.3233/JAD-171008

Shin M, Kim D, Heo YJ, Baek JW, Yun S, Jeong HW (2023) Gerstmann-Sträussler-Scheinker disease: a case report. J Korean Soc Radiol 84(3):745–749. https://doi.org/10.3348/jksr.2022.0089

Simon LV, Hashmi MF, Farrell MW (2025) Hyperkalemia. [Updated 2023 Sep 4]. In: StatPearls [Internet]. StatPearls Publishing, Treasure Island (FL). Available from: https://www.ncbi.nlm.nih.gov/books/NBK470284/

Sitammagari KK, Masood W (2025) Creutzfeldt Jakob disease. [Updated 2024 Jan 30]. In: StatPearls [internet]. StatPearls Publishing, Treasure Island (FL). Available from: https://www.ncbi.nlm.nih.gov/books/NBK507860/

Syed YY (2020) Sodium Oligomannate: first approval. Drugs 80(4):441–444. https://doi.org/10.1007/s40265-020-01268-1

Trindade-Silva AE, Lim-Fong GE, Sharp KH, Haygood MG (2010) Bryostatins: biological context and biotechnological prospects. Curr Opin Biotechnol 21(6):834–842. https://doi.org/10.1016/j.copbio.2010.09.018

Videnovic A (2013) Treatment of Huntington disease. Curr Treat Options Neurol 15(4):424–438. https://doi.org/10.1007/s11940-013-0219-8

Vincent B, Paitel E, Frobert Y, Lehmann S, Grassi J, Checler F (2000) Phorbol ester-regulated cleavage of normal prion protein in HEK293 human cells and murine neurons. J Biol Chem 275(45):35612–35616. https://doi.org/10.1074/jbc.M004628200

Yang LG, March ZM, Stephenson RA, Narayan PS (2023) Apolipoprotein E in lipid metabolism and neurodegenerative disease. Trends Endocrinol Metab 34(8):430–445. https://doi.org/10.1016/j.tem.2023.05.002

Yero T, Rey JA (2008) Tetrabenazine (Xenazine), an FDA-approved treatment option for Huntington's disease-related chorea. P & T 33(12):690–694

Yuksel M, Tacal O (2019) Trafficking and proteolytic processing of amyloid precursor protein and secretases in Alzheimer's disease development: an up-to-date review. Eur J Pharmacol 856:172415. https://doi.org/10.1016/j.ejphar.2019.172415

Marine-Derived Natural Bioactives to Benefit in the Prevention and Treatment of Neurological and Neurodegenerative Disorders

5

James Stoddard, Suhrud Pathak, Rishi Nadar, Muralikrishnan Dhanasekaran, Jaufar Palekkodan, Raakin S. Ameen, Husni Ahmed Abdullah Al-Goshae, and Wael Mahmoud Searan

Abstract

Bioactive substances found in marine natural products have shown great promise in biomedicine, especially in the prevention and treatment of neurological and neurodegenerative diseases. A wide variety of structurally varied and physiologically powerful chemicals have emerged as a result of the particular environmental stresses and biodiversity of marine ecosystems. These marine bioactives are potential options for treating complicated neurological diseases because they have a variety of therapeutic and preventive qualities, such as neuroprotective,

J. Stoddard · S. Pathak · R. Nadar
Drug Discovery and Development, Harrison College of Pharmacy, Auburn University, Auburn, AL, USA

M. Dhanasekaran (✉)
Drug Discovery and Development, Harrison College of Pharmacy, Auburn University, Auburn, AL, USA

Graduate School, Auburn University, Auburn, AL, USA
e-mail: dhanamu@auburn.edu

J. Palekkodan
Illham House, Changuvetti kund, Kottakkal, Kerala, India

R. S. Ameen
Senior at Thomas S Wootton High School, Rockville, MD, USA

H. A. A. Al-Goshae
International Medical School, Management and Science University, Shah Alam, Malaysia

W. M. Searan
Diabetic Foot Center, Center Diabetes & Endocrinology, KFSH, Buraidha Al-Qassim, Saudi Arabia

M. Bhupathyraaj et al. (eds.), *Neuroprotective Marine Bioactives*, Nutritional Neurosciences, https://doi.org/10.1007/978-981-95-5791-2_5

anti-inflammatory, and antioxidant activities. Notably, because of overlapping biological pathways, substances like fucoidan, omega-3 polyunsaturated fatty acids (PUFAs), and different marine polysaccharides have demonstrated benefit across a range of diseases. Pharmaceuticals generated from marine sources have previously helped treat conditions like Parkinson's disease, Alzheimer's disease, Alpers disease, and Sandhoff disease. Marine bioactives present a potent pathway for the development of novel treatments as life expectancy rises worldwide and the prevalence of neurological illnesses increases. A revolutionary breakthrough in world healthcare might be achieved if these marine substances are further investigated and incorporated into therapeutic approaches. This could lead to the management or perhaps cure of diseases that are presently thought to be incurable.

5.1 Introduction

Thanks to the incredible diversity of life on display within our oceans, marine organisms have proven to be an excellent source for the development of new bioactives. The value of marine natural products has become particularly evident in the world of medicine. Marine bioactives have been used in a number of ways medicinally, demonstrating preventative, therapeutic and diagnostic capabilities. Additionally, the incredible size and volume of life present in the marine environment means that the ability for future discoveries is nearly unlimited, making marine natural products an exceptionally exciting source for further advances.

5.1.1 Preventative Properties

The importance of one's diet in maintaining optimal health and preventing disease is well established, and the adverse effects of poor dietary habits can be significant. Although seafood consumption is associated with a number of benefits, arguably the most well-known are those provided by the fatty acids offered. Polyunsaturated fatty acids, including Omega-3 fatty acids, can work to improve cardiovascular function, improve brain function, and lower the risk of chronic ailments such as obesity and diabetes (Hamad 2015). Omega-3's are seen abundantly in seafood, with examples including the eicosapentaenoic and docosahexaenoic acids found in fish such as salmon and herring (Hamad 2015). Beyond the fatty acids, seafood is rich in protein and boasts an impressive amino acid profile when compared to other dietary sources. In addition, protein-pigment complexes, a subset of algae proteins found in cyanobacteria and red algae, demonstrate antioxidative and anti-inflammatory properties, lending them significant value in preventing cancer, neurodegenerative conditions, and more (Hamad 2015). Further boosting the value of seafood consumption are the nutrients on offer, including vitamin A, vitamin D, zinc, and magnesium. When considering all these factors, it comes as no surprise that increased seafood consumption is recommended by many established health and dietary organizations, such as the American Heart Association (Reames 2012). Even more, seafood is widely available and often reasonably affordable, making it a practical choice for many looking to find a sustainable yet healthy diet.

More generally, marine natural products have shown preventative value by working to combat and reduce risk factors associated with many common diseases. Oxidative stress, which can cause damage to cellular macromolecules, contributes to cancer, diabetes, neurodegenerative disorders, as well as cardiovascular conditions. Supplementation of chitosan, derived from the marine product chitosan, has shown great success in preventing oxidative stress by decreasing uremic toxins and lipid hydroperoxides (Šimat 2020). Other derivatives of chitosan have been shown to prevent hypertension, a massive driving factor behind cardiovascular conditions, along with demonstrating anti-inflammatory, anti-obesity, and antidiabetic abilities. Pigments derived from marine algae offer similar capabilities to chitosan derivatives, while polysaccharides sourced from carrageenan and agar take preventative action against cancer, tumor development, and offer antibacterial properties (Šimat 2020).

Overall, the preventative capabilities possessed by marine products are wide-ranging and backed by substantial evidence. These capabilities are arguably growing in importance, as many conditions associated with lifestyle factors are becoming progressively more common, especially in first-world countries. As recently as 2016, over half a million deaths in the United States were attributed directly to poor dietary habits, while the development of over 90% of cancer has been tied to lifestyle and environmental factors (Nova 2020). Further, data collected by the National Health and Nutrition Examination survey found that as of 2010, nearly 40% of American adults were classified as obese, and this number has unfortunately only grown since (Fryar 2012). As the data below illustrates, the prevalence of obesity in the United States has been steadily rising over the past few decades, which has also led to increases in the prevalence of conditions associated with obesity, such as cardiovascular disorders (Fryar 2012).

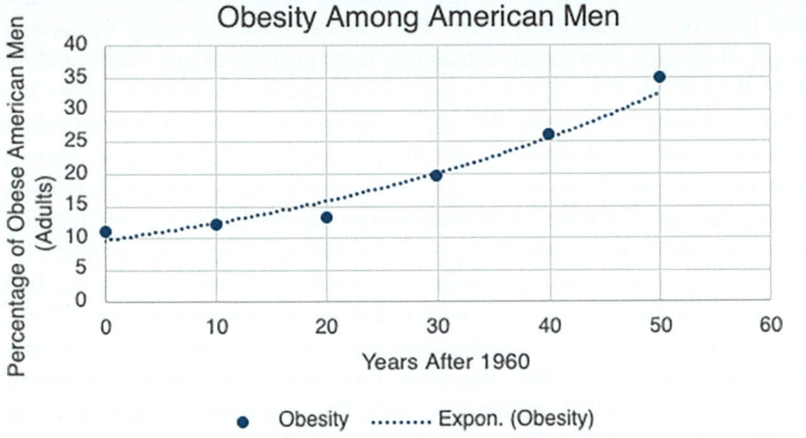

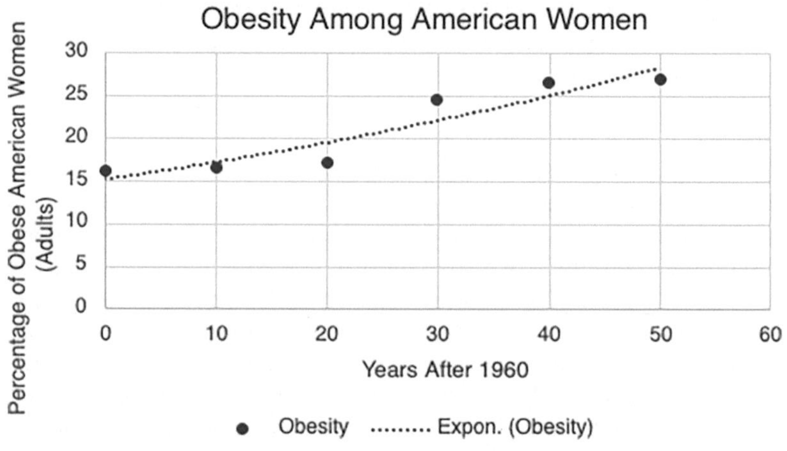

Although there are certainly other contributing factors, the fact that this remarkable rise in obesity prevalence has directly coincided with the rise of ultra-processed and fast food confirms just how important dietary habits are in ensuring optimal health.

5.1.2 Therapeutic Purposes

Although preventative measures are able to ward off a number of medical conditions, illness and disease will persist regardless of efforts to prevent them, and therapeutic measures must be taken. Once again, marine-derived bioactives have played a significant role within this realm. The beginnings of marine therapeutics can be traced back decades, and marine-based drugs have been used to successfully treat cancer, cardiovascular conditions, neurodegenerative diseases, and more.

Cancer is known to be one of the most pressing issues facing medicine and impacts the lives of millions every year. The disease comes in many forms and affects people of all ages, two factors attributing to its position as one of the leading causes of death globally. Cancer is associated with a number of hallmarks, but the primary driving factor behind the disease's catastrophic effects is uncontrolled growth and metastasis of improperly functioning cells, which prevents normal bodily function and contributes to a rapid decline in health if untreated (Bhatia 2022). The beginnings of marine-based therapeutics can be traced to a drug battling this disorder, as cytarabine became the first marine-derived drug to be approved by the FDA. Finding its origins in the sponge *Cryptotethya crypta,* the drug has had significant success in the treatment of many variations of leukemia (Pereria 2019). After being demonstrated to have superior treatment capabilities to alternative medications available in the 1960s, this drug altered the landscape of cancer treatment and became a staple in polychemotherapy, including common regimens such as MEC and DHAP (Pereria 2019). This proved to be only the beginning of anticancer marine-derived drugs, as a number of new drugs have been approved since. Trabectedin, isolated from *Ecteinascidia turbinate (*an ascidian organism found in the waters of the Caribbean), has become a widely used pharmaceutical for cancer patients (Bhatia 2022). Trabectedin's anticancer properties come from its ability to stop the growth of abnormal cells by preventing transcription of DNA, along with its capabilities of stopping metastasis. Further examples of marine-based anticancer agents include Eribulin mesylate and brentuximab vedotin (Bhatia 2022).

Much like cancer, cardiovascular disorders have been well established as one of the biggest threats to global health, impacting millions every year. More commonly known as CVDs, cardiovascular disorders encompass a number of conditions, from heart disease to stroke. Further, they have been consistently recognized as the leading cause of death worldwide for a number of years (Mensah 2019). Fortunately, much progress has been made with respect to treatment, and marine-derived substances have once again made a substantial positive impact. A number of marine drugs have been used to battle CVDs, demonstrating many courses of biological action, including anti-inflammatory, anticoagulation and lipid lowering abilities. This variety is particularly useful in the treatment of CVDs, as these conditions are attributed to a massive range of risk factors (Akram 2023). Some examples of successfully used marine products include saringosterol and asperlin, prescribed to treat symptoms of atherosclerosis, along with histochrome, which is used to treat ischemic heart disease. Even more, these drugs, as well as many other marine-derived pharmaceuticals, offer a lower risk of adverse complications when compared to competing drugs (Akram 2023). The value of marine natural products in fighting against cardiovascular disorders is only magnified when considering that lifestyle and dietary factors play a substantial role in the development of the disease. As illustrated earlier, marine products have shown great preventative capabilities against cardiovascular disorders and their underlying risk factors.

Thanks primarily to advances in medicine, the global average lifespan has seen a significant increase over the past decades, and in all likelihood will only continue to increase further. Although this is a positive change from almost all possible angles, one drawback is an increase in the number of disorders associated with aging, specifically neurodegenerative disorders. Characterized by neurodegeneration due to the death of neuronal cells, Alzheimer's disease is the most common form of dementia. This condition is seen substantially more frequently in elderly individuals and contributes to decreased cognitive function and memory decline (Kumar 2024). Alzheimer's is quickly proving to be an issue of significant priority, and researchers estimate that the frequency of new cases each year will increase to one million by 2050 (Martins 2020a, b). Currently, there are two major hallmarks associated with Alzheimer's. These include neuritic plaques of amyloid-β (Aβ) peptide, attributing to the death of neuronal cells, alongside hyper-phosphorylation of tau proteins, which leads to neurodegenerative conditions as a result of synaptic dysfunctions (Lakshmi 2018). Due to the complexity of this disorder, Alzheimer's has proved to be an exceptionally tough puzzle to solve, and current treatment primarily attempts to slow the progression of the disorder and manage symptoms, while preventative action is also of significant importance. As expected, marine-derived products have been used successfully to help manage Alzheimer's and offer protective capabilities against the disease. Bryostatin, sourced from the marine organism *Bugula neritina,* is commonly prescribed and contributes to overall cognitive repair, significantly improving quality of life. Gracilins and manzamine, both isolated from sponges, demonstrate an ability to reduce tau phosphorylation, giving them preventative capabilities against the disease (Lakshmi 2018).

As the second most prevalent neurodegenerative disorder, Parkinson's disease impacts millions globally. Dopamine, a chemical crucial in regulating bodily movements, is at the core of Parkinson's. The disease is attributed to decreased amounts of dopamine-producing cells and abnormally low dopamine levels as a result (Wong 2014). This causes a myriad of muscular issues, most notably tremors, slowed movement, and worsened reflexes. Nonmuscular symptoms result as well after the onset of the disorder, such as depression and cognitive impairment (Wong 2014). Although no cure currently exists, a number of drugs are offered to help alleviate symptoms and fight the disorder. From a marine standpoint, fungi have shown great potential in treatment. Neochinulin A, a fungal-derived alkaloid, is able to decrease neurotoxicity induced by MPP+ toxin and prevent mitochondrial dysfunction (Silva 2023). Also derived from fungi, Xyloketal B provides antioxidant and anti-inflammatory abilities, vital in maintaining adequate dopamine levels by preventing neuronal death, while alternative marine sources such as seaweeds and fungi possess similar capabilities (Catanesi 2021).

An often overlooked, yet vital aspect of effective treatment is pain management. Chronic pain is a common symptom associated with a plethora of ailments, and many common operations can force patients to accept discomfort as a tradeoff to correct the condition they are battling. Although a number of painkillers are in circulation, addiction and overreliance on these medications have become a massive problem due to withdrawal symptoms. However, marine drugs may present the answer to solving this dilemma. Marine-derived analgesics, namely, peptides, alkaloids, and terpenes, offer a significantly lower risk of addiction development compared to other options and present a wider spectrum of capabilities as they work to target certain pathways attributed to pain that other drugs cannot (Moovendhan 2024).

A more specific issue faced with respect to pain treatment and management is that of neuropathic pain. Neuropathic pain is often attributed to diabetes, trauma, and chemotherapy treatment, and most commonly used treatments have shown limited success (Mishra 2025). Fortunately, marine derivatives such as peptides, fucoidan, and docosahexaenoic acid may be able to provide better results. Thanks to a wider range of biological actions, these substances are able to treat complications and symptoms of neuropathic pain by decreasing oxidative stress and neuroinflammation (Mishra 2025).

5.1.3 Diagnostic Capabilities

The first step in the treatment of any disorder is diagnosis and identification, but this process is often not as straightforward as it might seem. Underlying processes behind disease can be exceptionally complex, making the identification of these hallmarks difficult. Further, quick and accurate diagnosis can be especially crucial, especially with respect to fast-acting conditions that must be identified and treated quickly. Fucoidan, a marine-sourced polysaccharide, has proved quite useful in this regard, as it has been used to help identify myocardial infarction as well as thrombosis, two conditions in which quick diagnosis is paramount. Fucoidan has demonstrated other diagnostic capabilities, such as in identifying aortic aneurysms (Majee 2017). Similarly, marine products have also been used as experimental tools, helping researchers to gain valuable insights into the inner workings of many complex medical conditions. Isolated from various dinoflagellates, okadaic acid can be used to study neurological disorders, such as Alzheimer's, associated with improper phosphorylation of proteins (Lindequist 2016). When considering just how challenging neurodegenerative conditions have been with respect to treatment, the ability to provide opportunities for further examination of these disorders is particularly crucial. Marine derivatives have also been utilized in the polymerase chain reaction (PCR). PCR is used to create up to millions of copies of DNA by amplifying a small amount of DNA. This capability gives it great utility in disease diagnosis, genetic research, and within the world of forensics (Garibyan 2013). The marine enzymes Taq polymerase and Pfu have both played crucial roles in the advancement of this technique (Lindequist 2016).

5.1.4 Summary

Marine natural products are incredibly diverse and have been used successfully in the treatment, prevention, and diagnosis of many common diseases and illnesses, such as cardiovascular conditions, neurological disorders, and cancer. In addition, many marine-based therapeutics offer advantages over competing drugs, such as more diverse courses of action (contributing to better treatment) as well as reduced risks of adverse side effects. The accessibility and sheer volume of marine products give them significant potential as sources for the synthesis of new and improved bioactives.

5.2 Neurological Pathologies and Marine Natural Products

5.2.1 Main Pathologies

As illustrated previously, marine natural products have shown tremendous value across many realms of medicine, but a specific area of interest and possibility is in treating neurological and neurodegenerative disorders. Although neurological disorders cover a very broad spectrum, conditions of particular importance include the following: Absence of the Septum Pellucidum Acid Lipase Disease, Acute Disseminated Encephalomyelitis, Adrenoleukodystrophy, Agenesis of the Corpus Callosum, Agnosia, Aicardi Syndrome, Aicardi-Goutières Syndrome, AIDS and

HIV, Alexander Disease, Alpers Disease, Alzheimer's Disease, Amyotrophic Lateral Sclerosis (ALS), Anencephaly, Angelman Syndrome, Antiphospholipid Syndrome, Aphasia, Apraxia, Arachnoid Cysts, Arachnoiditis, Arteriovenous Malformations (AVMs), Ataxia and Cerebellar or Spinocerebellar Degeneration, Ataxia Telangiectasia, Atrial Fibrillation and Stroke, Attention Deficit Hyperactivity Disorder, Autism Spectrum Disorder, Back Pain, Barth Syndrome, Batten Disease, Behcet's Disease; Bell's Palsy, Benign Essential Blepharospasm, Binswanger's Disease, Brachial Plexus Injury, Brain and Spinal Cord Tumors, Brown-Sequard Syndrome, CADASIL, Canavan Disease, Carpal Tunnel Syndrome, Central Cord Syndrome, Central Pain Syndrome, Central Pontine Myelinolysis, Cerebellar Degeneration, Cerebellar Hypoplasia, Cerebral Aneurysms, Cerebral Arteriosclerosis, Cerebral Atrophy, Cerebral Cavernous Malformations, Cerebral Hypoxia, Cerebral Palsy, Cerebro-Oculo-Fascio-Skeletal (COFS) Syndrome, Charcot-Marie-Tooth Disease, Chiari Malformations, Chorea, Chronic Inflammatory Demyelinating Polyneuropathy (CIDP), Coffin-Lowry Syndrome, Colpocephaly, Coma, Complex Regional Pain Syndrome, Congenital Myasthenia, Congenital Myopathy, Corticobasal Degeneration, Craniosynostosis, Creutzfeldt-Jakob Disease, Cushing's Syndrome, Cytomegalovirus, Dandy-Walker Syndrome, Deep Brain Stimulation for Movement Disorders, Dementia, Multi-Infarct, Dementias, Dermatomyositis, Developmental Dyspraxia, Diabetic Neuropathy, Dravet Syndrome, Dysautonomia, Dysgraphia, Dyslexia, Dyssynergia Cerebellaris Myoclonica, Dystonia, Empty Sella Syndrome, Encephalitis, Encephalitis Lethargica, Encephaloceles, Epilepsy and Seizures, Essential Tremor, Fabry Disease, Fahr's Syndrome, Familial Periodic Paralyses, Farber's Disease, Febrile Seizures, Fibromuscular Dysplasia, Foot Drop Syndrome, Friedreich Ataxia, Frontotemporal Disorders, Functional Neurologic Disorder, Gangliosidoses, Gaucher Disease, Gerstmann's Syndrome, Gerstmann-Straussler-Scheinker Disease, Giant Axonal Neuropathy, Guillain-Barré Syndrome, Headache, Hemicrania Continua, Hemifacial Spasm, Hereditary Neuropathies, Hereditary Spastic Paraplegia, Herpes Zoster Oticus, Holmes-Adie Syndrome, Holoprosencephaly, HTLV-1-Associated Myelopathy (HAM)/Tropical Spastic Paraparesis (TSP), Huntington's Disease, Hydranencephaly, Hydrocephalus, Hydromyelia, Hypersomnia, Hypertonia, Hypotonia, Hypoxic Ischemic Encephalopathy, Inclusion Body Myositis, Incontinentia Pigmenti, Infantile Neuroaxonal Dystrophy, Infantile Spasms, Inflammatory Myopathies, Iniencephaly, Isaacs Syndrome, Joubert Syndrome, Kearns-Sayre Syndrome, Kennedy's Disease, Kleine-Levin Syndrome, Klippel-Feil Syndrome, Klippel-Trenaunay Syndrome, Klüver-Bucy Syndrome, Krabbe Disease, Kuru, Lambert-Eaton Myasthenic Syndrome, Landau-Kleffner Syndrome, Learning Disabilities, Leigh Syndrome, Lennox-Gastaut Syndrome, Lesch-Nyhan Syndrome, Leukodystrophy, Lewy Body Dementia, Lipid Storage Diseases, Lipoid Proteinosis, Lissencephaly, Locked-In Syndrome, Lupus, Lyme Disease, Megalencephaly, Melkersson-Rosenthal Syndrome, Meningitis; Menkes Disease, Meralgia Paresthetica, Metachromatic Leukodystrophy, Microcephaly Migraine, Miller Fisher Syndrome, Mitochondrial Myopathies, Moebius Syndrome, Monomelic Amyotrophy, Motor Neuron Diseases,

Moyamoya Disease, Mucolipidoses, Mucopolysaccharidoses, Multifocal Motor Neuropathy, Multiple Sclerosis, Multiple System Atrophy, Muscular Dystrophy, Myasthenia Gravis, Myoclonus, Myotonia, Myotonia Congenita, Narcolepsy, Neuroacanthocytosis, Neurodegeneration with Brain Iron Accumulation, Neurofibromatosis, Neuroleptic Malignant Syndrome, Neuromyelitis Optica, Neuronal Migration Disorders, Neurosarcoidosis, Neurosyphilis, Neurotoxicity, Niemann-Pick Disease, Normal Pressure Hydrocephalus, Occipital Neuralgia, Ohtahara Syndrome, Olivopontocerebellar Atrophy, Opsoclonus Myoclonus, Orthostatic Hypotension, Pain, Paraneoplastic Syndromes, Paresthesia, Parkinson's Disease, Paroxysmal Choreoathetosis, Paroxysmal Hemicrania, Parry-Romberg Syndrome, Pelizaeus Merzbacher Disease, Peripheral Neuropathy, Periventricular Leukomalacia, Pervasive Developmental Disorders, Pinched Nerve, Piriformis Syndrome, Pituitary Tumors, Polymyositis, Pompe Disease, Porencephaly, Post-Polio Syndrome, Postural Tachycardia Syndrome, Primary Lateral Sclerosis, Progressive Multifocal Leukoencephalopathy, Progressive Supranuclear Palsy (PSP), Prosopagnosia, Pseudotumor Cerebri, Rasmussen's Encephalitis, Refsum Disease, Repetitive Motion Disorders, Restless Legs Syndrome, Rett Syndrome, Reye's Syndrome, Sandhoff Disease, Schilder's Disease, Schizencephaly, Septo-Optic Dysplasia, Shaken Baby Syndrome, Shingles, Sjögren's Syndrome, Sleep Apnea, Sotos Syndrome, Spasticity, Spina Bifida, Spinal Cord Infarction, Spinal Cord Injury, Spinal Muscular Atrophy, Spinocerebellar Ataxias including Machado-Joseph Disease, Stiff-Person Syndrome, Striatonigral Degeneration, Stroke, Sturge-Weber Syndrome, Subacute Sclerosing Panencephalitis, SUNCT Headache, Swallowing Disorders, Sydenham Chorea, Syncope, Syringomyelia, Tabes Dorsalis, Tardive Dyskinesia, Tarlov Cysts, Tay-Sachs Disease, Tethered Spinal CordSyndrome, Thoracic Outlet Syndrome, Thyrotoxic Myopathy, Todd's Paralysis, Tourette Syndrome, Transient Ischemic Attack (TIA), Transmissible Spongiform Encephalopathies, Transverse Myelitis, Traumatic Brain Injury (TBI), Tremor, Trigeminal Neuralgia, Troyer Syndrome, Tuberous Sclerosis Complex, Vasculitis in the Nervous System, Von Hippel-Lindau Disease (VHL), Wallenberg's Syndrome, Wernicke-Korsakoff Syndrome, Whiplash, Whipple's Disease, Williams Syndrome, Wilson Disease, Zellweger Syndrome (Pathak 2024). Marine natural products have served to help treat and prevent a good amount of these conditions, as listed in the table below.

Condition	Marine bioactives used
Alpers disease	Echinochrome A
Alzheimer's	Chitosan, fucoidan
AIDS and HIV	Lectins, sulfated polysaccharides
Amyotrophic lateral sclerosis	Coibamide A, papuamine
Attention deficit hyperactivity disorder and autism	Docosahexaenoic acid
Cerebral arteriosclerosis	Fucoidan, alginate, ulvan
Cerebral palsy	Xyloketal B

(continued)

Condition	Marine bioactives used
Congenital myasthenia	Makaluvamine G
Dementia	Docosahexaenoic acid
Epilepsy and seizures	2,5-diketopiperazines
Friedreich ataxia	Algal metabolites
Headache	Ziconotide
Huntington's disease	Polysaccharides, polyphenols
Inflammatory myopathy	DPHC
Lyme disease	Aureobasidin
Meningitis	m2–9
Multiple sclerosis	MS14
Myasthenia gravis	Bacteria, fungi, algae
Neurotoxicity	Fucoidan
Parkinson's disease	NP7, algae
Peripheral neuropathy	Polyphenols
Sandhoff disease	*Aeromonas caviae* CHZ306
Sleep apnea	Vitamin D, magnesium
Spinocerebellar ataxia	Amino acids, zinc
Stroke	Omega-3 PUFAs
Traumatic brain injury	Fucoxanthin

5.2.2 Bioactive Course of Action

1. *Alpers Disease*: Most commonly seen in children, Alpers disease is a progressive neurodegenerative disorder that can contribute to serious complications, including epilepsy, liver disease, neurological decline, and early death (Wiltshire 2008). This condition is attributed to mutations in polymerase γ (*POLG1*), a mtDNA polymerase gene (Wiltshire 2008). Most commonly known for its antioxidant and cardioprotective abilities, Echinochrome A is a pigment derived from sea urchins that could also play a role in the treatment of Alpers disease (Jeong 2014). Echinochrome A has been shown to have a positive effect on proper mtDNA transcription and regulation, as evidenced by its ability to increase mitochondrial number and mtDNA mass. These effects can help prevent mutations in POLG1, which means Echinochrome A can be a potential treatment for Alpers disease (Jeong 2014).
2. *Alzheimer's Disease*: Alongside the earlier discussed bioactives, many other marine natural products have exhibited potential in the treatment and prevention of Alzheimer's disease. One example of this can be found in chitosan, a polysaccharide derived from fungi and crustaceans, which works to ward off excessive amyloid beta (Aβ) plaque aggregation (a significant hallmark of the condition) by preventing neuronal cell death due to oxidative stress (Hu 2023). Another marine-derived polysaccharide, fucoidan, also works to prevent the death of neuronal cells. Fucoidan's biological course of action is to stop

apoptosis by preventing caspase-9 and caspase-3 enzymes from being activated by Aβ protein (Hu 2023). The wide-ranging biological courses of action offered by marine bioactives make them excellent candidates for future advances in the fight against Alzheimer's disease.

3. *HIV/AIDS*: AIDS and HIV have become a pressing issue over the past decades, and the potentially fatal effects of the condition are well known. Fortunately, a few marine-sourced bioactives offer hope for further advances in the battle against the virus. Firstly, lectins (found in a number of algae) have significant potential as a preventative agent thanks to their ability to prevent cells from being infected by HIV, as well as stop the transfer of the virus from infected cells (Besednova 2019). Also found abundantly in algae, sulfated polysaccharides have shown great antiviral potential, as they can stop viruses from attaching to target cell molecules, a potentially crucial application in stopping the progression of HIV (Besednova 2019).
4. *ALS*: Despite its place as one of the most well-known and researched neurodegenerative conditions, ALS has proven to be an exceptionally challenging disorder to treat. With no cure currently available, current treatment serves only to manage symptoms and slow the progression of the disorder. However, some marine bioactives offer protective capabilities against ALS, making them exciting options in preventing the onset of this condition. A common hallmark seen in many neurological diseases, including ALS, is failure of the autophagy-lysosome pathway. This failure leads to protein aggregation, specifically that of superoxide dismutase in ALS (Kotimchenko 2022). The marine compounds papuamine and coibamide A, isolated from sea sponges and cyanobacteria, respectively, have demonstrated an ability to restore proper function of the autophagy-lysosome pathway and prevent protein aggregation (Kotimchenko 2022). This capability gives both papuamine and coibamide A significant potential as ALS preventative agents.
5. *Attention Deficit Hyperactivity Disorder and Autism*: Impacting around 5–7% and 1.5% of American children, respectively, ADHD and autism spectrum disorder are two of the most well-known and researched neurodevelopmental disorders, and both can be tied to genetic and environmental factors. One particular environmental factor contributing to the development of these conditions is a lack of docosahexaenoic acid. This omega-3 fatty acid, found abundantly in a variety of seafood, is crucial in neurological development (DiNicolantonio 2020). Further, inadequate intake of DHA has been shown to increase the risk of neurodevelopmental conditions, such as autism spectrum disorder and ADHD (DiNicolantonio 2020). In addition, the increasing prevalence of ADHD and autism spectrum disorder in Western nations can potentially be attributed to a higher intake of omega-6 PUFAs. When compared to our ancestors, who consumed around a 2:1 ratio of omega-6 PUFAs to omega-3s, modern Western diets have seen this ratio rise to an astounding 30:1 (Martins 2019). Similar to the observed increases in prevalence of obesity and other lifestyle-related conditions, other factors are almost certainly in play. However, the potential value of taking dietary preventative measures is quite apparent, and increased intake

of DHA offers great value in the prevention of these two neurodevelopmental disorders, alongside improving general cognitive development and function.

6. *Cerebral Arteriosclerosis*: Arteriosclerosis is a significant risk factor contributing to the development and onset of many cardiovascular disorders. As discussed previously, fucoidan has shown protective capabilities against Alzheimer's disease and offers the same benefits with respect to arteriosclerosis. Fucoidan has demonstrated an ability to reduce lipid buildup in the arteries as well as correct abnormal levels of lipids in the blood, both of which are considered to be biological driving factors behind the development of arteriosclerosis (Cao 2020). Similarly, alginate is commonly isolated from brown seaweed and is capable of lowering the probability of developing cardiovascular disorders (including arteriosclerosis) by treating hyperlipidemia (Cao 2020). Another polysaccharide, ulvan, has demonstrated great value in the treatment of arteriosclerosis by means of hyperlipidemia correction as well (Cao 2020). With respect to cardiovascular conditions, the value of preventative action is enormous, making Fucoidan's protective properties all the more valuable.
7. *Cerebral Palsy*: Seen in roughly 1 in 500 newborns, neonatal hypoxic-ischemic encephalopathy is an extremely damaging neurological condition and directly contributes to the development of cerebral palsy. This condition currently has no known cure, meaning preventative action is vital (Xiao 2014). Xyloketal B, derived from fungi in the South China Sea, has shown a number of protective abilities against hypoxic-ischemic encephalopathy both in vitro and in vivo. One driving factor behind the onset of HIE is oxygen-glucose deprivation (Xiao 2014). Attributed to excessive calcium levels, oxygen-glucose deprivation can lead to cellular death. Upon testing in primary cortical neurons, Xyloketal B was shown to reduce calcium entry and reduce neuronal cell death from oxygen-glucose deprivation as a result (Xiao 2014). In addition, this compound also demonstrated abilities to reduce brain damage from hypoxic ischemia by lowering infarct volume, alongside offering general antioxidative benefits (Xiao 2014). Overall, Xyloketal B is an exciting option for the prevention of cerebral palsy thanks to the number of protective capabilities it offers against HIE. However, the potential value of Xyloketal B is not limited to cerebral palsy, as a number of other neurological conditions and defects (such as epilepsy and learning disorders) can arise as a result of neonatal hypoxic-ischemic encephalopathy (Xiao 2014).
8. *Congenital Myasthenia*: Caused by abnormalities in neuromuscular signal transmission at the motor end plate, congenital myasthenia constitutes a wide variety of disorders. Although there are other contributing factors to this faulty neuromuscular signal transmission, it is most commonly attributed to defects in nicotinic acetylcholine receptors, better known as nAChRs (Engel 2015). One specific kind of these ion channels is muscle nAChR. This nicotinic acetylcholine receptor is found on the motor end plate and is crucial in transmitting signals from motor neurons to muscle fibers (Kudryatsev 2018). The significance of muscle nAChR in the development of congenital myasthenia is best explained by allosteric transitions. Namely, deregulation of desensitization (transformation

from an open ion channel to a closed one) directly causes slow-channel congenital myasthenic syndrome, a rare and currently incurable form of congenital myasthenia (Kudryatsev 2018). Fortunately, there may be a potential marine solution to the problems posed by this condition in the form of Makaluvamine G, a pyrroloiminoquinone alkaloid isolated from an Indonesian sponge (Kudryatsev 2018). Makaluvamine G has shown an ability to increase the rate of desensitization in muscle nAChR through both competitive and noncompetitive inhibition, in addition to capabilities of both orthosteric and allosteric binding, making it a crucial source for future advancements in the prevention of slow-channel myasthenic syndrome (Kudryatsev 2018).

9. *Dementia*: As shown earlier in discussing the preventative capabilities of marine natural products, omega-3 polyunsaturated fatty acids offer substantial health benefits and can help prevent many common illnesses and disorders. Much like in the case of ADHD and Autism spectrum disorder, the omega-3 PUFA DHA is of particular importance with respect to dementia prevention. Only supplied via dietary means, DHA is a crucial type of omega-3 PUFA, and a loss of DHA is associated with dementia and overall cognitive decline (Cardoso 2016). Further, DHA has been shown to offer neuroprotective and overall cognitive benefits, helping to prevent dementia as well as other neurodegenerative conditions, including Alzheimer's disease (Cardoso 2016). In addition to its abilities in preventing the onset of dementia, DHA consumption has a number of other neurological benefits (most notably in the prevention of depression and mood disorders) alongside cardiovascular benefits (Cardoso 2016).
10. *Epilepsy and Seizures*: Affecting close to 100 million globally, epilepsy is a particularly dangerous neurological disorder that can contribute to premature death. The severity of this condition is only heightened when considering that current anti-seizure drugs only demonstrate success in around 65% of patients (Copmans 2022). This means that there is a massive need for the development of new and improved therapies for the treatment of epilepsy, and a few marine derivatives have shown potential in meeting this need. Firstly, the marine bioactive halimide (a 2,5-diketopiperazine) is sourced from *Aspergillus insuetus* and has demonstrated both anti-seizure as well as anti-epileptiform capabilities (Copmans 2022). Further, another 2,5-diketopiperazine, plinabulin, possesses these capabilities as well, making both excellent options for the development of new and improved antiseizure drugs (Copmans 2022).
11. *Friedreich Ataxia*: Better known as FA, this autosomal recessive condition leads to severe complications, such as heart failure, diabetes, as well as difficulty speaking and walking (Naik 2024). Friedreich's ataxia can be caused by damage to the cerebellum, spinal cord, or due to defects in the FXN gene. This gene mutation leads to reduced FXN transcription as a result of GAA trinucleotide repeats, depleting energy production of cardiac cells and neurons (Naik 2024). Much like many other neurological conditions, the complexity behind FA makes it quite challenging to treat and prevent. One potential solution comes in the form of metabolites derived from marine algae. Compounds such as alkaloids, phenols, phytosterols, and carbohydrates demonstrate a range of

neuroprotective abilities, making them a source of interest in working to prevent FXN defects and Friedreich's ataxia as a result (Naik 2024).

12. *Headache*: Approved in 2004 by the USFDA, ziconotide is a marine-derived analgesic. Functioning by blocking calcium channels, which regulate transmission of pain stimulation (McGivern 2006) on the nociceptive nerves present on the spinal cord, ziconotide has proved to be an effective agent to combat generalized pain associated with headaches or other ailments (Malve 2016). Additionally, as mentioned in the introduction, analgesics sourced from marine compounds (including alkaloids and peptides) have not only proven to be effective painkillers but carry a lower risk of addiction development when compared to competing drugs (Moovendhan 2024).
13. *Huntington's Disease*: Alongside other neurological conditions including Alzheimer's and Parkinson's, Huntington's disease is directly linked to neuroinflammation, which contributes to the onset of this condition as well as further neurodegeneration (Elbandy 2022). Current therapies are unable to stop the progression of this condition, meaning that the anti-inflammatory properties offered by marine bioactives are of particular value (Elbandy 2022). One marine source with anti-neuroinflammatory properties is found in sulfated polysaccharides (including fucans, fucoidans, and carrageenans), which are found abundantly in seaweed. Other examples include polyphenols (seen substantially in algae) as well as chitin/chitosan (found in crustaceans such as lobster, crayfish, and shrimp) (Elbandy 2022). When considering the wide availability of marine products offering anti-neuroinflammatory properties, marine natural products are an excellent source both for the prevention and management of Huntington's disease.
14. *Inflammatory Myopathy*: Although inflammatory myopathies encompass a large group of disorders with varying causes, one common driving factor behind this condition is an inflammatory cytokine, tumor necrosis factor (Kim 2020). Better known as TNF-α, tumor necrosis factor is capable of causing premature activation of atrogin-1 gene expression, worsening the rehabilitative effects of satellite cells, and causing protein degradation of IGF-1 as a result (insulin-like growth factor 1). These effects mean that excessive release of this cytokine is directly correlated with muscle loss and the onset of inflammatory myopathy (Kim 2020). Diphlorethohydroxycarmalol, abbreviated DPHC, is a marine natural product sourced from *Ishige okamurae.* Upon testing, this compound has shown an ability to lower the expression levels of the key protein behind muscular atrophy (MuRF-1 and MAFbx/Atrogin-1) stimulated by TNF-α, making it an exciting source for the development of future drugs to combat inflammatory myopathy (Kim 2020).
15. *Lyme Disease*: *Borrelia garinii,* the bacterium causing Lyme Disease, is also capable of causing headache, fatigue, as well as more severe symptoms such as meningitis (Basharat 2024). Arginine deminase, commonly known as ADI, is an enzyme found in *Borrelia garinii.* This enzyme plays an important role in protein synthesis within the bacteria, synthesizing citrulline and ammonia from arginine (Basharat 2024). When considering how valuable arginine deminase is

for the survival of *Borrelia garinii,* one potential course of action against Lyme disease would be to disrupt the proper function of this enzyme. Aureobasidin is a marine-based metabolite that has been shown to prevent proper function of ADI and inhibit protein synthesis, stopping bacterial growth as a result (Basharat 2024). This capability makes aureobasidin a particularly interesting source for future developments in preventing Lyme Disease and other conditions associated with *Borrelia garinii.*

16. *Meningitis*: *Angiostrongylus cantonensis* is the parasitic agent behind the majority of eosinophilic meningitis and meningoencephalitis cases observed in areas of Asia and islands of the Pacific Ocean. This worm directly leads to eosinophilic meningitis following ingestion, and the inflammation caused by the worm's death can lead to meningoencephalitis (Li 2011). Although anthelmintics like albendazole can be used to kill the worm, the complications associated with developing meningoencephalitis as a result make this a risky endeavor. However, the marine compound m2–9 may offer a solution. This fungal derivative has been shown to significantly reduce leukocyte numbers when used in combination with albendazole (a commonly used anthelmintic), as compared to the usage of albendazole alone (Li 2011). These results indicate that m2–9 can be used to improve on current therapeutic methods, as this marine compound helps to prevent the inflammatory effects that result from anthelmintic administration (Li 2011).
17. *Multiple Sclerosis*: Multiple sclerosis is a rare neurological condition that causes overall neurological decline due to the destruction of myelin and axons within the central nervous system (Goldenberg 2012). The disease has a complex pathology, with oxidative stress and inflammation playing a significant role in the development and onset of this condition (Goldenberg 2012). MS14 is a natural drug derived from both marine and herbal sources, with the majority of its composition being sourced from the marine organism *Penaeus latisculatus* (Ahmadi 2010). The makeup of this drug means that it possesses a number of trace elements, including bromine and zinc, that offer antioxidative properties. Further, this drug also possesses anti-inflammatory capabilities, which, combined with its antioxidative properties, help to slow the progression of multiple sclerosis (Ahmadi 2010), serving to improve the quality of life of those afflicted by this condition. Considering the number of driving factors behind multiple sclerosis, MS14 is particularly useful as its biological course of action offers very broad neuroprotective effects.
18. *Myasthenia Gravis*: With common symptoms including weakness and fatigue of skeletal muscles, myasthenia gravis is a neurological condition most commonly caused by a reduction in the number of functional AChR (which play an important role in muscle contraction and nerve impulse transmission), and this deficiency leads to the neuromuscular transmission blockage seen in myasthenia gravis (Moodie 2019). Much like many other neurological diseases, myasthenia gravis has been treated through the usage of cholinesterase inhibitors, which serve to improve neuromuscular communication. Including bacteria, algae, and fungi, a number of marine organisms possess compounds with

cholinesterase-inhibiting abilities, meaning they have the potential to be used in the treatment and prevention of myasthenia gravis (Moodie 2019).

19. *Neurotoxicity*: Neurotoxicity is a significant driving factor behind a number of neurological conditions, thanks to its neurodegenerative effects and capability of causing neuronal damage (Pangestuti 2011). As was the case with Alzheimer's disease and cerebral arteriosclerosis, the marine derivative Fucoidan can play a vital role in battling neurotoxicity. Fucoidan is able to prevent β-amyloid neurotoxicity by stopping the inhibition of phosphorylation of PKC (protein kinase C) (Pangestuti and Kim 2011). This biological course of action serves to stop neuronal death, giving fucoidan great capabilities in preventing a significant risk factor behind some of the most dangerous and prevalent neurological conditions (Pangestuti 2011).
20. *Parkinson's Disease*: As discussed earlier, Parkinson's is one of the most common neurodegenerative conditions globally and is caused by a lack of dopamine due to loss of dopamine-producing neurons. The loss of these dopaminergic neurons is mainly attributed to oxidative stress, neuroinflammation, abnormal mitochondrial function, and accumulation of the protein α-synuclein (Huang 2019). One marine-based compound with significant potential in Parkinson's treatment is NP7, sourced from the bacterium *Streptomyces* sp. This compound could be useful in reducing oxidative stress by preventing neuronal cell death and necrosis, as well as stopping excessive activation of microglial cells (Huang 2019). In addition, marine algae possess many of these same capabilities, making them another compound of interest in the prevention of Parkinson's (Huang 2019).

    However, the benefits of marine bioactives in battling Parkinson's disease extend far beyond preventive properties, as there are substantial amounts of marine-based drugs in trial for the treatment of Parkinson's, with two examples including pramipexole and inosine (Huang 2019). Although Parkinson's is considered incurable as of now, a breakthrough could be looming, and marine derivatives may very well prove to be the source.
21. *Peripheral Neuropathy*: A common side effect of diabetes, neuropathy results due to stimulation of the aldose reductase enzyme as a result of chronic hyperglycaemia (Samaddar 2019). Polyphenols isolated from the red microalga *Symphocladia latiuscula* have demonstrated an ability to prevent peripheral neuropathy by reducing the activity of the aldose reductase enzyme. Further, these polyphenols showed a capability to inhibit manifestation of hallmarks associated with aldose reductase stimulation, including sorbitol aggregation and faulty activity of the sodium potassium pump in the sciatic nerve (Samaddar 2019). As discussed earlier, lifestyle-associated conditions such as diabetes have seen a sharp rise in prevalence, further increasing the value of these marine bioactives.
22. *Sandhoff Disease*: A member of the chitinase family, N-acetyl-glucosaminidases (GlcNAcases) are enzymes playing a crucial role in intracellular protein function and in catalyzation of *N*-acetyl-D-glucosamine (GlcNAc) hydrolysis (Cardozo 2023). Thanks to these functions, GlcNAcases have significant impor-

tance in the treatment of Sandhoff disease (as deficiencies in intracellular protein function and GlcNAc hydrolysis can contribute to the development of Sandhoff disease, as well as some other neurological disorders) (Cardozo 2023). A Brazilian marine bacterium, *Aeromonas caviae* CHZ306, has demonstrated great ability in extracellular chitinase synthesis. This capability means this bacterium offers a potential source for large-scale production of GlcNAcases, crucial for more effective treatment of Sandhoff disease (Cardozo 2023).

23. *Sleep Apnea*: Beyond causing general fatigue, the inability to achieve sound, restful sleep due to sleep apnea can lead to more serious complications such as heart disease and diabetes (Hall 2024). Primarily, sleep apnea treatment consists of dietary and lifestyle changes. With respect to dietary preventative measures, adequate intake of vitamin D and magnesium is considered especially crucial in treating sleep apnea (Hall 2024). While direct supplementation is frequently used, marine-based options present a great alternative. In particular, fatty fish are considered to be a great source of magnesium, while oily fish can provide vitamin D (Larsen 2011). These simple dietary changes can pay dividends in preventing sleep apnea and improving overall health and quality of life as a result.
24. *Spinocerebellar Ataxia*: Affecting around 1 in 100,000 individuals globally, spinocerebellar ataxias are a group of rare, progressive neurodegenerative diseases, with common symptoms including dementia, limb ataxia, epilepsy, and cerebellar gait (Wright 2023). Due to the progressive nature of these disorders, early intervention is important in ensuring the neurological condition is maintained as well as possible. Those impacted by a spinocerebellar ataxia will have higher demands for nutrients as a result of increased brain activity (Wright 2023), and many marine organisms are great sources of these required nutrients. Zinc, an antioxidant associated with cognitive health, can be found in a variety of fish. Alongside this, salmon and trout are two excellent sources of serine, leucine, and other important amino acids with respect to neurological function (Wright 2023). Overall, marine-based organisms can provide a number of compounds key to preventing cognitive decline in individuals with SCA.
25. *Stroke*: Due to the severe and potentially fatal results of a stroke, preventive action is of the utmost importance. Once again, the wide-ranging benefits of omega-3 PUFAs apply in stroke prevention, and many marine organisms are a fantastic source of these compounds, as discussed previously. Studies have shown that increased intake and supplementation of the omega-3 PUFAs DHA and DPA is correlated with lowering the likelihood of stroke (Saber 2017).
26. *Traumatic Brain Injuries*: More commonly known as TBI, traumatic brain injuries can be exceptionally damaging, directly contributing to brain contusions, ruptured blood vessels, and other significant physical damage (Hannan 2020). Further, outside of these primary impacts, neuronal degeneration can occur later, leading to a host of other neurological problems (Hannan 2020). Fortunately, a marine solution may be available to help prevent further neurological issues. Fucoxanthin, a carotenoid derivative, has shown an ability to

prevent TBI by inhibiting the neuroprotective effects of the Nrf2-ARE and Nrf2-autophagy pathway, which can prevent the development of neurodegenerative conditions as a consequence of TBI. (Hannan 2020).

5.3 Conclusion

Although natural bioactives can be found from a nearly endless range of sources, marine natural products have proven to be truly unique, thanks to the incredible diversity of life along with the unique environmental demands placed upon marine organisms. As a result, marine bioactives have contributed to advances in nearly every aspect of medicine, offering preventative, therapeutic, and diagnostic abilities. Marine natural products have demonstrated great value in battling some of the most damaging conditions faced globally, including cardiovascular disorders, cancer, diabetes, and more. However, these bioactives are of particular importance in the advancement of treatment for neurological and neurodegenerative conditions. Firstly, preventative measures are absolutely crucial with respect to these conditions, and marine bioactives possess an almost endless reservoir of important protective properties against neurological conditions. From neuroprotective capabilities to anti-inflammatory properties, marine-sourced bioactives offer great value in the prevention of neurological and neurodegenerative diseases. Beyond this, the complex pathologies behind these disorders have made them exceptionally challenging to treat, and marine natural products could very well be the solution to this problem. Further, many of these conditions share overlap with respect to biological driving factors, meaning that individual bioactives can be used in the treatment of multiple disorders (as illustrated by compounds such as fucoidan, omega-3 PUFAs, polysaccharides, etc). Marine-derived pharmaceuticals have shown great variety in regards to biological course of action, as evidenced by their contributions to the advancement in treatment for conditions such as Alzheimer's disease, Parkinson's disease, as well as other less known disorders including Alpers disease, Sandhoff disease, and more. Overall, marine bioactives have already helped to make great strides in improving existing therapies as well and in the creation of new pharmaceuticals for the treatment of neurodegenerative and neurological disorders, and the potential for future advances is substantial.

As the average global lifespan continues to steadily climb, the prevalence of neurological diseases will only continue to increase as well, and the effects could be exceptionally damaging. However, all hope is not lost, as these developments have led to a greater focus on improving treatment protocols, and much progress has already been made. Marine bioactives have been at the forefront of this progress and may be the key to conquering diseases currently considered incurable, saving millions of lives globally in the process.

References

Ahmadi A (2010) MS14, an Iranian herbal-marine compound for the treatment of multiple sclerosis. Chin J Integr Med 16:270

Akram W (2023) Marine-derived compounds applied in cardiovascular diseases: submerged medicinal industry. Mar Drugs 21(3):193

Basharat Z (2024) Screening marine microbial metabolites as promising inhibitors of Borrelia garinii: a structural docking approach towards developing novel Lyme disease treatment. BioMed Res Int 2024:9997082

Besednova NN (2019) Marine algae metabolites as promising therapeutics for the prevention and treatment of HIV/AIDS. Metabolites 9(5):87

Bhatia S (2022) Biotechnological innovations from ocean: transpiring role of marine drugs in management of chronic disorders. Molecules 27(5):1539

Cao Q (2020) Current research landscape of marine-derived anti-atherosclerotic substances. Mar Drugs 18(9):440

Cardoso C (2016) Dietary DHA and health: cognitive function aging. Nutr Res Rev 29(2):281

Cardozo FA (2023) Enhanced production of N-acetyl-glucosaminidase by marine Aeromonas caviae CHZ306 in bioreactor. Braz J Microbiol 54:1533–1545

Catanesi M (2021) Benefits under the sea: the role of marine compounds in neurodegenerative disorders. Mar Drugs 19(1):24

Copmans D (2022) From the North Sea to drug repurposing, the Antiseizure activity of Halimide and Plinabulin. Pharmaceuticals 15(2):247

DiNicolantonio JJ (2020) The importance of marine omega-3s for brain development and the prevention and treatment of behavior, mood, and other brain disorders. Nutrients 12(8):2333

Elbandy M (2022) Anti-inflammatory effects of marine bioactive compounds and their potential as functional food ingredients in the prevention and treatment of Neuroinflammatory disorders. Molecules 28(1):2

Engel AG (2015) Congenital myasthenic syndromes: pathogenesis, diagnosis, and treatment. Lancet Neurol 14(4):420

Fryar CD (2012) Prevalence of overweight, obesity, and extreme obesity among adults: United States, trends 1960–1962 through 2009–2010. CDC/NCHS

Garibyan L (2013) Research techniques made simple: polymerase chain reaction (PCR). J Invest Dermatol 133(3):e6

Goldenberg MM (2012) Multiple sclerosis review. Pharm Ther 37(3):175

Hamad I (2015) Marine bioactive compounds and their health benefits: a review. Compr Rev Food Sci Food Saf 14:446–465

Hannan A (2020) Neuroprotective potentials of marine algae and their bioactive metabolites: pharmacological insights and therapeutic advances. Mar Drugs 18(7):347

Hu D (2023) Application of marine natural products against Alzheimer's disease: past, present and future. Mar Drugs 21(1):43

Huang C (2019) Marine-derived natural compounds for the treatment of Parkinson's disease. Mar Drugs 17(4):221

Jeong SH (2014) Echinochrome a increases mitochondrial mass and function by modulating mitochondrial biogenesis regulatory genes. Mar Drugs 12(8):4602

Kim S-Y (2020) Diphlorethohydroxycarmalol (DPHC) isolated from the Brown alga Ishige okamurae acts on inflammatory myopathy as an inhibitory agent of TNF-? Mar Drugs 18(11):529

Kotimchenko YS (2022) Marine natural products from the Russian Pacific as sources of drugs for neurodegenerative diseases. Mar Drugs 20(11):708

Kudryatsev DS (2018) Makaluvamine G from the marine sponge Zyzzia fuliginosa inhibits muscle nAChR by binding at the Orthosteric and allosteric sites. Mar Drugs 16(4):109

Kumar A (2024) Alzheimer Disease. StatPearls Publishing

Lakshmi S (2018) Marine derived bioactive compounds for treatment of Alzheimer's disease. Front Biosci 10:537

Larsen R (2011) Health benefits of marine foods and ingredients. Biotechnol Adv 29(5):508
Lindequist U (2016) Marine-derived pharmaceuticals—challenges and opportunities. Biomol Ther 24(6):561
Majee SB (2017) Pharmacological, pharmaceutical, cosmetic and diagnostic applications of sulfated polysaccharides from marine algae and bacteria. Afr J Pharm Pharmacol 11(5):68
Malve H (2016) Exploring the ocean for new drug developments. J Pharm Bioallied Sci 8(2):83
Martins M (2020a) Marine natural products, multitarget therapy and repurposed agents in Alzheimer's disease. Pharmaceuticals 13(9):242
Martins BP (2020b) The role of marine omega-3 in human neurodevelopment, including autism Spectrum disorders and attention-deficit/hyperactivity disorder—a review. Crit Rev Food Sci Nutr 60(9):1431
McGivern JG (2006) Targeting N-type and T-type calcium channels for the treatment of pain. Drug Discov Today 11(5–6):245
Mensah GA (2019) The global burden of cardiovascular diseases and risk factors: 2020 and beyond. J Am Coll Cardiol 74(20):2529
Mishra S (2025) Marine-derived bioactive compounds for neuropathic pain: pharmacology and therapeutic potential. Naunyn-Schmiedeberg's Arch Pharmacol 398:6193
Moodie LWK (2019) Natural cholinesterase inhibitors from marine organisms. Nat Prod Rep 36:1053
Moovendhan M (2024) Natural pain killers from marine sources: a new frontier in neurosurgical pain management. Neurosurg Rev 47:498
Naik B (2024) Utilizing marine algal metabolites to fight neurodegenerative diseases. Mar Biotechnol Bioprod 11:1370839
Nova P (2020) Health benefits and bioavailability of marine resources components that contribute to health – what's new? Crit Rev Food Sci Nutr 60(21):3680
Pangestuti R (2011) Neuroprotective effects of marine algae. Mar Drugs 9(5):803
Pathak S (2024) Assessing effects of cannabis on various neuropathologies: a systematic review. J Ayurveda Integr Med 15(3):100911
Pereria RB (2019) Marine-derived anticancer agents: clinical benefits, innovative mechanisms, and new targets. Mar Drugs 17(6):329
Reames E (2012) Nutritional benefits of seafood. Southern Regional Aquaculture Center
Saber H (2017) Omega-3 fatty acids and incident ischemic stroke and its Atherothrombotic and Cardioembolic subtypes in 3 US cohorts. Stroke 48(10):2678
Samaddar S (2019) Polyphenols of marine red macroalga Symphyocladia latiuscula ameliorate diabetic peripheral neuropathy in experimental animals. Heliyon 5(5):e01781
Silva J (2023) Marine-derived components: can they be a potential therapeutic approach to Parkinson's disease? Mar Drugs 21(8):451
Šimat V (2020) Recent advances in marine-based nutraceuticals and their health benefits. Mar Drugs 18(12):627
Wiltshire E (2008) Juvenile Alpers disease. Arch Neurol 65(1):121
Wong SL (2014) Parkinson's disease: prevalence, diagnosis and impact. Health Rep 25(11):10–14
Wright DA (2023) Nutritional care and intervention in spinocerebellar ataxia. In: Ataxia – practice essentials and interventions. Intech Open
Xiao A-J (2014) Marine compound Xyloketal B reduces neonatal hypoxic-ischemic brain injury. Mar Drugs 13(1):29

Marine-Derived Natural Bioactives to Alleviate and Benefit Dementia

6

Jack Carlisle, Sophie Mae Talbert, Keyi Liu, Suhrud Pathak, Rishi Nadar, Muralikrishnan Dhanasekaran, G. Jeyabalan, Jaufar Palekkodan, Manha S. Ameen, Yaman Walid Kassab, and Muath Fahmi Najjar

Abstract

Dementia is a complex syndrome that is caused by multifaceted neurodegenerative disorders and cerebrovascular diseases, which refers to cognitive loss. The common dementia disorders include Alzheimer's disease (AD), dementia with Lewy bodies (DLB), frontotemporal dementia (FTD), Parkinson's disease (PD), limbic-predominant age-related TDP-43 encephalopathy (LATE), and vascular

J. Carlisle · S. M. Talbert · K. Liu · S. Pathak · R. Nadar
Drug Discovery and Development, Harrison College of Pharmacy, Auburn University, Auburn, AL, USA

M. Dhanasekaran (✉)
Drug Discovery and Development, Harrison College of Pharmacy, Auburn University, Auburn, AL, USA

Graduate School, Auburn University, Auburn, AL, USA
e-mail: dhanamu@auburn.edu

G. Jeyabalan
Alwar Pharmacy College, Rajasthan University of Health Sciences, Alwar, Jaipur, India

J. Palekkodan
Illham House, Changuvetti Kund, Kottakkal, Kerala, India

M. S. Ameen
Sophomore at Thomas S Wootton High School, Rockville, MD, USA

Y. W. Kassab
College of Pharmacy, National University of Science and Technology, Muscat, Oman

College of Pharmacy, Al-Bayan University, Baghdad, Iraq

M. F. Najjar
Clinical Pharmacy Department, Al Rayan National College of Health Sciences and Nursing, Al Rayan National Colleges, Al-Madinah al-Munawwarah, Saudi Arabia

M. Bhupathyraaj et al. (eds.), *Neuroprotective Marine Bioactives*, Nutritional Neurosciences, https://doi.org/10.1007/978-981-95-5791-2_6

dementia. Modern research shows that dementia often exhibits as the coexistence of multiple protein pathologies (cerebral multimorbidity), which further increases the complexity of disease diagnosis and treatment. Current conventional clinical therapies, such as cholinesterase inhibitors, NMDA receptor antagonists, and monoclonal antibodies, have been indicated to slow the progression of the disease. However, those therapies are also concerning due to their limited efficacy and ignorable side effects. Nowadays, marine natural products have become a trendy hot spot with the development of gathering and exploration technology. In this article, we systematically reviewed the recent advances in the treatment, prevention, and diagnosis of dementia, especially Alzheimer's disease, through the application of marine natural products, including marine-derived alkaloids, marine polyunsaturated fatty acids, astaxanthin, and fucoidan. The development of the exploration of marine biological resources reveals its considerable potential in the discovery of new anti-dementia natural medicines. And it is also expected to provide a safer and effective multi-target intervention strategy for dementia and also open a new direction for precision medicine and drug development for neurodegenerative diseases.

Keywords

Dementia · Alzheimer's disease · Marine natural products · Neurodegenerative diseases · Multi-target therapy

6.1 Dementia: Classification, Pathology, History, and Current Therapy

6.1.1 Introduction of Dementia

Dementia pathologies can be defined in a complex array of causes, studies, and treatments (Schneider 2022). Most commonly, neurodegenerative pathologies that tend to be the root causes of dementia and dementia-related disease in aging are amyloid-beta plaques, as well as paired helical filament (PHF)-tau tangles regarding Alzheimer's disease. Lewy bodies are highly associated with dementia with Lewy bodies (DLB) as well as dementia as a result of Parkinson's disease (PDD). TDP-43 and FTLD-tau or FTLD-TDP are most commonly associated with Limbic-related Age-Associated TDP-43 Encephalopathy (LATE) and frontotemporal dementia, respectively. Studies show that the location of the brain atrophy or brain damage is indicative and highly correspondent to the presentation and clinical phenotype of the specific dementia.

Dementia in history was commonly understood as a cover-all term for major neurocognitive disorders that resulted in psychosocial consequences (Assal 2019). It wasn't until the nineteenth century that dementia became a medical disease as a result of progress in neuropathology mapping. This progress allowed for the fragmentation of the disease into different neuropathological conditions, with many of the various types of dementias being linked to a vascular origin. Three main types of dementia were typically used to differentiate the disease: acute, chronic, and senile {Anonymous, 2025 #3305}{Anonymous, 2023 #3306}. Acute dementia is characterized by the rapid onset of symptoms that lead to a swift decline. Chronic dementia, which is most commonly observed today, displays a more gradual and progressive cognitive decline. Senile dementia or dementia as a result of the regular aging process is generally referred to as outdated. Despite specific classification, scientific advances reveal that a major cause of dementia is attributed to a loss of brain tissue following vascular events as opposed to plaques due to arteriosclerosis (Boller 2008).

6.1.2 Pathophysiology of Dementia

The pathophysiology of dementia is broadly represented as the aggregation of misfolded proteins and cerebrovascular disease (Sin 2023). Amyloid-beta plaques are neurofibrillary tangles that are viewed as prime examples of Alzheimer-related dementia. Though dementias across the board have characteristic symptoms such as cognitive decline and impairment, behavioral inconsistencies, and psychosocial defects, many mixed neuropathologies are found within the brains of those who have been diagnosed with dementia. This has opened the door to many clinical implications.

Dementia cannot be reduced to one cause, but it can be beneficial to observe the most common factors and leading contributors to the onset of dementia. Alzheimer's disease (AD) is the most common cause of late-onset dementia (Scheltens et al. 2021; Soria Lopez et al. 2019). The defining pathological characteristic of Alzheimer's disease (AD), as previously mentioned, is the accumulation of amyloid-beta plaques and neurofibrillary tangles between neurons. Deemed the amyloid cascade hypothesis, a further evaluation into the cause of AD attributed the abnormal accumulation of amyloid-beta plaques to an immense imbalance in amyloid-beta production and amyloid-beta clearance. Studies that investigated this hypothesis found that inherited mutations in one of three different genes led to the overproduction and eventual aggregation of amyloid beta. Additionally, neurofibrillary tangles due to phosphorylated tau proteins, another hallmark feature of AD according to the amyloid cascade hypothesis, play a crucial role in the pathological process. Hyperphosphorylation of these tau proteins results in the destabilization of microtubules. The downstream effects of this process create severe neurodegenerative damage. However, evidence has been increasingly compiled that supports the multicausality of pathophysiology.

Dementia with Lewy bodies (DLB) is the next most common form of neurodegenerative dementia (McKeith et al. 2017). DLB has four core clinical features: spontaneous motor parkinsonism, recurrent well-formed visual hallucinations, cognitive fluctuations, and REM sleep behavior disorder. DLB typically presents with at least one of these features alongside downgraded cognitive function. The pathology of DLB is understood to be the accumulation of alpha-synuclein, a synaptic protein. This accumulation occurs within Lewy bodies and Lewy neurites within the brain. Neuronal loss in the substantia nigra can also be attributed to the neuropathology of DLB. Parkinson's disease dementia (PDD) is a subtype of Lewy body dementia. Classification yields to the results of the onset of the specific disease, but they are both regarded as alpha-synucleinopathies. Furthering the idea of the multicausality of dementia as a disease, studies found that comorbid AD-related proteinopathies were frequently detected in people with DLB.

Vascular dementia is a concept that has evolved in recent years that attempts to connect CVD with cognitive impairment {Anonymous, 2019 #3307} (O'Brien and Thomas 2015). There is much overlap between neuroimaging and neuropathology between those with vascular dementia and older, healthy individuals who present with no cognitive symptoms. Thus, atherothromboembolic disease and small vessel disease are typically common among neuropathological findings in vascular dementia and can be used as markers for a diagnosis. Variability in the interpretation of vascular dementia furthers the multicausality theme of dementia as a disease and offers an explanation for various complexities of the disease.

Frontotemporal dementia is featured as a selective degeneration of the frontal and temporal lobes. It is the common cause of early-onset dementia, and it is found to be highly correlated with genetic factors.

Cerebral multimorbidity becomes the term used to describe dementia, as it is commonly displayed as a coexistence of multiple neuropathologies (Attems and Jellinger 2013). The synergistic interactions of these neurodegenerative processes contribute to both clinical diagnoses (specifically their thresholds) as well as disease trajectories. Future approaches must aim to identify strategies to combat a myriad of interworking pathologies to reduce the cognitive defects produced by these interactions.

6.1.3 Clinical Manifestations and Etiological Factors of Dementia

Common risk factors that contribute to dementia are age, family history, past medical history, diet, exercise, cardiovascular factors, substance use, physical and emotional trauma, and many more. Some of these factors can be altered by lifestyle choices, while others cannot be changed.

The symptoms of dementia can vary depending on the type of dementia and the leading cause of the disease. However, there is much overlap, including many cognitive changes and psychological changes that occur {Anonymous, 2024 #3308} {Anonymous, 2024 #3309}. These include memory loss, trouble communicating, trouble performing complex tasks, problems with reasoning or problem-solving, confusion, disorientation, personality changes, depression, anxiety, agitation, and many more. Such symptoms can intensify over time as the disease continues to progress.

6.1.4 Conventional and Complementary Dementia Therapies

Treatment is still an ever-growing topic of research with regard to dementia. Synthetic and natural treatments are both at the forefront of research and development. Some of the most common synthetic, FDA-approved treatments that are on the market now include cholinesterase inhibitors, NMDA receptor antagonists, combination therapy, aducanumab, and lecanemab.

The main cause of AD is thought to be due to the reduction of acetylcholine synthesis (Sharma 2019). Thus, the deterioration of the cholinergic neurons, which corresponds to loss of neurotransmission, is the major cause of cognitive decline. Cholinergic deficiency is highly correlated with Alzheimer's disease (AD), which makes cholinesterase inhibitors a prime target for therapeutic intervention. Examples of these inhibitors include donepezil, rivastigmine, and galantamine, which work by increasing levels of acetylcholine. This is done indirectly by inhibiting cholinesterase, enzymes that hydrolyze acetylcholine. Donepezil exhibits unique neurotransmitter level, molecular level, and chemical level activity to selectively inhibit acetylcholinesterase (AChE). Similarly, galantamine binds nicotinic cholinergic receptors and has been shown to be effective in the treatment of the cognitive symptoms of AD. Rivastigmine, which targets both AChE and butyrylcholinesterase (P. PubChem 2004), is typically used for the treatment of Parkinson's disease-associated dementia. However, the efficacy of these drugs has proven to be somewhat limited by adverse side effects and incomplete arrest of disease progression. Thus, these cholinesterase inhibitors are better viewed as a slowing or delaying of the disease rather than a true treatment of the disease.

N-methyl-d-aspartate receptors (NMDAR), as a neurotransmission pathway, play a critical role in synaptic plasticity as well as the survival of neurons (Wang and Reddy 2017). However, overuse and excessive NMDAR activity can lead to cell death, a primary potential cause of the neurodegeneration that occurs in AD. Studies have shown that the activation of extrasynaptic NMDARs is the major contributing factor that promotes cell death. Thus, an NMDAR antagonist that selectively blocks extrasynaptic NMDAR function is the goal of treatment. Memantine takes this role. Memantine functions to regulate the activity of glutamate. Glutamate is an abundant primary excitatory neurotransmitter within the central nervous system. It activates extrasynaptic NMDA receptors when too much is released. Memantine functions in limiting the activation of these extrasynaptic NMDARs by regulating glutamate.

Memantine can also be used in combination with cholinesterase inhibitors, specifically donepezil. Namzaric is the combination of memantine and donepezil and represents the idea of combination therapy—the use of multiple medications simultaneously to improve the efficacy of treatment.

Aducanumab (2021) is an anti-amyloid-beta monoclonal antibody. It has been formulated and intricately designed to specifically target AD (Rahman et al. 2023). It specifically targets amyloid-beta aggregates with the intention of breaking down the plaques. It has proven to be effective, but new studies show the occurrence of a higher incidence of amyloid-related imaging abnormalities, a serious adverse side effect. Lecanemab (2023) is the newest monoclonal antibody designed to clear amyloid-beta plaques (van Dyck et al. 2023). Similarly to Aducanumab, it binds to the toxic amyloid-beta plaques and promotes their removal from the brain, slowing the progression of AD and AD-related dementia. It also carries amyloid-related imaging abnormalities, but at a more limited rate compared to previous treatments. The long-term effects of both drugs are still underway.

Natural treatments tend to be less versed in the realm of dementia treatment. Some of these natural interventions look into dietary strategies, herbal remedies, physical and mental exercise, stress reduction, and even certain vitamins and supplements. These treatments tend to focus more on preventative measures.

The dietary approach involves diets that are rich in fruits, vegetables, whole grains, lean proteins, etc. (Dominguez et al. 2021). Popular diets that fall into this category include the ketogenic diet, which serves to provide an alternative energy source for the brain, as well as the DASH diet, which aims to lower blood pressure, a health defect that has shown a positive correlation with the development of dementia. Fish oil and omega-3 fatty acids also take part in lowering the risk of all-cause dementia and cognitive decline (Wei et al. 2023).

Similarly, herbal remedies have shown possible cognitive-enhancing effects, and there are many studies currently being conducted to evaluate herbal remedies as part of or minor contributors within an overall treatment program focused on the prevention and treatment of cognitive decline. Some of the remedies that have gained the most traction include ginkgo biloba, bacopa monnieri, ashwagandha, and rosemary. However, some of these remedies have shown significant improvement only within individuals with preexisting cognitive impairment and were not effective in the prevention of disease or accelerating the cognitive function of healthy individuals (Tan et al. 2014).

Physical and mental exercise continues to prove its continuous benefits (De la Rosa et al. 2020). Aerobic exercises improve adequate blood flow to the brain while puzzles and memory games supply the brain with long-term stimulation. Both physical and mental exercise are thought to play a role in neurogenesis within brain areas crucial for memory, such as the hippocampus. Nonetheless, as observed in many of the synthetic and natural treatments, further studies need to take place.

There has been growing interest in the idea of the utilization of marine natural products for therapeutic, preventative, and diagnostic purposes for dementia. This is an area of science and research that is relatively untapped and has large potential benefits and upside for the treatment of many diseases. The focus will be placed on marine natural products that have been at the core of novel research studies to alleviate the detrimental effects and cognitive decline related to dementia.

6.2 Marine Natural Products: Therapeutic, Preventive, and Diagnostic Purposes

6.2.1 Introduction of Marine Natural Products

Marine ecosystems harbor an immense biodiversity of organisms that produce a wide array of bioactive compounds with significant medical potential. These compounds, known as marine natural products (MNPs), exhibit unique structural diversity and potent biological activities that have made them valuable candidates for pharmaceutical development. Due to their diverse chemotypes, MNPs have emerged as promising templates for novel drugs aimed at treating a broad spectrum of diseases, including cancer, infectious diseases, and inflammatory conditions. Over the past few decades, MNPs have significantly impacted modern medicine through therapeutic, preventative, and diagnostic applications, leading to the discovery of numerous compounds now used clinically or in advanced stages of research.

Marine natural products have been developed into drugs for therapeutic use. These marine-derived drugs have been successfully employed as anticancer agents and antimicrobial agents. Macromolecules are expressed by cancer cells, activating oncogenic signaling pathways, which lead to uncontrolled cell division, proliferation, growth, and apoptosis (Yun et al. 2019). However, natural bioactive molecules exhibit cytotoxic effects by attacking cancerous macromolecules. Marine sponges and fungi have been found to have prolific bioactive secondary metabolites, able to generate a marine-derived compound, *salinosporamide A*. S*alinosporamide A* has been found to bind irreversibly to the catalytic sites of the proteasome, halting the degradation of misfolded proteins and triggering apoptosis in cancerous cells. This mechanism has proven especially effective against certain hematological malignancies, such as multiple myeloma and some leukemias, where rapid protein turnover is essential for cell survival.

Marine natural products have also shown broad antimicrobial activity. They are effective against a variety of pathogens, including bacteria, viruses, fungi, and parasites. Additionally, a large diversity of MNPs has shown efficacy or potential in drug-resistant infections (Abdelmohsen et al. 2017). *Salinosporamide A* also exhibits antibiotic properties as proteasome inhibition interferes with the degradation of proteins in bacterial cells, leading to cell death. In addition, many marine polysaccharides extracted from seaweed, such as *fucoidans and carrageenan*, display broad-spectrum antiviral activity. The antiviral activity targets a wide range of viruses, including herpes simplex virus (HSV), influenza viruses, and HIV, by preventing viral entry into host cells. Their broad-spectrum activity and low toxicity make them promising candidates for topical microbicides and antiviral drugs.

In terms of antiparasitic applications, compounds like *avrainvillamide* and *marinopyrrole A* have shown potential against parasites such as *Plasmodium falciparum,* malaria, and *Leishmania* species, offering new treatments in regions burdened by parasitic diseases (Zhang et al. 2023). Marine fungi and cyanobacteria have also yielded compounds capable of combating fungal infections caused by *Candida* and *Aspergillus* species, which are especially dangerous to immunocompromised individuals (Kumar 2022; Arendrup 2014). This advancement is promising due to the need for new antibiotics in the face of antibiotic resistance.

In addition to their therapeutic benefits, marine natural products have a wide range of preventative health applications. Marine natural products contain rich sources of anti-inflammatory, antioxidant, and nutraceutical compounds, which contribute to disease prevention. Marine algae and cyanobacteria have significant anti-inflammatory and antioxidant properties. Found in brown seaweed, *fucoxanthin* displays free-radical scavenging abilities that have been proven to reduce inflammatory symptoms associated with diabetes, obesity, and cardiovascular disease. *Fucoxanthin* also modulates gene expressions involved in lipid metabolism and insulin sensitivity, offering potential benefits for obesity management and metabolic syndrome.

Additionally, brown algae produce *phlorotannin*, which modulates inflammatory pathways by inhibiting key enzymes such as cyclooxygenase (COX), lipoxygenase (LOX), and nuclear factor kappa b (NF-kB) (Rajan et al. 2021). Also, *phlorotannin* acts as a natural scavenger of reactive oxygen species (ROS), thereby protecting cells from oxidative stress. Oxidative stress is a major contributor to aging and chronic diseases; therefore, phlorotannin has preventative effects against cancer, neurodegenerative conditions, and metabolic disorders. Their ability to neutralize reactive oxygen species (ROS) makes them effective in preventing cellular damage associated with Alzheimer's and Parkinson's.

One of the most notable marine-derived preventative agents is omega-3 polyunsaturated fatty acids extracted from fish oil. Found abundantly in cold-water marine fish, omega-3 s are crucial in cardiovascular health (Rohman et al. 2023). These fatty acids reduce blood triglyceride levels, lower blood pressure, and stabilize heart rhythms. *Docosahexaenoic acid* (DHA), a form of omega-3, is a vital structural component of neural and retinal membranes. DHA supports synaptic functions, neuroplasticity, and neurotransmission. These processes are required for functional learning, memory, visual acuity, and cognitive performance. Omega-3 nutrients help prevent and manage conditions such as rheumatoid arthritis, Alzheimer's, and bowel disease (Shahidi and Ambigaipalan 2018). These nutraceuticals are mainly given by dietary supplements in forms such as capsules, gels, or liquids.

Beyond therapeutics and prevention, marine natural products have revolutionized diagnostics and molecular biology tools. *Aequorea victoria* jellyfish produce a *green fluorescent protein (GFP), which is used as a genetically encoded fluorescent label* (Chudakov et al. 2010). These fluorescent proteins allow for the visualization of structural organization, protein localization, and the processes of living cells and organisms. Also, these proteins have the ability to provide super-resolution imaging and monitor the activity of enzymes and tumors.

Similarly, corals produce another fluorescent protein, DsRed and mCherry, expanding the color palette in multichannel imaging (Shaner et al. 2004). The wider variety of fluorescent proteins allows simultaneous visualization of multiple biological target organisms. These tools are especially valuable in cancer diagnostics, where they can be used to monitor tumor progression, metastasis, and responses to treatment. The fluorescence is able to track tumor progression and treatment responses. Also, marine-derived oxidases and electroactive compounds are used in biosensors to detect pathogens, toxins, and biomarkers with high sensitivity. These sensors offer rapid, specialized, and noninvasive diagnostic capabilities, which are crucial for early disease detection and effective treatment planning.

6.2.2 Marine Natural Products for Alzheimer's Disease Intervention

Alzheimer's disease (AD) is currently the most common cause of dementia. Its hallmark characteristics pose considerable implications for any affected individuals, and it remains one of the fastest-growing global health concerns. Alzheimer's disease-related dementia (ADRD) is the result of a combination of biological changes and damage within the brain, making the pathogenesis of the disease complicated and unique for each case (Hafez Ghoran and Kijjoa 2021). However, this intricate combination of factors also opens the door for a wide variety of products to potentially target specific parts or specific combinations of disease progression. Marine sponges, marine algae, and marine fish oils present the possibility of significant positive results in therapeutic and preventative treatments.

The combination of biological changes and damage within the brain occurs throughout various parts of the brain. However, the hippocampus and the cerebral cortex tend to be some of the most highly targeted areas within the pathogenesis of AD. Marine natural products will induce neuroprotective effects on these areas of the brain, typically through indirect action such as chemical upregulation or inhibition of neurotransmitters or receptors in these areas. Some products play a role as a physical barrier of protection for the brain.

Marine alkaloids are natural bioactive compounds that contain nitrogenated structures and are utilized in diverse biological activities (Munekata et al. 2021). They can be found in various marine macroorganisms, such as sponges and algae, and can also be chemically modified upon extraction. These alkaloids work to treat the effects of ADRD by inhibiting key enzymes involved in AD. A few of these enzymes include acetylcholinesterase (AChE), butyrylcholinesterase (BuChE), beta-secretase (BACE-1), and other kinases. Grassystatin A analog, 9-methylfascaplysin fascaplysin analog, and *Caenorhabditis elegans* have all undergone in vivo studies in animals. Grassystatin A analog was shown to reduce levels of amyloid-beta peptides in the brains of treated animals. Additionally, 9-methylfascaplysin, a fascaplysin analog, reduced the impending effects of cognitive dysfunction and inhibited the phosphorylation of amyloid-beta peptides by tau protein in an in vivo experiment on mice. *Caenorhabditis elegans* contains circumdatin D, a secondary metabolite that produces neuroprotective effects in the AD pathway.

6.2.2.1 Marine-Derived Alkaloids

Cholinesterase enzymes have been identified as relevant targets in the inhibition of AD progression. Marinoquinoline A is an alkaloid derived from a marine bacterium. It exhibits potent AChE inhibitory activity as it contains an IC_{50} value of 4.9 μM. For reference, the IC_{50} value of galantamine, a current cognitive-enhancing medication used to treat ADRD, is 0.6 μM. Additionally, ethyl acetate extracts of Acrostalagmus luteoalbus obtained from marine green algae present methoxyindole diketopiperazine derivatives with AChE inhibition activity (Hafez Ghoran and Kijjoa 2021). Each IC_{50} value represents the amount of drug necessary to produce its effect. IC_{50} value and drug potency are inversely related (Table 6.1).

Beta-amyloid aggregates are created through the cleavage of the amyloid precursor protein (APP). This cleavage is done by beta-secretase (BACE-1) and gamma-secretase. Thus, they are prime targets for inhibition of the pathogenesis of AD, and some marine alkaloids have been shown to be effective in inhibiting BACE-1. Lodopyridone (A, B, C) was purified from a marine-derived actinomycete, *Saccharomonospora*. Each compound displayed a weak inhibition (42–60%) of BACE-1. Similar results were found for 3-[(2,4-dimethoxy)benzylidene]-anabaseine, a nicotinic acid-derived alkaloid, and Ianthellidone F, a pyrrolidone alkaloid. Lodopyridone A also acted in the reduction of the production of beta-amyloid aggregation by nearly 50%. Bis-indole alkaloids such as racemosins A, racemosins B, and caulerpin, derived from green algae, also displayed a neuroprotective effect, inducing neurotoxicity in beta-amyloid aggregates (Leclerc et al. 2001).

Inhibition of protein kinase has the potential to reduce beta-amyloid levels and reverse AD phenotypes. PKC inhibitors are valuable because protein kinase assists in the control of tau phosphorylation and, consequently, aggregation. Biscogniauxone obtained from a marine-derived fungus can inhibit glycogen synthase kinase 3-beta (GSK-3β). GSK-3β is essential in multiple signaling pathways, including cell proliferation, apoptosis, inflammation, and glycogen metabolism. Thus, this is a prime target for the treatment of AD as well as other degenerative disorders.

Many marine alkaloids are in the early stages of preclinical studies. The information that has been gathered so far demonstrates promise for the safety and efficacy of these marine alkaloids in terms of their therapeutic and preventative effects on ADRD. Their neuroprotective and anti-inflammatory properties hold immense upside potential for the field of medicine.

Table 6.1 IC_{50} value of marine-derived alkaloids

Compound	IC_{50} value
Galantamine	0.6 μM
Marinoquinoline A	4.9 μM
N-methoxyindolediketopiperazine A (racemic mixture)	9.5 μM
N-methoxyindolediketopiperazine B (racemic mixture)	60.7 μM
N-methoxyindolediketopiperazine C (racemic mixture)	130.5 μM

6.2.2.2 Marine Polyunsaturated Fatty Acids

Though marine alkaloids are one of the apparent heads of interest in marine-derived treatments studies for dementia, omega-3 fatty acids have also gained much attention in recent years. Fish-derived omega-3 polyunsaturated fatty acids, specifically EPA and DHA, have been shown to be highly associated with reducing and minimizing the onset and effects of AD.

EPA is eicosatetraenoic acid, a common type of omega-3 fatty acid. It plays a significant role in anti-inflammatory effects, and it acts as a base from which multiple derivatives are formed within the brain (von Schacky 2021). It is also known for its cardiovascular benefits. EPA is thought to have the main effects of improved mood and vascular cognitive performance, as opposed to its counterpart, DHA, and its structural performance. Nonetheless, studies and trials indicate that EPA supplementation aids in the slowing of the progression of the early stages of cognitive decline. DHA is docosahexaenoic acid, another common type of omega-3 fatty acid. DHA functions in major structural roles for the brain and nervous system components. It also plays a role in vision. Its effects are most crucial during infancy and aging. Results of a recent meta-analysis show that "a low Omega-3 index predisposes to a more rapid cognitive decline and an earlier onset of dementia compared to a higher Omega-3 index" (von Schacky). It was also found that omega-3 fatty acids are beneficial during the early-onset phase of AD.

DHA is commonly found within neuron membrane phospholipids. It aids the crucial structural component in maintaining the integral barrier, which, when compromised, can affect the proper function of the nervous system. Studies have shown that patients with AD and cognitive impairment displayed significantly low levels of EPA and DHA (Swanson et al. 2012). A second study evaluated a diet that contained many omega-3 fatty acid-rich foods combined with a small amount of foods low in omega-3 fatty acids, which was strongly associated with a lower risk for AD. Brain imaging scans of DHA-rich diets displayed a significantly reduced plaque, with the most noticeable decreases within the hippocampus and parietal cortex—regions most commonly associated with AD. Another study also found that EPA and DHA supplementation reduced the release of inflammatory factors IL-1β, IL-6, and granulocyte colony-stimulating factor, which, within AD pathogenesis, lead to dysfunction of the neurons in the brain.

DHA exhibits multiple neuroprotective effects by various mechanisms, with a focus on brain structure and function (Cole et al. 2009). First, it reduces arachidonic acids and their metabolites. Because DHA is preferentially taken in by the brain, it competes with arachidonic acid for esterification into phospholipids. AD is partially characterized by inflammation of some regions of the brain as well as excitotoxic activity, both of which result in increased arachidonic acid products. Increased ratios of DHA relative to arachidonic acid have been shown to have a positive response with some neuroprotective observations similar to NSAID-like activity. DHA is also involved in trophic factor signaling. It can increase the neuronal phosphatidylserine levels and the rate of the critical docking step (Akt membrane docking). This creates a neurotrophic signaling increase, which functions to downregulate proapoptotic regulators. Other mechanisms include an increase in the production of

BDNF, which is depleted in AD-affected hippocampus, as well as the promotion of antioxidant defenses and reduction of oxidative damage commonly seen in AD.

One specific study conducted a randomized controlled clinical trial to evaluate the effects of supplemental DHA on the prevention of AD and, as a result, ADRD. Participants with a positive family history of dementia were chosen, and a placebo treatment was also administered. The study specifically focused on the status of APOE4 carriers in the human brain. APOE4 carriers inhibit the delivery of DHA to the brain, and they have been identified as high-risk individuals for AD development. The results of this study showed a lower brain uptake of DHA compared to noncarriers, indicating that the APOE4 genotype is a limiting factor in DHA delivery to the brain (Arellanes et al. 2020). This result was found within the placebo group as well as the supplemental group. This study further demonstrates that DHA supplementation increases general brain DHA delivery within healthy adults and contributes to maintaining cognitive health among healthy adults.

Omega-3 polyunsaturated fatty acids have multiple potential action mechanisms on brain functions. These include modifications to membrane fluidity, membrane-bound enzyme activity, receptor affinity, ion channel function, neurotransmitter production and activity levels, and signal transduction.

Nearly 50% of the neuronal membrane is composed of fatty acids (Yehuda et al. 2005). The ability of polyunsaturated fatty acids to change the fluidity index of the membrane allows them to potentially monitor and alter neuronal functions in major components of the brain. Additionally, these fatty acids play a contributing role in myelin sheath composition. The proper function of axons within the nervous system is crucial for optimal cognitive and physical function.

Most of these polyunsaturated fatty acids are supplied by the diet. Thus, they must be able to cross the blood-brain barrier. The blood-brain barrier is crucial within the aging process as its selective permeability keeps the integrity of the sensitive area. The buildup of beta-amyloid plaques within the brain due to a permeable blood-brain barrier or the lack of proper amyloid clearance is a known pathway for the progression of ADRD. It is also crucial that the blood-brain barrier be porous enough to allow the passage of these fatty acids.

6.2.2.3 Astaxanthin

Astaxanthin (AST) shows strong potential as a natural neuroprotective compound for Alzheimer's disease. Astaxanthin exhibits potent antioxidant and anti-inflammatory effects, showing protective action against brain aging and cognitive decline. Astaxanthin is a red-orange carotenoid pigment found naturally in marine organisms, including microalgae, krill, salmon, shrimp, and lobsters. AST is structurally unique from other carotenoids in that it displays a polar region on each end (Chang and Xiong 2020). Astaxanthin features polar ionone rings at each end of its structure, enabling it to anchor into cell membranes and effectively neutralize free radicals. Also, the nonpolar zone is made up of 13 carbon-carbon double polyunsaturated bonds, allowing high-energy electrons to be removed. These distinctive structural features of astaxanthin contribute to its superior antioxidant capacity compared to other carotenoids (Donoso et al. 2021).

Overproduction of reactive oxygen species (ROS) results in oxidative stress, which plays a critical role in the etiology of Alzheimer's disease (Si and Zhu 2022). ROS damages lipids, membranes, proteins, and nucleic acids in neurons, leading to neuronal deficits and apoptosis. However, astaxanthin plays a crucial role in maintaining cellular integrity by effectively quenching molecular oxygen and scavenging free radicals. This antioxidant action not only preserves the structural integrity of cellular membranes but also supports immune system function and gene expression regulation, thereby contributing to cellular homeostasis and reducing the oxidative damage associated with neurodegenerative diseases such as Alzheimer's. AST neutralizes ROS both in the cytosol and within mitochondrial membranes, where oxidative damage is most critical in AD. Studies have shown that rats fed AST have enhanced activity of endogenous antioxidant enzymes such as superoxide dismutase (SOD), catalase (CAT), and thiobarbituric acid reactive substances in the liver and plasma (Ranga Rao et al. 2010).

Alzheimer's disease is highly associated with neuroinflammation due to activated microglia and astrocytes releasing pro-inflammatory cytokines, resulting in neural dysfunction (Netea et al. 2017). The NF-κB pathway is a key regulatory pathway for inflammation and immune responses. Excessive activation of the NF-kB pathway has been linked to deleterious inflammation related to neurodegenerative diseases such as AD (Si and Zhu 2022). When exposed to external stimuli, the NF-kB pathway is activated by the dissociation of IkB phosphorylated by the IkB kinase complex (IKKa and IKKb). Once detached from IkB and activated, NF-kB enters the nucleus and binds to specific DNA sites, triggering the production of inflammatory cytokines, including IL-1β, IL-6, and TNF-α. Therefore, inhibiting the activation of the NF-κB pathway is seen as an effective strategy for reducing excessive inflammation. Astaxanthin (AST) can help block this overactivation by either downregulating the phosphorylation of IκB-α or increasing the production of IκB-α. In addition, astaxanthin demonstrates significant anti-inflammatory activity by suppressing cyclooxygenase-1 (COX-1) expression and reducing nitric oxide production in lipopolysaccharide (LPS)-stimulated BV2 microglial cells, indicating its broader regulatory role in inflammation-associated gene expression (Kohandel et al. 2022).

Astaxanthin (AST) has the ability to cross the blood-brain barrier (BBB), a selectively permeable membrane that restricts most therapeutic compounds from entering the central nervous system (CNS). This barrier is efficient in protecting neuronal function while maintaining a stable and safe environment. In AD, the BBB becomes disrupted, permitting harmful substances to cross the barrier, contributing to inflammation and neurodegeneration. Astaxanthin is a lipid-soluble pigment that enables efficient penetration into brain tissue. This allows astaxanthin to exert localized antioxidant and anti-inflammatory actions directly within neuronal and glial cells, offering therapeutic potential that many other antioxidants lack.

In a recent study, astaxanthin (AST) accumulated in the hippocampus and cerebral cortex of rat brains following oral administration (Si and Zhu 2022; El-Agamy et al. 2018). This accumulation in key brain regions involved in memory and cognition suggests that AST contains the potential to preserve or enhance cognitive function. A growing body of research has demonstrated that astaxanthin (AST) may play a neuroprotective and neurorestorative role in the brain by promoting neurogenesis and enhancing the expression of key genes involved in neuronal repair, structural maintenance, and functional recovery. AST's capacity to modulate these genes suggests its therapeutic potential in conditions characterized by neuronal damage, such as Alzheimer's disease, stroke, and traumatic brain injury. Among the genes influenced by AST are glial fibrillary acidic protein (GFAP), microtubule-associated protein 2 (MAP-2), brain-derived neurotrophic factor (BDNF), and growth-associated protein 43 (GAP-43). Together, the modulation of these genes by astaxanthin underscores its potential as a multi-target neurotherapeutic agent, capable of not only protecting neurons from oxidative and inflammatory damage but also promoting structural and functional recovery in the injured or aging brain.

Amyloid-β (Aβ) is a protein fragment formed when amyloid precursor protein (APP) is broken down by enzymes beta-secretase and gamma-secretase. In normal blood-brain barriers (BBB), these protein fragments clear; however, they accumulate and clump together in Alzheimer's disease. The accumulation of Aβ from extracellular amyloid plaques is toxic to neurons. These amyloid plaques disrupt cell-cell communications, neuronal function, and transport across the BBB (Babalola et al. 2023). In Alzheimer's, Aβ42 aggregation is one of the first detectable symptoms in the brain.

In clinical studies, astaxanthin (AST) displays the ability to improve Aβ plaques and impaired clearance. AST binds to Aβ peptides, altering their structure, which decreases the probability of toxic plaque formation. In a clinical study, astaxanthin was seen to directly bind to Aβ 42, the most common Aβ protein associated with AD, and inhibits the formation of fibrils (Alghazwi et al. 2019). Also, AST inhibits the formation of β-sheets, which formation needed for the amyloid-beta plaques. Astaxanthin's ability to reduce the likelihood of the formation of Aβ plaques and clearance of the BBB makes it a promising candidate for prevention and protection as a therapeutic strategy for AD.

Overall, astaxanthin, a naturally occurring carotenoid found in marine bioactive organisms, has shown promising potential in the prevention and management of Alzheimer's disease. Due to its potent antioxidant, anti-inflammatory, and anti-amyloid properties, astaxanthin can directly affect processes involved in Alzheimer's progression. Astaxanthin helps reduce oxidative stress by neutralizing harmful reactive oxygen species (ROS), which are elevated in Alzheimer's and contribute to neuronal damage. One of its key roles is inhibiting the aggregation of amyloid-beta (Aβ) peptides, which are toxic protein fragments that accumulate in the brain and form plaques characteristic of the disease. Additionally, it appears to suppress neuroinflammation by downregulating inflammatory cytokines and supporting the health of brain immune cells like microglia. Astaxanthin may also protect mitochondrial function and enhance autophagy, aiding the brain in clearing harmful protein buildups due to its ability to cross the blood-brain barrier. While most research so far has been conducted in animal models and cell cultures, early findings suggest that astaxanthin could be a valuable neuroprotective supplement. However, more human clinical trials are needed to confirm its therapeutic effects in Alzheimer's patients.

6.2.2.4 Fucoidan

Fucoidan (FUC) exhibits promising neuroprotective effects against Alzheimer's disease (AD) due to its anti-inflammatory, antioxidant, and anti-amyloid properties. Fucoidan is a class of polysaccharides extracted from the extracellular matrix of bioactive brown seaweed, such as *Fucus vesiculosus* and *Undaria pinnatifida*. Chemically, it is primarily composed of L-fucose residues, often accompanied by sulfate groups, and occasionally other monosaccharides such as galactose, mannose, and xylose, depending on the seaweed source and extraction method.

Chronic neuroinflammation exacerbates Alzheimer's disease. Fucoidan has been seen to mediate neuroinflammatory symptoms. In a clinical study, lipopolysaccharide (LPS) was used to induce neuronal and cognitive damage in the hippocampus of mice (Wang et al. 2022). Fucoidan was introduced as a neurorestorative compound to HT22 neural cells in the hippocampus. FUC protected the HT22 cells through anti-inflammation and antioxidative effects. Also, fucoidan ameliorated the LPS injury through counteracting the induced oxidative stress. FUC inhibited the release of inflammatory cytokines and the NLRP3 inflammasome activation of HT22 cells. Also, it repressed AChE activity, which activates the hydrolysis of acetylcholine and aggravates neuronal loss. The alleviation of neuroinflammatory effects in LPS-induced injury promotes cognitive and memory improvements.

One of the key mechanisms through which fucoidan exerts neuroprotective effects in the context of Alzheimer's disease (AD) is its ability to inhibit the toxicity of amyloid-beta (Aβ) peptides, which are heavily implicated in the initiation and progression of the disease. Aβ peptides aggregate to form extracellular plaques in the brain, disrupting synaptic communication and triggering a cascade of neurodegenerative events. Fucoidan has been shown in neuronal culture studies to significantly reduce the neurotoxic impact of Aβ, preserving neuronal activity and preventing the typical decline in synaptic function caused by Aβ accumulation (Jhamandas et al. 2005). This neuroprotective effect suggests that fucoidan may interfere with either the aggregation process of Aβ or its binding to neuronal receptors, ultimately safeguarding neurons from Aβ-induced apoptosis and functional impairment.

In addition to combating Aβ toxicity, fucoidan also addresses another major pathological hallmark of AD, oxidative stress. Oxidative stress arises when there is an imbalance between reactive oxygen species (ROS) production and the brain's antioxidant defense systems. In AD, elevated ROS levels contribute to lipid peroxidation, DNA damage, mitochondrial dysfunction, and neuronal death (Wang et al. 2018). Fucoidan's antioxidant capabilities have been demonstrated through its ability to lower intracellular ROS levels, reduce lipid peroxidation markers such as malondialdehyde (MDA), and preserve mitochondrial integrity. By enhancing mitochondrial function and overall cell viability, fucoidan not only helps prevent oxidative damage but may also improve energy metabolism and neuronal resilience. This dual action, blocking Aβ toxicity and suppressing oxidative stress, positions fucoidan as a particularly compelling candidate for multi-targeted therapeutic intervention in Alzheimer's disease (Wolfe 2021).

Fucoidan also shows promise in mitigating tau pathology, another central feature of Alzheimer's disease. Tau protein, under normal physiological conditions, plays a vital role in maintaining microtubule stability within neurons. However, in AD, tau becomes abnormally hyperphosphorylated, causing it to detach from microtubules and aggregate into insoluble neurofibrillary tangles (NFTs), structures strongly associated with neurodegeneration and cognitive decline. The NFTs are misfolded and insoluble, leading to a buildup in the hippocampus. The pathological spread of tau across brain regions contributes significantly to disease progression. Recent research suggests that fucoidan reduces tau hyperphosphorylation through the upregulation of the kt/GSK-3β pathway. The kt/GSK-3β pathway is known for regulating the phosphorylation of the tau proteins (Park et al. 2019). Although these findings are largely based on in vitro and animal model studies, they provide a compelling rationale for further investigation into fucoidan as a multi-target therapeutic agent capable of addressing both amyloid and tau pathologies in Alzheimer's disease (Mandal 2017) (Fig. 6.1).

Overall, fucoidan (FUC), a sulfated polysaccharide derived from brown seaweed, demonstrates significant neuroprotective potential in the treatment of Alzheimer's disease (AD). Its therapeutic effects are attributed to its anti-inflammatory, antioxidant, anti-amyloid, and anti-tau properties. Fucoidan protects neural cells by reducing neuroinflammation, as seen in studies using LPS-induced hippocampal injury models, where it suppressed pro-inflammatory cytokines, inhibited NLRP3 inflammasome activation, and preserved acetylcholine levels by repressing AChE activity. It also counters amyloid-beta (Aβ) toxicity by preventing its aggregation and interaction with neuronal receptors, thereby preserving synaptic function and neuronal survival. Furthermore, fucoidan combats oxidative stress by lowering reactive oxygen species (ROS), minimizing lipid peroxidation, and preserving mitochondrial function, enhancing neuronal resilience. In addition to these effects, fucoidan mitigates tau pathology by reducing tau hyperphosphorylation through the regulation of the Akt/GSK-3β signaling pathway, which is crucial in maintaining tau stability. Collectively, these multi-targeted actions position

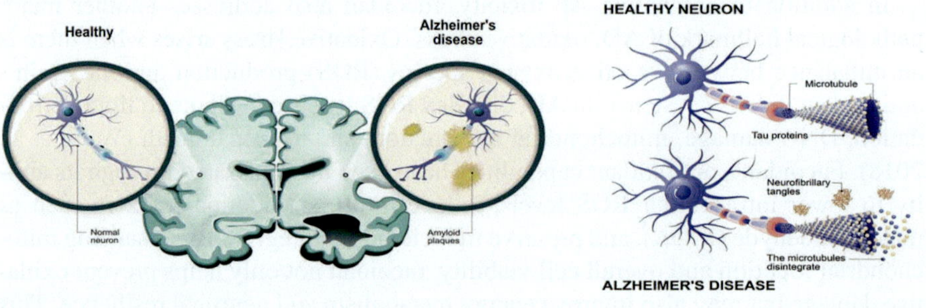

Fig. 6.1 Healthy vs Alzheimer's disease

fucoidan as a promising natural compound for the prevention and treatment of Alzheimer's disease, warranting further exploration in clinical settings.

6.3 Conclusion and Future Prospects

Marine natural products' potential benefits extend beyond their therapeutic and preventative effects on dementia. Current studies also demonstrate antitumor effects, cardioprotective activity, antiseizure activity, antidepressant-like activity, anti-inflammatory activity, antiparasitic activity, antimicrobial activity, and more. Cancer research has also begun to tap into the marine natural product world (Khalifa et al. 2019; Kumar and Adki 2018). Being one of the world's most lethal diseases, the need for novel treatments and action plans is at the forefront of the research world. These products possess many new possible mechanisms and biological effects that have the potential to be the breakthrough researchers are hoping for. Marine natural products' potential for pharmaceutical and medical advancements is seemingly endless.

Additionally, marine natural products are considered to be a plentiful and sustainable resource. Advancements in sustainable harvesting and aquaculture are already beginning to gain traction. Research also shows that some of these common marine derivatives can be chemically replicated and modified. The potential upside of these marine resources creates immense research potential. Though an opportunity and a challenge, marine compounds are already showing promise in the field of treating an array of neurodegenerative diseases.

Additionally, marine natural products are being investigated within the context of cosmetics. Their natural hydrating and antioxidant properties give them an "anti-aging" effect that many cosmeceutical companies are looking to capitalize on (Fonseca et al. 2023). The same goes for nutraceutical companies that look to supply the demand for sources of nutrients to benefit leading minor diseases worldwide. Most notably, fish oils, containing the well-known omega-3 fatty acids, have entered the field of health and fitness (Tvrzicka et al. 2011). The idea and interest in "functional foods" and supplements derived from marine natural products will continue to rise.

Marine natural products provide a promising, hopeful future. The demand is present, the technological advancements and the biotechnologies necessary are present, and the research is underway. Problems such as cost, scalability, access, and regulation are the factors that threaten the success of implementing these bioactive compounds. To address these concerns is to invest in the promising future of the relatively untapped world of marine natural products.

References

Abdelmohsen UR, Balasubramanian S, Oelschlaeger TA, Grkovic T, Pham NB, Quinn RJ, Hentschel U (2017) Potential of marine natural products against drug-resistant fungal, viral, and parasitic infections. Lancet Infect Dis 17(2):e30–e41

Alghazwi M, Smid S, Musgrave I, Zhang W (2019) In vitro studies of the neuroprotective activities of astaxanthin and fucoxanthin against amyloid beta (Aβ1-42) toxicity and aggregation. Neurochem Int 124:215–224

Arellanes IC, Choe N, Solomon V, He X, Kavin B, Martinez AE et al (2020) Brain delivery of supplemental docosahexaenoic acid (DHA): a randomized placebo-controlled clinical trial. EBioMedicine 59:102883

Arendrup MC (2014) Update on antifungal resistance in *Aspergillus* and *Candida*. Clin Microbiol Infect 20:42–48

Assal F (2019) History of dementia. Front Neurol Neurosci 44:118–126

Attems J, Jellinger K (2013) Neuropathological correlates of cerebral multimorbidity. Curr Alzheimer Res 10(6):569–577

Babalola JA, Lang M, George M, Stracke A, Tam-Amersdorfer C, Itxaso I et al (2023) Astaxanthin enhances autophagy, amyloid beta clearance and exerts anti-inflammatory effects in in vitro models of Alzheimer's disease-related blood brain barrier dysfunction and inflammation. Brain Res 1819:148518

Boller F (2008) History of dementia. In: Handbook of clinical neurology, vol 89. Elsevier, pp 3–13

Chang MX, Xiong F (2020) Astaxanthin and its effects in inflammatory responses and inflammation-associated diseases: recent advances and future directions. Molecules 25(22):5342

Chudakov DM, Matz MV, Lukyanov S, Lukyanov KA (2010) Fluorescent proteins and their applications in imaging living cells and tissues. Physiol Rev 90(3):1103–1163

Cole GM, Ma Q-L, Frautschy SA (2009) Omega-3 fatty acids and dementia. Prostaglandins Leukot Essent Fat Acids 81(2):213–221

De la Rosa A, Olaso-Gonzalez G, Arc-Chagnaud C, Millan F, Salvador-Pascual A, García-Lucerga C et al (2020) Physical exercise in the prevention and treatment of Alzheimer's disease. J Sport Health Sci 9(5):394–404

Dominguez LJ, Veronese N, Vernuccio L, Catanese G, Inzerillo F, Salemi G, Barbagallo M (2021) Nutrition, Physical activity, and other lifestyle factors in the prevention of cognitive decline and dementia. Nutrients 13(11):4080

Donoso A, González-Durán J, Muñoz AA, González PA, Agurto-Muñoz C (2021) Therapeutic uses of natural astaxanthin: an evidence-based review focused on human clinical trials. Pharmacol Res 166:105479

El-Agamy SE, Abdel-Aziz AK, Wahdan S, Esmat A, Azab SS (2018) Astaxanthin ameliorates doxorubicin-induced cognitive impairment (Chemobrain) in experimental rat model: impact on oxidative, inflammatory, and apoptotic machineries. Mol Neurobiol 55(7):5727–5740

Fonseca S, Amaral MN, Reis CP, Custódio L (2023) Marine natural products as innovative cosmetic ingredients. Mar Drugs 21(3):170

Hafez Ghoran S, Kijjoa A (2021) Marine-derived compounds with anti-Alzheimer's disease activities. Mar Drugs 19(8):410

Jhamandas JH, Wie MB, Harris K, MacTavish D, Kar S (2005) Fucoidan inhibits cellular and neurotoxic effects of beta-amyloid (A beta) in rat cholinergic basal forebrain neurons. Eur J Neurosci 21(10):2649–2659

Khalifa SAM, Elias N, Farag MA, Chen L, Saeed A, Hegazy M-EF et al (2019) Marine natural products: a source of novel anticancer drugs. Mar Drugs 17(9):491

Kohandel Z, Farkhondeh T, Aschner M, Pourbagher-Shahri AM, Samarghandian S (2022) Anti-inflammatory action of astaxanthin and its use in the treatment of various diseases. Biomed Pharmacother 145:112179

Kumar MS (2022) Antifungal and antibacterial activity of marine sponges from Ratnagiri coast of India. J Med Mycol 32(4):101305

Kumar MS, Adki KM (2018) Marine natural products for multi-targeted cancer treatment: a future insight. Biomed Pharmacother 105:233–245

Leclerc S, Garnier M, Hoessel R, Marko D, Bibb JA, Snyder GL et al (2001) Indirubins inhibit glycogen synthase kinase-3 beta and CDK5/p25, two protein kinases involved in abnormal tau phosphorylation in Alzheimer's disease. A property common to most cyclin-dependent kinase inhibitors? J Biol Chem 276(1):251–260

Mandal A (2017) Alzheimer's disease specific brain deposits photographed for the first time in vivid detail. News Med Life Sci

McKeith IG, Boeve BF, Dickson DW, Halliday G, Taylor JP, Weintraub D et al (2017) Diagnosis and management of dementia with Lewy bodies: fourth consensus report of the DLB consortium. Neurology 89(1):88–100

Munekata PES, Pateiro M, Conte-Junior CA, Domínguez R, Nawaz A, Walayat N et al (2021) Marine alkaloids: compounds with in vivo activity and chemical synthesis. Mar Drugs 19(7):374

Netea MG, Balkwill F, Chonchol M, Cominelli F, Donath MY, Giamarellos-Bourboulis EJ et al (2017) A guiding map for inflammation. Nat Immunol 18(8):826–831

O'Brien JT, Thomas A (2015) Vascular dementia. Lancet 386(10004):1698–1706

P. PubChem Compound Summary for CID 6253, Cytarabine Bethesda (MD): National Library of Medicine (US), National Center for Biotechnology Information; 2004. Available from: https://pubchem.ncbi.nlm.nih.gov/compound/Cytarabine

Park SK, Kang JY, Kim JM, Yoo SK, Han HJ, Chung DH et al (2019) Fucoidan-rich substances from Ecklonia cava improve Trimethyltin-induced cognitive dysfunction via down-regulation of amyloid β production/tau hyperphosphorylation. Mar Drugs 17(10):591

Rahman A, Hossen MA, Chowdhury MFI, Bari S, Tamanna N, Sultana SS et al (2023) Aducanumab for the treatment of Alzheimer's disease: a systematic review. Psychogeriatrics 23(3):512–522

Rajan DK, Mohan K, Zhang S, Ganesan AR (2021) Dieckol: a brown algal phlorotannin with biological potential. Biomed Pharmacother 142:111988

Ranga Rao A, Raghunath Reddy RL, Baskaran V, Sarada R, Ravishankar GA (2010) Characterization of microalgal carotenoids by mass spectrometry and their bioavailability and antioxidant properties elucidated in rat model. J Agric Food Chem 58(15):8553–8559

Rohman A, Irnawati, Windarsih A, Riswanto FDO, Indrayanto G, Fadzillah NA et al (2023) Application of chromatographic and spectroscopic-based methods for analysis of Omega-3 (ω-3 FAs) and Omega-6 (ω-6 FAs) fatty acids in marine natural products. Molecules 28(14):5524

Scheltens P, De Strooper B, Kivipelto M, Holstege H, Chételat G, Teunissen CE et al (2021) Alzheimer's disease. Lancet 397(10284):1577–1590

Schneider JA (2022) Neuropathology of dementia disorders. Continuum 28(3):834–851

Shahidi F, Ambigaipalan P (2018) Omega-3 polyunsaturated fatty acids and their health benefits. Annu Rev Food Sci Technol 9:345–381

Shaner NC, Campbell RE, Steinbach PA, Giepmans BNG, Palmer AE, Tsien RY (2004) Improved monomeric red, orange and yellow fluorescent proteins derived from Discosoma sp. red fluorescent protein. Nat Biotechnol 22(12):1567–1572

Sharma K (2019) Cholinesterase inhibitors as Alzheimer's therapeutics (Review). Mol Med Rep 20(2):1479–1487

Si P, Zhu C (2022) Biological and neurological activities of astaxanthin (review). Mol Med Rep 26(4):300

Sin CK (2023) Pathophysiology of dementia. Aust J Gen Pract 52:516–521

Soria Lopez JA, González HM, Léger GC (2019) Chapter 13: Alzheimer's disease. In: Dekosky ST, Asthana S (eds) Handbook of clinical neurology, vol 167. Elsevier, pp 231–255

Swanson D, Block R, Mousa SA (2012) Omega-3 fatty acids EPA and DHA: health benefits throughout life. Adv Nutr 3(1):1–7

Tan M-S, Yu J-T, Tan C-C, Wang H-F, Meng X-F, Wang C et al (2014) Efficacy and adverse effects of ginkgo Biloba for cognitive impairment and dementia: a systematic review and meta-analysis. J Alzheimer's Dis 43(2):589–603

Tvrzicka E, Kremmyda L-S, Stankova B, Zak A (2011) Fatty acids as biocompounds: their role in human metabolism, health and disease – a review. Part 1: classification, dietary sources and biological functions. Biomed Pap 155(2):117–130

van Dyck Christopher H, Swanson CJ, Aisen P, Bateman RJ, Chen C, Gee M et al (2023) Lecanemab in early Alzheimer's disease. N Engl J Med 388(1):9–21

von Schacky C (2021) Importance of EPA and DHA blood levels in brain structure and function. Nutrients 13(4):1074

Wang R, Reddy PH (2017) Role of glutamate and NMDA receptors in Alzheimer's disease. J Alzheimer's Dis 57(4):1041–1048

Wang X, Yi K, Zhao Y (2018) Fucoidan inhibits amyloid-β-induced toxicity in transgenic Caenorhabditis elegans by reducing the accumulation of amyloid-β and decreasing the production of reactive oxygen species. Food Funct 9(1):552–560

Wang Y, Wang Q, Duan L, Li X, Yang W, Huang T et al (2022) Fucoidan ameliorates LPS-induced neuronal cell damage and cognitive impairment in mice. Int J Biol Macromol 222:759–771

Wei B-Z, Li L, Dong C-W, Tan C-C, Xu W (2023) The relationship of Omega-3 fatty acids with dementia and cognitive decline: evidence from prospective cohort studies of supplementation, dietary intake, and blood markers. Am J Clin Nutr 117(6):1096–1109

Wolfe MS (2021) Disruption of amyloid β-protein processing drives Alzheimer's. Research Outreach

Yehuda S, Rabinovitz S, Mostofsky DI (2005) Essential fatty acids and the brain: from infancy to aging. Neurobiol Aging 26(1, Supplement):98–102

Yun CW, Kim HJ, Lee SH (2019) Therapeutic application of diverse marine-derived natural products in cancer therapy. Anticancer Res 39(10):5261

Zhang M, Zhang Q, Zhang Q, Cui X, Zhu L (2023) Promising Antiparasitic natural and synthetic products from marine invertebrates and microorganisms. Mar Drugs 21(2):84

Viewing Marine Natural Products' Role in Neuronal Hyperarousal

7

Cammi Griffith, Keyi Liu, Suhrud Pathak, Rishi Nadar, Muralikrishnan Dhanasekaran, K. Reeta Vijayarani, and Kirubakaran Narayanan

Abstract

Neuronal hyperarousal is a pathological condition related to various neuropsychiatric diseases, which manifests with abnormal, excessive, and prolonged activation of the nervous system. This physiological phenomenon is modulated by mechanisms involving cortisol and adrenergic signaling and has been indicated to play an important role in Alzheimer's disease (AD), Parkinson's disease (Ong W et al. Cancers (Basel) 14(16):4025, 2022), anxiety disorders, difficulty falling asleep, and post-traumatic stress disorder (PTSD). However, neuronal hyperarousal has also been considered a lack of efficient treatment because of its complex mechanism. Recently, with the development of the marine natural products (MNPs) explosion, more and more marine bioactives have been brought to the public's attention due to their unique chemical structure and potential neuromodulatory activity. This article has a systematic review of the introduction, history, clinical symptoms, and current therapy of neuronal hyperarousal and brings

C. Griffith · K. Liu · S. Pathak · R. Nadar
Drug Discovery and Development, Harrison College of Pharmacy, Auburn University, Auburn, AL, USA

M. Dhanasekaran (✉)
Drug Discovery and Development, Harrison College of Pharmacy, Auburn University, Auburn, AL, USA

Graduate School, Auburn University, Auburn, AL, USA
e-mail: dhanamu@auburn.edu

K. R. Vijayarani
Department of Pharmaceutics, Surya School of Pharmacy, Vikravandi, Villupuram. Affiliated to The TN Dr. M.G.R. Medical University, Chennai, Tamil Nadu, India

K. Narayanan
Drugs Testing Laboratories, Directorate of Drugs Control, DMS Teynampet, Chennai, Tamil Nadu, India

M. Bhupathyraaj et al. (eds.), *Neuroprotective Marine Bioactives*, Nutritional Neurosciences, https://doi.org/10.1007/978-981-95-5791-2_7

a foresight of the potential application of marine natural products in this field. A rising number of studies have evidenced that the anti-inflammatory, antioxidant, and neuroplasticity effects of the marine natural products provide a valuable, novel, and safer strategy for both diagnosis and treatment in neuronal hyperarousal. In summary, neurological hyperarousal is a key target in the study of neurological diseases. In-depth exploration of its pathological mechanisms and the combination with marine natural products research is of great significance for the treatment of neuropsychiatric diseases.

Keywords

Neuronal hyperarousal · Adrenergic signaling · Post-traumatic stress disorder (PTSD) · Marine natural products

7.1 Introduction to Neuronal Hyperarousal

Arousal, a key component of consciousness, is controlled by a network involving the brainstem, thalamus, hypothalamus, and cortex (Xie et al. 2024). It influences cognition, attention, and perception. Arousal is a key part of everyday function, so if it becomes unregulated, it will have dire consequences on an individual's health and subjective well-being. Neuronal hyperarousal is a state of excessive and prolonged activation of the nervous system, leading to heightened alertness, anxiety, and difficulty relaxing. It is a collection of symptoms that can manifest when the sympathetic nervous system is in a constant state of "fight-or-flight." It can often be triggered by a traumatic event and continue to persist due to unstable hormones such as cortisol, epinephrine, and norepinephrine. Individuals experiencing neuronal hyperarousal often report symptoms such as restlessness, irritability, inability to sleep, and inability to focus. One of the primary triggers of neuronal hyperarousal is exposure to traumatic events. The brain releases cortisol, epinephrine, and norepinephrine when it perceives a threat. These are stress hormones that have distinct functions in regulation. The steroid hormone cortisol is released by cells of the adrenal cortex. It is associated with long-term stress, which is the chronic stress that comes from repeated exposure to stressful situations. In a normal stress response, these hormones would help the body prepare for action and then recalibrate to normal levels once the threat has passed. However, in cases of hyperarousal, these hormones remain elevated, preventing the nervous system from returning to a calm state. This prolonged stress response can contribute to anxiety disorders, post-traumatic stress disorder (PTSD), and chronic stress-related illnesses. The effects of neuronal hyperarousal extend beyond psychological distress. Physiological consequences include increased heart rate, elevated blood pressure, and a weakened immune system. Over time, chronic hyperarousal can lead to metabolic imbalances, cardiovascular disease, and cognitive impairments. The inability to regulate arousal also disrupts sleep patterns, exacerbating fatigue and further impairing cognitive function. Addressing neuronal hyperarousal requires a multifaceted approach. Techniques such as meditation, deep breathing exercises, and progressive muscle relaxation can help regulate the nervous system. Additionally, cognitive behavioral therapy (CBT) and other therapeutic interventions can assist individuals in managing stress responses and retraining the brain's reaction to perceived threats. Pharmacological treatments, including selective serotonin reuptake inhibitors (SSRIs) and beta-blockers, may also be beneficial in cases of severe hyperarousal. Understanding and managing neuronal hyperarousal is crucial for maintaining overall well-being. By implementing strategies to balance the autonomic nervous system and reduce chronic stress, individuals can improve their mental health, enhance cognitive function, and promote long-term physiological stability.

7.2 History of Treatment and Diagnosis in Neuronal Hyperarousal

Hyperarousal, as a symptom within the context of post-traumatic stress disorder (PTSD), has been recognized and included in diagnosis since the 1980s, when PTSD was first defined as a mental disorder. However, it was not always correctly identified in patients. The American Civil War created a necessity for addressing a seemingly unidentifiable but prevalent set of symptoms (Friedman 2025). At first, literature called the condition a state of nostalgia, diagnosed it as "soldier's heart" or "irritable heart," referring to the symptom of a rapid heart rate, and often wrote it off as shellshock or classified it as "war neurosis." In European hospitals, hydrotherapy or electrotherapy was used as treatment along with hypnosis. In World War II, the diagnosis of shell shock was replaced by combat stress reaction (CSR), commonly known as "battle fatigue." This diagnosis was often looked down upon by generals or deemed not real. Eventually, this symptom set was classified as PTSD, and the idea of neuronal hyperarousal was solidified.

Many chronic psychiatric disorders come from changes in the development of neural circuits. Chronic hyperarousal during key developmental periods can lead to long-term circadian and emotional dysregulation, contributing to mood disorders, PTSD, cardiovascular risk, and other psychiatric conditions (Xie et al. 2024). Hyperarousal influences many functions of the brain, including myelination, which is a key factor in neural plasticity. Myelination is the formation of the myelin sheath that wraps around a nerve to allow for improved conduction and quick memory recall. In hyperarousal, circuits like the amygdala, prefrontal cortex, and limbic structures are chronically activated, which can weaken emotion regulation. In a study using brain imaging (Chen et al. 2008), children with hyperarousal showed increased activity of fear and hostility. Monoamine depletion studies suggest that serotonin and dopamine imbalances contribute to a change in circadian rhythm. Patients with bipolar disorder, a type of hyperarousal, even in remission, exhibit abnormal circadian activity patterns, highlighting the long-term effects and difficulty of reversal. Circadian rhythm and chronic hyperarousal positively correlate. Chronic stress and hyperarousal lead to excessive release of cortisol, a hormone that helps regulate the circadian cycle. Poor or irregular sleep, which is crucial to maintaining a balanced circadian rhythm, leads to increased emotional reactivity, irritability, and anxiety, which promotes hyperarousal. Delayed sleep phase syndrome (DSPS) results in difficulty falling asleep and waking up at socially acceptable times, further reinforcing stress and hyperarousal. Once in this toxic loop of stress and hyperarousal, many patients find it difficult to work toward treatment. Genetic continuums of homeostatic stability may predispose to specific multivariable-influenced, endophenotypic pathologies, reflecting central nervous system hyperarousal and/or distress. Chronic hyperarousal in developmentally plastic neural circuits contributes to the development of chronically dysregulated circadian and emotional states. Many chronic psychiatric disorders stem from developmental variations in specific neural circuits. Inherited genetic variations and a loss of homeostasis contribute to chronic hyperarousal, which then disrupts circadian rhythms and the regulation of emotions. Circadian dynamics are influenced by genetic and neurochemical factors such as norepinephrine, dopamine, and endorphins. Glucocorticoid hormones also play a key role in stress response and brain aging and can be affected by an offset circadian rhythm. Impaired stress regulation can lead to disruptions in circadian rhythms as well (Duke 2008). These serve as markers of neurophysiological function and arousal. Hyperarousal also has potential long-term effects when it is chronic or persists, including altered myelination patterns and limbic hyperactivation as seen in conditions like PTSD and pediatric bipolar disorder. Irritability is linked to circadian dysregulation and limbic dysfunction, with implications for mood disorders, dementia, and chronic pain. Autographic studies show that circadian disruptions persist even in bipolar patients outside acute episodes.

7.3 Symptoms of Neuronal Hyperarousal

Abnormal arousal patterns are seen in various disorders, including depression, autism spectrum disorder (Tasdemir et al. 2002), and PTSD. Depression, in particular, is linked to persistent hyperarousal and difficulties with sleep regulation. Physiological markers include increased heart rate and decreased heart rate variability, dilated pupils and increased sweat production, hyperactive HPA axis, and overactive locus coeruleus. In depressed individuals, a hyperstable EEG vigilance pattern is seen. This indicates sustained wakefulness and difficulty transitioning to lower arousal states, permitting them to stay awake. The thalamus and its cortical connections are hyperactive as well, with increased connectivity to the default mode network (DMN), sensory cortex, and prefrontal regions. FMRI studies suggest that thalamocortical hyperconnectivity may underlie both hyperarousal and depressive symptoms (Riemann et al. 2010). Hyperarousal is one of the four main PTSD symptom clusters, and it shares many common causes. It is a dominant symptom subtype, distinct from dissociative PTSD. It strongly influences other PTSD symptoms but is not significantly influenced by them. It is automated by the amygdala, which processes peritraumatic hyperarousal and transforms it into persistent PTSD symptoms. The amygdala receives hyperarousal inputs from the cardiovascular, respiratory, and autonomic nervous systems and then regulates and determines the severity of the input. The amygdala's signals to the brainstem, hippocampus, anterior cingulate cortex, and orbitofrontal cortex contribute to PTSD symptoms as well. These connections drive exaggerated startle responses, sleep disturbances, hypervigilance, and trauma-related sensory intrusions. Amygdala hyperactivity is automatic, is involuntary, and continues even in the absence of current danger, making it a daily hindrance for the individuals affected by it (Weston 2014). Biological mechanisms also play a role in hyperarousal. PTSD patients show elevated heart rate and respiration rate after trauma, which often predicts PTSD onset and warns patients of an upcoming episode, colloquially known as a panic attack. The amygdala modulates brainstem regions responsible for arousal, contributing to exaggerated startle and wakefulness. The amygdala also influences visual cortex activity, potentially intensifying trauma-related intrusions and flashbacks. This is commonly seen in episodes with patients who complain of "blacking out," referencing their visual symptoms. It encodes hyperarousal experiences on "grounded cognition" principles, meaning it reactivates the same brain regions involved in the initial trauma when PTSD symptoms arise. This neural reactivation contributes to persistent, autonomic hyperarousal symptoms. Amygdala projections to the visual cortex heighten perceptual sensitivity to threat-related cues, leading to exaggerated attention to potential dangers in the environment. Its strong connection to the hippocampus leads to enhanced consolidation of traumatic memories, making them vivid and intrusive, and its suppression of mOFC contributes to anger, irritability, and impulsive aggression in PTSD. Impaired mOFC function leads to poor impulse control, which manifests as reckless or self-destructive behaviors. The amygdala processes both physical and emotional pain, integrating them into PTSD symptomatology. Chronic pain and distress may be processed similarly to hyperarousal, reinforcing PTSD symptoms. Hyperarousal drives the development of other PTSD symptoms, including avoidance, negative mood, and intrusive reexperiencing. The model suggests hyperarousal is the central mechanism in PTSD, with amygdala hyperactivity as its primary driver. Previously, it was held that insomnia was a condition related to acute stress (Doghramji 2014). However, after reviewing previous cases and revising the causes of insomnia, it can be better understood as a direct result of central nervous system hyperarousal. Insomnia, viewed now as a disorder of hyperarousal, is a key feature of neuropsychiatric and neurodegenerative conditions. Experimental studies were conducted to show that stress primarily exists in the central nervous system. In this sleep study, all subjects were exposed to the same stressor. Instead of permitting the same response, stress had two overwhelming effects. One group was able to maintain proper sleep habits and mood regulation, whereas the other was found to have higher heart rates and increased sympathetic nervous system activity. The conclusion of the study was that given a predisposition to physiological activation, any number of physiological stressors will increase the likelihood of an insomnia response. This brings into question whether there is causality between increased arousal and insomnia. A study was conducted in which patients were given doses of caffeine, attempting to mirror chronic insomnia. The chronic use of caffeine increased metabolic rate, and sleep efficiency declined significantly. These patients had increased multiple sleep latency test (MSLT) scores and began to show signs of anxiety closer to psychopathology, as measured by the anxiety scale of the Minnesota Multiphasic Personality Inventory (Pt) (Krupnik 2021). This displayed that cognitive and personality components frequently seen in patients with insomnia could be significantly influenced by a chronically increased level of CNS arousal.

7.4 Treatment of Neuronal Hyperarousal

Since hyperarousal can be thought of as a set of symptoms, treatment can be difficult, and very rarely is hyperarousal determined to be cured. Instead, clinicians focus on teaching or helping patients to manage and improve symptoms through a variety of treatments. In a study using a mouse model (Ruat et al. 2021), the effects of trauma on brain morphology were studied using MRI scans taken before and after a traumatic stressor was inflicted via foot shocks. Hyperarousal symptoms were assessed through responses to various sensory threats and integrated into a single Hyperarousal Score. The findings demonstrated that a brief traumatic incident, such as a shock, can change the volume of the brain. Increased volume suggests a heightened neural activity, inflammation, or structural reorganization, while a decreased volume suggests neuronal loss, shrinkage, or reduced synaptic connections. The findings also suggested that the globus pallidus is an important center for the regulation of fear responses to threatening stimuli. The globus pallidus is a triangular-shaped, subcortical structure within the basal ganglia and can be treated with deep-brain stimulation (Damier et al. 2007). Deep-brain stimulation (DBS) is a surgical therapy where electrodes are placed on specific areas of the brain and deliver electrical stimulation (Ayaz et al. 2024). DBS works by modulating abnormal brain activity. Electrical stimulation disrupts or alters the signals that cause movement disorders, improving symptoms. Other nonsurgical therapies can be implemented, such as prolonged exposure therapy, eye movement desensitization and reprocessing, and cognitive behavioral therapy. These therapies focus on having the patient change their thought process and learn to cope with and avoid triggers that worsen symptoms (Kaczkurkin and Foa 2015). There are other treatments that use medications that inhibit or encourage hormone reuptake (Sutherland and Davidson 1994). Pharmacotherapy is an important aspect of treatment during the early stages of the illness and may become necessary on a long-term basis for a large number of patients. There is not one treatment for hyperarousal as its symptoms may be very specific and require a specified medication. For example, in addressing the symptoms of poor impulse control, lithium, beta-blocking drugs, and carbamazepine are used. There are also natural approaches to treatment, which include lifestyle and behavioral changes, such as dietary and herbal remedies, and a change of sleep habits. These have been shown to improve symptoms and help patients manage the condition (Song et al. 2020).

7.5 The Role of Marine Natural Products in Neuronal Hyperarousal

It is understood and accepted that marine invertebrates produce bioactive natural products that appear to be useful for developing new drugs. By exploring untapped geographical sources such as the ocean and novel groups of organisms, such as those found in the ocean, scientists can maximize the search for new marine drugs to treat human diseases (Leal et al. 2012). Natural products have made large strides in the context of medicine. The use of natural products as medicines has been utilized for centuries in the form of traditional medicines, remedies, potions, and oils, with many of these natural bioactive products still remaining unidentified (Dias et al. 2012). The main source of understanding of natural product uses from medicinal plants is a result of scientists experimenting and observing trial and error for hundreds of centuries through edibility trials or premature deaths, searching for available cuisine for the treatment of diseases. While humanity has been trying to tap into the use of natural products as a source of treatment for all of time, marine natural products (MNPs) had never been approved for human use until 2004 (Molinski et al. 2009). This is partially because of a lack of knowledge of the ocean due to a lack of tools, such as SCUBA gear. Oceans, due to the vast area of the earth they take up and the diverse ecosystem they provide, are fundamental to the earth. They harbor the majority of the biodiversity of life that exists on earth. As scientific tools have advanced, natural products from the ocean have begun to be discovered and considered for treatments and research. Marine natural products are bioactive compounds derived from marine organisms, such as sponges, corals, algae, and bacteria, that exhibit diverse biological activities, including potential applications in medicine, agriculture, and biotechnology. They are effective for treatment without needing a large quantity of material (Haque et al. 2022) and promising sources of drugs among other natural products of plant, animal, and microbial origin. Marine natural products are often potent because the unique and extreme conditions of the ocean force the evolution of diverse and specialized defenses in marine organisms in the form of chemical adaptations if the organism wants to survive, leading to the production of novel and potent bioactive compounds. In 1999, marine organisms had already surpassed terrestrial organisms in providing larger percentages of bioactive natural products. Twenty drugs from marine sources are in clinical use as of 2022. FDA-approved marine-derived drugs for clinical use include treatments for cancer, pain management, and antiviral and anti-inflammatory drugs. For example, *Ecteinascidia turbinata* is used as a cancer drug that binds to DNA and prevents cancer cell proliferation (Cappello and Nieri 2021). Marine natural products demonstrated their efficacy against a vast array of diseases: Some possess new and specific mechanisms of action, and others are the most potent among their inhibitor families. This allows them to have a role in helping find a long-lasting, holistic cure for hyperarousal. Sponges are a particularly rich source of physically diverse and bioactive natural products, with approximately one-third of all known marine-derived compounds coming from this group (Lackner et al. 2017). A study on uncultured filamentous bacteria in a sponge, *Candidatus Entotheonella* factor and

Candidatus Entotheonella gemina, suggests that there is untapped potential lying in marine natural products. Some MNPs interact with the central nervous system (CNS), attuning neurotransmitter activity and decreasing stress-related symptoms like hyperarousal. For example, bryostatin-1 enhances synaptic plasticity, showing promise for PTSD treatment. A 2020 study demonstrated that bryostatin-1 increases synapse formation and decreases the density of dendrites in the spine in a protein kinase C (PKC)-dependent manner. These effects suggest that bryostatin-1 may improve cognitive functions by modulating neuronal structures involved in synaptic transmission (Ly et al. 2020). Some MNPs act on gamma-aminobutyric acid (GABA), serotonin, and dopamine receptors. These are all involved in the body's stress response and arousal levels. By modulating these systems, MNPs can potentially help reduce hyperarousal, which is a hallmark of PTSD and anxiety disorders.

As mentioned earlier, MNPs can have a role in pain suppression, such as ziconotide (Tian et al. 2023). This is derived from *Conus magus* (marine cone snail), which works by blocking calcium channels in neurons. These channels are crucial for pain transmission, and if successfully blocked, they could potentially modulate hyperarousal by calming overactive neural pathways associated with stress responses. Another symptom of hyperarousal, neuroinflammation (Chen et al. 2021), is a key factor in PTSD, anxiety, and depression. Marine natural products possess anti-inflammatory properties that may help modulate the immune system and reduce inflammation in the brain, potentially alleviating hyperarousal symptoms. Fucoidans, derived from brown seaweed, have demonstrated anti-inflammatory effects and have the potential to play a role in calming the neuroimmune system, thus reducing hyperarousal associated with stress. MNPs can influence neuroplasticity, which refers to the brain's ability to adapt and reorganize in response to new stimuli. This can be particularly beneficial in conditions like PTSD, where synaptic remodeling may help reregulate dysfunctional stress pathways. As mentioned, bryostatin-1 (from marine bryozoans) enhances synaptic plasticity and neurogenesis by modulating protein kinase C (PKC) activity. This enhancement of synaptic connections is promising for PTSD treatment, as it may help the brain recover from the trauma-induced disruption of neural networks involved in stress regulation and emotional responses. Some marine natural products specifically target neuroreceptors involved in regulating arousal levels. Marine natural products may work in tandem with existing therapeutic approaches to enhance the treatment of hyperarousal and related symptoms. If MNPs, pharmacotherapy, and cognitive behavioral therapy can be used together, the highest success in treatment may be seen. Marine natural products benefits do not end at treatment; rather, they start at diagnosis. MNPs have shown promise in various diagnostic purposes across several fields, including disease detection, biomarker identification, and imaging. Some marine natural products, particularly those with strong fluorescent properties, are being studied as potential imaging agents for the diagnosis of diseases, including cancer and neurological disorders (Kumar and Adki 2018). Their autofluorescence and specific binding affinities to molecular targets make them excellent candidates for noninvasive imaging techniques. Coelenterazine, a luciferin-derived compound from marine organisms like jellyfish, is a bioluminescent marker that has been explored for use in diagnostic imaging (Ellen Williams 2023). This molecule can be fused with an antibody or peptide that attacks tumors and then imaged, which allows radiologists to identify the specific location of the tumor. For hyperarousal, MNPs that influence neurotransmitter systems (like dopamine and serotonin) could be used to measure the neurochemical imbalance typically seen in hyperarousal. For instance, certain marine algae-derived compounds may interact with serotonergic pathways, helping to diagnose disorders where hyperarousal and stress responses are prominent. Marine natural products are often synthesized using a one-pot strategy. The one-pot strategy uses the conjugation of iminosugars with lipophilic moieties, like cholesterol, through a breakable succinic acid conjugator that would promote substance movement and release within the bacterial cell. The inorganic procedure, favored by the combined use of the polymer-supported triphenylphosphine, molecular iodine, and imidazole in a one-pot strategy with cholesterol, succinic acid, and unprotected iminosugars, allowed for the final compounds to be made in high yields (De Fenza et al. 2019; Manzo 2021). The approach involves joining bioactive iminosugars with lipophilic moieties, like cholesterol, to give the compound a higher susceptibility to transfer across the cell membrane of bacteria and eventually be released inside cells. This synthesis method allows researchers to efficiently produce MNP-related compounds in large quantities, which is essential for studying their biological activity, especially for potential therapeutic uses. Essentially, this strategy enables easier and more efficient creation of MNPs that can be studied for their bioactive properties, including those that might help with hyperarousal in the nervous system.

7.6 Future of Marine Natural Products

Even though new technologies have provided great advances, especially in the last two decades, for gathering and examining the function of marine samples, marine chemical ecology is yet to catch up with its terrestrial counterpart due to years of delay. The future of marine natural products (MNPs) in addressing neuronal hyperarousal is promising, with ongoing research exploring their therapeutic potential in neurological and psychiatric disorders. While many marine resources have already been fiercely exploited and utilized as they hold great monetary potential, several groups of marine invertebrates are still greatly undervalued. Still, most of those marine invertebrates hold an untapped potential for the development of new products from the sea. As biotechnology and marine pharmacology continue to be studied and techniques are streamlined, they will likely lead to the identification of novel MNPs that target hyperarousal mechanisms, such as dysregulated neurotransmitter systems and neuroinflammation. Marine-derived compounds, like bryostatin-1 and ziconotide, have already demonstrated neuroprotective and neuromodulator properties, which show promise for future advancements. Due to advancing subsea technology, increased exploration of deep-sea ecosystems and uncultured marine microbes may uncover new bioactive molecules with unique mechanisms of action. Future research will likely focus on MNPs that modulate key neurotransmitters (e.g., GABA, serotonin, and dopamine) to regulate hyperarousal symptoms. These neurotransmitters are directly involved in the body's stress response as the brain regulates when to send hormones like GABA to calm down the body or glutamate to increase neuronal activity. Enhanced synaptic plasticity and neurogenesis can help with hyperarousal by promoting brain reorganization and strengthening neural pathways involved in emotional regulation and stress response, potentially leading to improved emotional processing and a reduced hyperactive state. Compounds that enhance synaptic plasticity and neurogenesis, like bryostatin-1, may be further developed to repair neural networks disrupted by stress and trauma. Advances in neuropharmacology and genomics may enable personalized treatments using MNPs tailored to individuals with PTSD, anxiety, or other hyperarousal-related disorders. This is particularly helpful due to the hyper-specific set of symptoms each patient displays in neuronal hyperarousal. Future treatments may combine MNPs with existing pharmacological agents, cognitive behavioral therapy (CBT), or neurostimulation techniques to enhance efficacy and further the ability to renormalize these patients' everyday lives. MNPs with fluorescent properties, such as coelenterazine, could aid in imaging and biomarker detection, helping diagnose hyperarousal-related disorders at earlier stages. Future research may explore MNP-based biosensors for detecting neurochemical imbalances associated with stress and PTSD. Efforts to develop synthetic biology and biotechnological methods will be crucial for sustainable MNP production, reducing reliance on harvesting marine organisms. Genetically engineered bacteria or yeast may be used to synthesize marine-derived compounds efficiently for therapeutic use. In contrast to land plants, which all shared a single common ancestor around 470 million years ago, algal diversity encompasses several distantly related groups of mainly photoautotrophic organisms predominantly inhabiting aquatic environments. Despite the significant global progress that has further outlined the potential of algal biomass as a source of natural products, some challenges remain, and further research needs can be recognized. Significant gaps exist concerning biomass supply, but also in the taming, enhancement, and simulation of compound biosynthesis, applying some of the biotechnological approaches (Walker 2015).

References

Ayaz A et al (2024) Brain MR image simulation for deep learning based medical image analysis networks. Comput Methods Prog Biomed 248:108115

Cappello E, Nieri P (2021) From life in the sea to the clinic: the marine drugs approved and under clinical trial. Life (Basel) 11(12):1390

Chen B et al (2008) Would pancreas duct-epithelium-derived stem/progenitor cells enhance islet allograft survival by means of islets recruitment and tolerance induction in Edmonton protocol era? Med Hypotheses 70(3):661–664

Chen X et al (2021) Neuroprotective natural products for Alzheimer's disease. Cells 10(6):1309

Damier P et al (2007) Bilateral deep brain stimulation of the globus pallidus to treat tardive dyskinesia. Arch Gen Psychiatry 64(2):170–176

De Fenza M et al (2019) Exploring the effect of chirality on the therapeutic potential of N-alkyl-deoxyiminosugars: anti-inflammatory response to Pseudomonas aeruginosa infections for application in CF lung disease. Eur J Med Chem 175:63–71
Dias DA, Urban S, Roessner U (2012) A historical overview of natural products in drug discovery. Meta 2(2):303–336
Doghramji PP (2014) Integrating modern concepts of insomnia and its contemporary treatment into primary care. Postgrad Med 126(5):82–101
Duke BJ (2008) Pathogenic effects of central nervous system hyperarousal. Med Hypotheses 71(2):212–217
Ellen Williams TF (2023) Biomarkers as a diagnostic tool for PTSD. Available from: https://www.bioanalysis-zone.com/biomarkers-as-a-diagnostic-tool-for-ptsd/?utm_
Friedman MJ (2025) PTSD: National Center for PTSD 2025. Available from: https://www.ptsd.va.gov/understand/what/history_ptsd.asp
Haque N et al (2022) Marine natural products in clinical use. Mar Drugs 20(8):528
Kaczkurkin AN, Foa EB (2015) Cognitive-behavioral therapy for anxiety disorders: an update on the empirical evidence. Dialogues Clin Neurosci 17(3):337–346
Krupnik V (2021) Tackling hyperarousal: an integrative multimodal approach. Cogn Neuropsychiatry 26(3):199–212
Kumar MS, Adki KM (2018) Marine natural products for multi-targeted cancer treatment: a future insight. Biomed Pharmacother 105:233–245
Lackner G et al (2017) Insights into the lifestyle of uncultured bacterial natural product factories associated with marine sponges. Proc Natl Acad Sci USA 114(3):E347–E356
Leal MC et al (2012) Trends in the discovery of new marine natural products from invertebrates over the last two decades—where and what are we bioprospecting? PLoS One 7(1):e30580
Ly C et al (2020) Bryostatin 1 promotes synaptogenesis and reduces dendritic spine density in cortical cultures through a PKC-dependent mechanism. ACS Chem Neurosci 11(11):1545–1554
Manzo E (2021) Synthesis of marine natural products and molecules inspired by marine substances. Mar Drugs 19(4):208
Molinski TF et al (2009) Drug development from marine natural products. Nat Rev Drug Discov 8(1):69–85
Ong W et al (2022) Application of artificial intelligence methods for imaging of spinal metastasis. Cancers (Basel) 14(16):4025
Riemann D et al (2010) The hyperarousal model of insomnia: a review of the concept and its evidence. Sleep Med Rev 14(1):19–31
Ruat J et al (2021) Structural correlates of trauma-induced hyperarousal in mice. Prog Neuro-Psychopharmacol Biol Psychiatry 111:110404
Song K et al (2020) Complementary and alternative therapies for post-traumatic stress disorder: a protocol for systematic review and network meta-analysis. Medicine (Baltimore) 99(28):e21142
Sutherland SM, Davidson JR (1994) Pharmacotherapy for post-traumatic stress disorder. Psychiatr Clin North Am 17(2):409–423
Tasdemir D et al (2002) Cytotoxic bromoindole derivatives and terpenes from the Philippine marine sponge Smenospongia sp. Z Naturforsch C 57(9–10):914–922
Tian Z et al (2023) Bryostatin-1: a promising compound for neurological disorders. Front Pharmacol 14:1187411
Walker JM (ed) (2015) Natural products from marine algae. Springer
Weston CS (2014) Post-traumatic stress disorder: a theoretical model of the hyperarousal subtype. Front Psych 5:37
Xie M et al (2024) Neurobiological underpinnings of hyperarousal in depression: a comprehensive review. Brain Sci 14(1):50

Marine-Derived Natural Bioactives to Identify and Improve Depression

8

Bennett Schrimpf, Suhrud Pathak, Rishi Nadar, Muralikrishnan Dhanasekaran, K. Ilango, and S. Daisy Chella Kumari

Abstract

Depression persists as a major cause of disability worldwide, and it is frequently characterized by considerable side effects from traditional antidepressants, delayed therapeutic start, and limited treatment success. With their diverse modes of action that target oxidative stress, inflammation, neurogenesis, and neurotransmitter regulation—all key pathways linked to depressive disorders—marine-derived natural bioactives provide a new and potential therapy option for depression. Despite present issues with pharmacokinetics, bioavailability, and mechanistic comprehension, developments in drug delivery technologies and marine biotechnology are opening the door to more efficient use of these substances. The therapeutic potential of marine bioactives in reducing depression is examined in this review, which also highlights important bioactive classes, new research, and prospective avenues for clinical translation. The results of mental health care might be changed by creating safer, more effective, and more com-

B. Schrimpf · S. Pathak · R. Nadar
Drug Discovery and Development, Harrison College of Pharmacy, Auburn University, Auburn, AL, USA

M. Dhanasekaran (✉)
Drug Discovery and Development, Harrison College of Pharmacy, Auburn University, Auburn, AL, USA

Graduate School, Auburn University, Auburn, AL, USA
e-mail: dhanamu@auburn.edu

K. Ilango
Crescent School of Pharmacy, B.S. Abdhur Rahman Institute of Science and Technology, Chennai, India

S. D. C. Kumari
College of Pharmacy, Madras Medical College, Affiliated to The TN Dr. M.G.R. Medical University, Chennai, Tamil Nadu, India

M. Bhupathyraaj et al. (eds.), *Neuroprotective Marine Bioactives*, Nutritional Neurosciences, https://doi.org/10.1007/978-981-95-5791-2_8

prehensive substitutes for conventional antidepressants by incorporating marine-derived chemicals into individualized treatment plans.

8.1 Introduction

8.1.1 History of Depression

Depression, also known as major depressive disorder (MDD), is a severe mental disorder that affects 350 million people worldwide, with almost 800,000 people dying each year from depression-induced suicide (Zhao et al. 2024). Before the modern concept of depression existed, the affliction known today was referred to as melancholy in the late 1700s–1800s. This melancholy was defined as a disorder of intellect associated with sadness, also often referred to as partial insanity. In the mid-1800s, this definition evolved from a disorder of intellect to recognition of a mood-based origin, referring to it as melancholia with delusions (Kendler 2020). These delusions were thought to emerge from the irregular disposition. So this transfer of thought shifted the model of depression to what we understand it as today. First, it was thought to be a condition that affected cognition, which then, in turn, affected the individual's mood. After this realization, it was perceived as a disorder that affected mood and therefore altered thought processes, and the modern approaches to depression are based on this assumption as well. In the late nineteenth century, a criteria-based diagnostic approach was developed, clustering signs and symptoms of known individuals to diagnose others (Kendler 2020).

The first treatments for depression and melancholia were vastly different from what is seen today. The first evidence of this treatment came from the Middle East. In the eighth and ninth centuries, it was one of the first regions to establish specialized hospital wings for mental illnesses, and western regions soon followed this practice in the thirteenth century (Haenel 1986). During the Renaissance, this melancholy was treated mainly with nutrition, encouraging the integration of lean meats, fish, and grapes into the diet. Another treatment gaining favor in the seventeenth and eighteenth centuries was music therapy, and baths and massages were popular in the nineteenth century (Haenel 1986). Early drug treatments did not arise until the nineteenth century with the emergence of opium, ergot, ferrous iodine, arsenic, strophanthin, and cinchona. Following further research and experimentation, modern antidepressants—specifically lithium salts—were developed around 1948 (Haenel 1986).

As treatments evolved from early holistic approaches to modern antidepressants, researchers shifted their focus toward understanding physiological processes. Almost 60 years ago, the monoamine hypothesis of depression was constructed, which proposed that individuals with depression had depleted levels of the neurotransmitters dopamine, serotonin, and norepinephrine (Hillhouse and Porter 2015). One primary line of reasoning for this hypothesis at this time was the effect of reserpine on serotonin. Reserpine is a drug commonly used for hypertensive vascular disease in the 1950s, extracted from a shrub named Rauwolfia serpentina.

Taking this drug, however, was found to trigger or precede some cases of depression in those with this prescription, and reserpine was observed to induce depression-like symptoms in animals (Hillhouse and Porter 2015). This was supported by further evidence, which demonstrated that reserpine hindered a protein transporter called vesicular monoamine transporter (VMAT) (Hillhouse and Porter 2015). This inhibition resulted in reduced serotonin and catecholamines in the brain, illustrating the relationship between depressive symptoms and a deficiency of neurotransmitters in the brain.

8.1.2 Pathophysiology

The pathophysiology of depression is complex, and as a result, depression can affect multiple bodily symptoms. The monoamine hypothesis generated almost 60 years ago still stands and is gaining evidence currently. Modern studies have exceedingly confirmed that the monoamine neurotransmitters (serotonin, dopamine, and norepinephrine) have a substantial influence on cognition, emotion regulation, and psychological stress response in MDD, with serotonin regarded as the major contributor. Research has indicated that serotonin dysregulation is a potential precursor for MDD development, especially tryptophan metabolism impairment. Tryptophan is the biological precursor of both melatonin and serotonin, so impairments to this pathway can negatively impact serotonin synthesis (Robinson et al. 2020). In addition to monoamines, the neurotransmitters gamma-aminobutyric acid (GABA) and glutamate are also affected in patients with MDD, disrupting neural pathways and leading to an imbalance in their levels (Zhao et al. 2024). This hypothesis has been extensively tested and is strongly supported by substantial evidence.

Another prevalent pathophysiological mechanism of depression addresses dysfunction of the hypothalamic-pituitary-adrenal (HPA) axis. The HPA axis is the neuroendocrine system that relays signals from the hypothalamus to the pituitary gland to the adrenal gland to produce hormones in the fight-or-flight response. Specifically in response to environmental stress, the hypothalamus releases corticotropin-releasing hormone (CRH) (Menke 2024). This hormone binds to receptors in the anterior pituitary gland, inducing secretion of adrenocorticotropic hormone (ACTH) into circulation (Menke 2024). This hormone travels through the bloodstream to the adrenal cortex, subsequently activating the production of cortisol. Arginine-vasopressin (AVP) is another hormone influenced by stress in the HPA axis. AVP is produced both in the paraventricular nucleus of the hypothalamus and induces the production of ACTH, and locally in the adrenal cortex, which also activates the synthesis of cortisol (Menke 2024). Cortisol works in two discrete pathways. Firstly, when activated, intracellular cortisol receptors (GR) can migrate to the nucleus and regulate gene expression (Menke 2024). Secondly, membrane-bound cortisol receptors also work in protein kinase signaling cascades (Menke 2024). To reset homeostasis after a stress response, cortisol binds to GR located in the hypothalamus and the pituitary gland (Menke 2024). This causes a negative feedback loop, halting the secretion of CRH. The sensitivity of this feedback loop is significant for an adaptive stress response (Menke 2024) (Fig. 8.1).

One of the mediators of the HPA axis responsible for the regulation of the negative feedback loop is a gene called FKBP5. Individuals with variants of this gene, rendering a protein that is inadequate in its function within the HPA axis, have been linked to a higher risk of major depressive disorder, contributing to the almost 40% heritability of MDD (Menke 2024). This gene and other regulators of the HPA axis are greatly influenced by physical, sexual, and emotional childhood abuse, also leading to a higher incidence of MDD (Menke 2024). This trauma increases the sensitivity of the HPA axis, resulting in an intensification of stress responses. Furthermore, in later life, stressful events negatively impact the regulation of the negative feedback system of the HPA axis as well.

Growing research indicates that inflammation and oxidative stress play a key role in the neuroprogression of MDD, relating to cell death, reduced neural plasticity, and neurodegeneration, as evidenced by elevated biomarkers of both processes in affected individuals (Bakunina et al. 2015). Oxidative stress is an imbalance in the production of reactive oxygen species (ROS) or other free radicals and antioxidant defenses (Correia et al. 2023). This dysfunction results in impaired intracellular processes that rely on ROS. ROS act as secondary messengers in inflammatory mechanisms through protein kinase signaling cascades (Bakunina et al. 2015). Under standard conditions, cells protect themselves from the harmful effects of ROS through enzymatic antioxidant defenses to maintain or return to homeostasis (Correia et al. 2023). However, in states of MDD and other pathologies, the

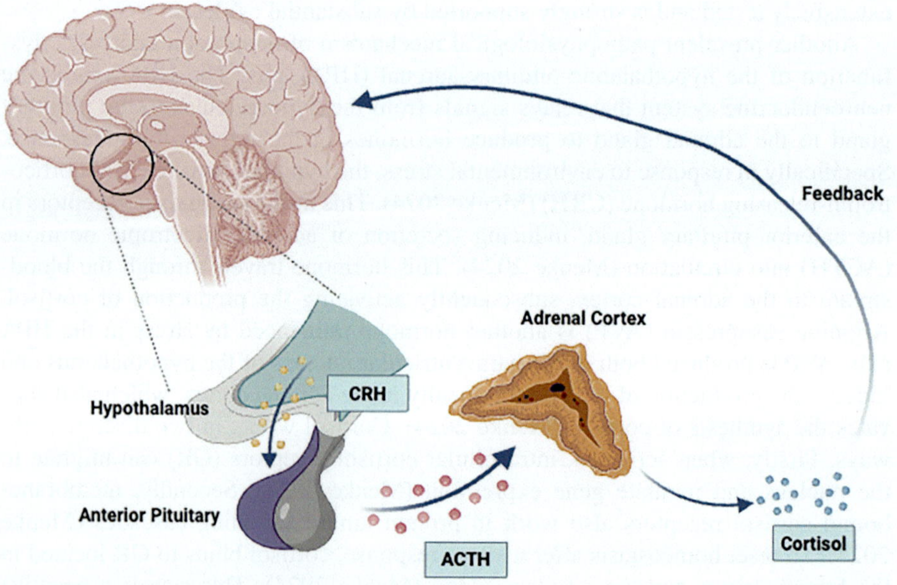

Fig. 8.1 HPA axis illustration created using BioRender

overproduction of free radicals overwhelms the antioxidant defenses, promoting cell damage. This deficit in antioxidant protection causes an increase in the synthesis of pro-inflammatory compounds which triggers immune response pathways, ultimately leading to cell death and disease neuroprogression.

Additionally, this excessive accumulation of ROS can induce heightened activity of the HPA axis (Correia et al. 2023). This relationship can also be reversed, as HPA axis activation and cortisol production may trigger the formation of free radicals. Studies have shown that in exposure to glucocorticoids, oxidative markers increased and enzymatic systems for antioxidant activity decreased, causing damage to the hippocampus and resulting in cognitive function decline (Correia et al. 2023). A similar study showed this relationship by first inducing depression-like symptoms and analyzing cortisol levels and oxidative stress markers, both of which were elevated (Correia et al. 2023).

One mechanism that is often overlooked in patients is the influence of depression on the microbiota-gut-brain axis. Patients with depression normally experience gastrointestinal conditions, interferences in appetite and metabolism, and irregularities in the composition of gut microbiota (Zhu et al. 2022). Studies examining human fecal matter showed that there are five phyla of microorganisms in the gut: Bacteroidetes, Firmicutes, Actinobacteria, Fusobacteria, and Proteobacteria, with the first two listed being the most prevalent and the most influenced by depression (Zhu et al. 2022). From all five phyla, nine bacteria had a larger population in patients with depression (Anaerostipes, Blautia, Clostridium, Klebsiella, Lachnospiraceae, Parabacteroides, Parasutterella, Phascolarctobacterium, and Streptococcus) and six had a smaller population (Bifidobacterium, Dialister, Escherichia, Shigella, Faecalibacterium, and Ruminococcus) (Zhu et al. 2022). The variation exhibited can result in gastrointestinal problems and further disorders, such as irritable bowel syndrome (IBS). Both depression and IBS coincide with an increase in Bacteroidetes and Firmicutes (Zhu et al. 2022). This relationship can also occur in the inverse, as IBS can indicate a higher risk for the development of depression (Zhu et al. 2022).

The effect of depression on microbiota is not unidirectional, as evidenced by a fecal matter transplant experiment performed on mice. This experiment demonstrated that fecal matter from a depressed mouse with the irregular microbiota composition transplanted into a nondepressed mouse caused depression-like symptoms, increased interferon γ (IFN-γ), and increased tumor necrosis factor-alpha (TNF-α) in the hippocampus (Zhu et al. 2022). The behavior and processes of the recipient mice mimicked that of their donors: increased anxiety, decreased happiness, and dysfunction of tryptophan metabolism, interfering with serotonin production (Zhu et al. 2022). Related experiments also showed that fecal matter transplants from depressed patients into unaffected mice caused these symptoms and a permanent change in microbiota composition to match the depressed patient (Zhu et al. 2022). Real-world application of this phenomenon explored by the above experiments can be seen in the finding that people with alcoholism have variations in their gut microbiota, which can result in their host manifesting symptoms of depression and anxiety (Zhu et al. 2022).

The mechanism of action of microorganisms in the gut may be linked to the release of metabolic molecules. Gut microbiota has simple communication with the host through only one layer of intestinal epithelial cells, at which molecules can pass through and enter circulation to act on the nervous system. Gamma-aminobutyric acid (GABA), a significant neurotransmitter in the brain, is one such metabolite that is released with the breakdown of foods that can directly impact the central nervous system (Zhu et al. 2022). In addition, CRH and the HPA axis play a role in intestinal motility and permeability. The permeability to inflammatory molecules was observed to be increased in those with depression (Zhu et al. 2022).

8.1.3 Symptoms and Diagnosis

According to the Diagnostic and Statistical Manual of Mental Disorders, Fifth Edition (DSM-5), written by the American Psychological Association (2013), major depressive disorder is characterized and diagnosed by the following criteria:

Five (or more) of the following symptoms have been present during the same 2-week period and represent a change from previous functioning; at least one of the symptoms is either (1) depressed mood or (2) loss of interest or pleasure: (1) Depressed most of the day, nearly every day as indicated by a subjective report (e.g., feels sad, empty, hopeless) or observation made by others (e.g., appears tearful). (2) Markedly diminished interest or pleasure in all, or almost all, activities most of the day, nearly every day (as indicated by subjective account or observation). (3) Significant weight loss when not dieting or weight gain (e.g., change of more than 5% of body weight in a month), or decrease or increase in appetite nearly every day. (4) Insomnia or hypersomnia nearly every day. (5) Psychomotor agitation or retardation nearly every day (observable by others, not merely subjective feelings of restlessness or being slowed down). (6) Fatigue or loss of energy nearly every day. (7) Feelings of worthlessness or excessive or inappropriate guilt (which may be delusional) nearly every day (not merely self-reproach or guilt about being sick). (8) Diminished ability to think or concentrate, or indecisiveness, nearly every day (either by subjective account or as observed by others). (9) Recurrent thoughts of death (not just fear of dying), recurrent suicidal ideation without a specific plan, or a suicide attempt or a specific plan for committing suicide. (American Psychiatric Association 2013, p. 161).

The above symptoms must not only be present, but they must also cause significant distress and dysfunction in social, emotional, or other areas of functioning (American Psychiatric Association 2013). To fully consider this instance a major depressive disorder, the clinician has to verify that these symptoms cannot be credited to another medical condition, psychotic disorder, or substance. Additionally, the patient would be assessed to determine if they have experienced a manic episode. If so, a diagnosis of MDD would not be appropriate, and instead, a mania-related disorder such as bipolar disorder (BPD) may be considered. Screening is generally recommended for all adults, regardless of risk, through the application of questionnaires and surveys that evaluate the above symptoms (Maurer et al. 2018).

Beyond criteria-based diagnostics, neuroimaging and other screening tools are utilized to determine a diagnosis of MDD. Structural MRI assesses brain volume, while functional MRI (fMRI) and positron emission tomography evaluate the activity and functionality of specific brain regions. Over time, numerous imaging studies have found that individuals with depression exhibit a significant reduction in brain volume in the medial prefrontal cortex, lateral prefrontal cortex, striatum, and hippocampus (Wise et al. 2014). Additionally, activity levels have been shown to increase in the medial prefrontal cortex, amygdala, and hippocampus, while decreased activity is observed in the lateral prefrontal cortex and striatum (Wise et al. 2014). These regions are associated with stress response, memory, higher-order cognition, and emotional processing. As a result, their impairment in depression leads to dysfunction in these processes.

It is important to note that men and women may have slight differences in the manifestation of MDD. In general, women diagnosed with depression report headaches, muscle pain, gastrointestinal problems, and other physical symptoms more than men (Maurer et al. 2018). Women also tend to exhibit more emotional sensitivity such as increased stress and crying. On the other hand, men diagnosed with depression report aggressive behavior, substance use problems, and higher risk-seeking behavior (Maurer et al. 2018).

There are many risk factors for developing MDD and other depression disorders, normally categorized into adverse life events, internal factors, and external factors. Adverse life events include experiences such as childhood sexual abuse, chronic health conditions, trauma, divorce, and limited social support. Internal factors encompass traits like neuroticism and low self-esteem, while external factors involve influences such as substance abuse (Maurer et al. 2018). In addition to various risk factors, depression is highly comorbid with not only other mental health disorders but also seemingly unrelated physical and neurological conditions. Some associated disorders include epilepsy, multiple sclerosis, cardiomyopathy, chronic liver disease, and HIV (Maurer et al. 2018).

8.1.4 Treatment Strategies

With regard to FDA-approved drugs for the treatment of depression, there are a few classes of drugs that mostly target monoamines (SSRIs, TCAs, MAOIs, etc.) that are generally used for the treatment of depression. Monoamine oxidase inhibitors (MAOIs) act to deactivate monoamine oxidase (MAO), an enzyme that breaks down monoamine neurotransmitters in the presynaptic terminal not packaged in vesicles. By inhibiting MAO, these neurotransmitters (serotonin, dopamine, epinephrine, norepinephrine) are more available for release to alleviate deficiencies that normally accompany depression (Hillhouse and Porter 2015). These drugs, however, are not first-line treatments anymore due to safety concerns. MAO works in a negative feedback loop to regulate serotonergic, adrenergic, and dopaminergic neurons; therefore, inhibition can cause side effects like hypertension, sweating, and increased heart rate.

Another class of drugs that can be implemented in depression treatment is tricyclic antidepressants (TCA). TCAs are unique in that they have diverse effects, causing inhibition at both ends of the synapse. It affects two reuptake transporters, presynaptic norepinephrine reuptake transporters and presynaptic serotonin reuptake transporters, and three receptor proteins, postsynaptic adrenergic α1 and α2 receptors, postsynaptic muscarinic receptors; and postsynaptic histamine H1 receptors (Hillhouse and Porter 2015). The inhibition of these reuptake transporters causes an excess of norepinephrine and serotonin synapse, prolonging the effect of these neurotransmitters on the postsynaptic receptors. However, due to the blocking of receptor proteins, there are some unintended side effects of dizziness, memory deficits, and lethargy.

Selective serotonin reuptake inhibitors (SSRI) are the most commonly prescribed drugs in depression treatment. SSRIs block serotonin reuptake transporters preferentially, with little to no affinity for postsynaptic receptors or norepinephrine reuptake transporters (Hillhouse and Porter 2015). Thus, there is an increased concentration of serotonin in the synapse, resulting in greater activity of postsynaptic serotonin receptors. Due to increased selectivity, there are generally fewer side effects. However, there are some, including nausea, trouble sleeping, and sexual dysfunction (Hillhouse and Porter 2015).

Some unorthodox antidepressants have very different mechanisms and binding properties, such as bupropion. Bupropion is a dopamine and norepinephrine reuptake inhibitor, with a higher affinity for dopamine transporters and no affinity for serotonin transporters (Hillhouse and Porter 2015). Despite this atypical nature, this approach is quite effective yet and decreases the side effects of sexual dysfunction that commonly accompanies SSRIs, not without some side effects of its own.

Apart from FDA-approved treatments, there are some natural treatment options for alleviation of depression symptoms. Many herb-derived bioactives function as antioxidants and promote anti-inflammatory properties. Examples include apigenin (from citrus fruits), baicalin (from *Scutellaria baicalensis*, a traditional Chinese medicinal herb), curcumin (from turmeric), luteolin (from celery and green bell peppers), quercetin (a flavonoid found in various fruits and vegetables), saffron, and piperine (from black pepper) (Noori et al. 2022). Additionally, folic acid, commonly known as vitamin B9, is a major vitamin that plays a significant role in the production of monoamine neurotransmitters and is found in reduced quantities in individuals with depression (Noori et al. 2022). Consequently, supplementation of vitamin B9 into the diets of patients could ameliorate symptoms of depression. Ultimately, both FDA-approved medications and natural treatments can be useful, offering distinct therapeutic benefits.

8.2 Marine Natural Products: Therapeutic Purposes

8.2.1 Marine Biodiversity as a Source of Bioactives

Before the 1950s, the application of marine-derived molecules in medicine was limited to folk medicine. Unbeknownst to scientists and medical professionals, these marine-derived bioactives could be used in actual, evidence-based, treatment. This realization was made around 1951 when researchers isolated C-nucleosides from Tethya, a Caribbean sponge (Bergmann and Feeney 1951). Currently, 30,000 natural products, with that number increasing by 1000 per year since 2018, have been identified in the marine ecosystem (Santhiravel et al. 2025). These derivatives have been the subject of many research experiments, as drug industries are striving to find new pharmaceuticals that can have applications in many treatments of diseases. Now, more than half of commercially available drugs are derived from natural sources or synthesized using natural products (Robinson et al. 2020). Additionally, there are marine-based natural drugs approved by the US FDA for therapeutic use. Some bioactive compounds of continued interest include polyunsaturated fatty acids, collagen, gelatin, enzymes, polyphenols, polysaccharides, pigments, vitamins, and minerals (Santhiravel et al. 2025).

One particular source of these bioactive compounds that has proven to include many benefits is seaweed. Seaweeds contain glycoproteins, cell wall-attached proteins, enzymes, lectins, peptides, mycosporine-like amino acids, and red algae phycobiliproteins (Echave et al. 2022). These molecules have been associated with antioxidant, anti-inflammatory, and even antitumoral properties (Echave et al. 2022). Seaweed contains high levels of essential amino acids, with percentages comparable to those found in foods typically recognized for their abundance of nutrients, around 30–40% of amino acid composition (Echave et al. 2022). Essential amino acids are those that cannot be produced by the body and must be obtained through food to carry out cellular processes, which is why they are important to note nutritionally. Besides their essential amino acid content, other seaweed proteins hold significant biological value. Seaweed protein hydrolysates have also been found to contain bioactive peptides (BAPs) with a variety of beneficial biological effects, as supported by extensive in vitro and in vivo research. In Japan, extracts from *Undaria pinnatifida* (wakame) and *Neopyropia yezoensis* (nori) have been designated as "foods for specified health uses" due to claims of helping lower blood pressure (Echave et al. 2022). Another application of seaweed is in brain health. Seaweed-derived bioactives have been shown to cross the blood-brain barrier, exerting neuro-immunomodulatory functions (Schepers et al. 2020). This discovery has led to multiple avenues of research, particularly in drug development for neurodegenerative disorders. Specifically, phytosterols derived from seaweed have demonstrated use in treatment of neurodegenerative disorders due to their effects of neuroplasticity, phagocytic clearance of neurotoxins, and anti-inflammatory properties (Schepers et al. 2020).

Another rich source of marine-derived bioactives is algae. Algae, overall, contains many bioactive compounds that are proven to have therapeutic effects, such as antioxidants, pigments, fatty acids, and specific minerals. There are differences in these molecules between species, however, due to abiotic factors (e.g., temperature, salinity, light intensity), which support the production of these compounds (Menaa et al. 2021). One emerging application of algae is the use of extracts in nanoparticle formation, where the extracts from algae can act as stabilizing and reducing agents for synthesis (Menaa et al. 2021). These nanoparticles can then be used for drug delivery, leading to optimized drug efficacy and safety. Another use is the application of photoprotective substances in therapeutic and preventative treatments. Photoprotective substances are compounds that help protect the skin and other tissues from the harmful effects of ultraviolet (UV) radiation from the sun, as well as serve antioxidant and immunomodulating activities. These substances consist of mycosporine-like amino acids (MAAs), sulfated polysaccharides, carotenoids, and polyphenols (Pangestuti et al. 2018). Therefore, photoprotective substances have uses in preventative skin care and pharmaceuticals. Additionally, the polysaccharides of various algae species have been proven to have anticoagulant, antithrombotic, antitumor, antilipemic, hypoglycemic, anti-inflammatory, and antioxidant activity (de Jesus Raposo et al. 2015). These properties have been linked to individual structure and chemical characteristics, leading to different applications of different species due to their biodiversity.

One reason why marine-derived bioactives are an ongoing and rapidly developing field of pharmaceutical research is due to their availability. There is a vast and diverse marine environment, encompassing oceans, seas, and freshwater ecosystems, which covers 70% of the earth's surface. This offers an extensive array of organisms, featuring algae, sponges, corals, and marine invertebrates that produce unique compounds with therapeutic potential. Moreover, the simplicity and sustainability of certain extraction techniques, such as solid-liquid and liquid-liquid extraction, make it easier for researchers to access and study the bioactive compounds in marine organisms. These methods are cost-effective and environmentally friendly, encouraging the exploration of marine sources for bioactive molecules. As marine biotechnology advances, the potential for discovering naturally derived treatments grows, promising new approaches to combating a range of diseases. In addition, the increasing awareness of environmental sustainability has prompted greater efforts to source marine organisms responsibly. Consequently, marine-derived bioactives remain an innovative and promising area of pharmaceutical research with significant implications for developing new therapeutic agents.

8.3 Marine Natural Products and Depression Pathology

8.3.1 Omega-3 Fatty Acids

Omega-3 long-chain polyunsaturated fatty acids (n-3 PUFAs), particularly eicosapentaenoic acid (EPA) and docosahexaenoic acid (DHA), are essential compounds primarily sourced from marine organisms such as deep-sea fish and shrimp (Wang et al. 2024b). Structurally, these fatty acids consist of long hydrocarbon chains with multiple double bonds, which contribute to their fluidity and biological activity. Their unique molecular structure allows them to integrate into cell membranes, affecting membrane fluidity, receptor function, and intracellular signaling pathways.

EPA and DHA are critical for brain development, neuronal growth, and cognitive health (Ahmmed et al. 2023). In various animal models, continuous oral administration of EPA or DHA has demonstrated significant reductions in depressive-like behavior induced by chronic unpredictable mild stress (CUMS) (Wang et al. 2024b). These effects were assessed using behavioral tests such as the sucrose preference test (SPT), open field test (OFT), and FST.

EPA and DHA modulate several key pathways involved in neuroinflammation and neuroprotection. They were shown to reduce the expression of pro-inflammatory cytokines like IL-1β, IL-6, and TNF-α and decrease microglial activation, as indicated by reduced microglial marker CD11b (Wang et al. 2024b). Simultaneously, they upregulate the expression of the astrocyte marker glial fibrillary acidic protein (GFAP), supporting glial health and brain homeostasis (Wang et al. 2024b). These effects are mediated through the regulation of the NF-κB/p38 signaling pathway. EPA and DHA also positively influence neurotrophic signaling by regulating the BDNF/TrkB pathway. This leads to increased production of neurotrophic factors such as brain-derived neurotrophic factor (BDNF), glial cell-derived neurotrophic factor (GDNF), and nerve growth factor (NGF) while reducing apoptotic markers and enhancing anti-apoptotic proteins (Wang et al. 2024b). This neurotrophic activity contributes to neuronal growth, repair, and plasticity while reducing neuronal apoptosis. Furthermore, these omega-3 fatty acids impact the HPA axis, reducing serum corticosterone levels and improving neurotransmitter balance in the hippocampus by enhancing serotonin and norepinephrine levels (Wang et al. 2024b).

Clinical studies have mostly supported these findings, with EPA showing greater efficacy than DHA in reducing depressive symptoms (Osher and Belmaker 2009). For instance, in patients with hepatitis C receiving IFN-α treatment, EPA supplementation significantly reduced the incidence of depression, whereas DHA did not produce the same effect (Wang et al. 2024b). Additionally, population studies among pregnant women have indicated that EPA/DHA supplementation may lower postpartum depression symptoms in individuals with a history of depression; however, the results have been mixed (Wang et al. 2024b).

The pharmacokinetic delivery of omega-3s can be further enhanced by lysophosphatidylcholine (LPC)-based systems, which utilize the major facilitator superfamily domain-containing protein 2 (Mfsd2a) to transport LPC-DHA across the blood-brain barrier (Ahmmed et al. 2023). This delivery mechanism improves brain bioavailability of DHA and EPA and supports the upregulation of serotonin receptors and BDNF expression in the brain, highlighting its potential in depression treatment (Ahmmed et al. 2023). Although many studies have demonstrated the importance of LPC as a carrier of DHA and EPA, there are some drawbacks associated with its use. LPC has been reported to exhibit pro-inflammatory, cytotoxic, and anti-hemostatic properties in some experimental models (Ahmmed et al. 2023). Despite this, these adverse effects appear to depend on both the concentration of LPC and the type of fatty acid attached to it, underscoring the need for careful consideration of its therapeutic applications (Ahmmed et al. 2023). Currently, the Food and Drug Administration (FDA) has approved omega-3 fatty acid ethyl ester (Wang et al. 2024b). However, the approved use is unrelated to depression. Yet this approval has made the compound widely accessible and relatively affordable for the average consumer, establishing omega-3 fatty acids as an ideal target for drug development.

PUFAs have also been shown to regulate the gut-brain axis by modulating gut microbiota composition and reducing intestinal inflammation in some studies (Costantini et al. 2017). Studies have demonstrated that omega-3 supplementation can increase the abundance of beneficial bacterial genera such as *Lactobacillus* and *Bifidobacterium* while reducing populations of pro-inflammatory bacteria (Costantini et al. 2017). This shift in gut microbiota supports the integrity of the intestinal barrier and reduces gut permeability, which prevents systemic inflammation that can negatively affect brain function. Moreover, omega-3s have been shown to influence the production of short-chain fatty acids (SCFAs), like butyrate and others, which exert anti-inflammatory effects and support neuronal health through neuroprotective signaling pathways (Costantini et al. 2017). Through these mechanisms, omega-3 fatty acids can potentially alleviate neuroinflammation and promote neurotransmitter balance, linking gut health with mental health outcomes. This emerging evidence highlights the potential for omega-3s to serve as modulators of the gut-brain axis.

In addition to their anti-inflammatory and neurotrophic effects, omega-3 fatty acids enhance synaptic function by integrating into neuronal membranes, increasing membrane fluidity, and improving receptor responsiveness and neurotransmitter release. This supports efficient signal transduction and synaptic plasticity, critical for mood regulation and cognitive function. Moreover, omega-3s may indirectly reduce oxidative stress by stabilizing mitochondrial membranes and enhancing cellular resilience against neuroinflammatory damage (Zirpoli et al. 2020). Emerging research also suggests that the therapeutic efficacy of omega-3 supplementation may depend on the ratio of EPA to DHA, with EPA-dominant formulations often showing greater antidepressant effects (Wang et al. 2024b). Despite promising clinical and preclinical findings, questions remain regarding optimal dosing strategies, bioavailability, and the influence of individual factors such as genetic variation and gut microbiota composition. These areas present important future directions for refining omega-3-based interventions in depression treatment.

The multiple potential actions of omega-3 fatty acids highlight their significant potential as therapeutic agents in brain health. Their ability to regulate neuroinflammation, neurotrophic support, neurotransmitter balance, and the gut-brain axis demonstrates a broad spectrum of benefits. While challenges remain regarding delivery methods and individual variability in response, ongoing research continues to strengthen the case for omega-3s as a valuable target in depression management research. As understanding deepens around their mechanisms and optimal dosing, omega-3 fatty acids represent a promising, well-tolerated, and accessible avenue for advancing mental health therapies (Table 8.1).

Table 8.1 ADME parameters of EPA and DHA by SwissADME

SwissADME		EPA	DHA
Physicochemical properties	Molecular formula	$C_{20}H_{30}O_2$	$C_{22}H_{32}O_2$
	Molecular weight (g/mol)	302.45	328.49
	# heavy atoms	22	24
	# aromatic heavy atoms	0	0
	Fraction Csp3	0.45	0.41
	Num. rotatable bonds	13	14
	Num. H-bond acceptors	2	2
	Num. H-bond donors	1	1
	Molar refractivity	97.66	106.8
	TPSA (Å^2)	37.3	37.3
Lipophilicity	Log Po/w (WLOGP)	5.99	6.55

There was no data for the other parameters

8.3.2 Indole Alkaloids

Indole alkaloids (IAs) are naturally occurring compounds built around a core indole ring structure, consisting of a six-membered benzene ring fused to a five-membered nitrogen-containing pyrrole ring. The nitrogen atom in this ring gives these molecules basic chemical properties, allowing them to interact with key biological targets. The indole structure plays a significant role in drug development due to its structural similarity to endogenous neurotransmitters such as serotonin. This similarity has driven research into the potential neurological effects of IAs. However, studies exploring the antidepressant potential of indole alkaloids from plants and marine sources remain limited.

IAs primarily exert their biological activity by interacting with G protein-coupled receptors (GPCRs), with particular emphasis on serotonin (5-HT) receptors found throughout the brain and body (Munir et al. 2020). There are seven families of 5-HT receptors, all but one of which are GPCRs (Munir et al. 2020). These receptors regulate mood, anxiety, sleep, appetite, and migraines. Some IAs are capable of mimicking serotonin and binding to these receptors, either activating or blocking them depending on their structure (Munir et al. 2020). This mechanism makes them valuable candidates in the development of antidepressants, antianxiety medications, and migraine therapies. Additionally, certain IAs can influence other biological pathways, including the inhibition of enzymes like MAO, which also play a role in mood regulation.

Modern migraine medications, known as triptans (e.g., sumatriptan, frovatriptan, and zolmitriptan), also rely on an indole ring structure. These compounds mimic serotonin and bind to 5-HT receptors to alleviate migraines by constricting blood vessels and inhibiting pain pathways (Munir et al. 2020). However, the ability to cross the blood-brain barrier and interact with other serotonin receptor families can sometimes result in side effects like drowsiness and dizziness.

A large number of IAs are found in marine organisms including sponges, seaweeds, mollusks, and marine bacteria. The ocean is a particularly rich source of these compounds due to specialized enzymes like haloperoxidases that facilitate the production of halogenated indole alkaloids (Munir et al. 2020). Several marine-derived IAs, such as 8,9-dihydrobarettin, sigma-conotoxin, barettin, and gelliusines A and B, have structures resembling serotonin and have been studied for their ability to interact with serotonin receptors (Munir et al. 2020). Other compounds, like methylaplysinopsin from *Aplysinopsis reticulata*, act as MAO inhibitors, preventing serotonin breakdown in presynaptic neurons and increasing serotonin availability (Munir et al. 2020). In addition to serotonergic pathways, IAs have demonstrated an affinity for dopamine and adrenergic receptors, suggesting potential for broader antidepressant effects (Munir et al. 2020).

Among the most notable marine-derived IAs are N-3′-ethylaplysinopsin, 6-bromoaplysinopsin, and 6-bromo-2′-de-N-methylaplysinopsin, isolated from the sponge *Smenospongia aurea*. These compounds have been shown to bind to serotonin receptors (5-HT2A and 5-HT2C), displacing receptor antagonists and influencing receptor activity (Hu et al. 2002). The receptor selectivity varies between these compounds; N-3′-ethylaplysinopsin showed no clear preference for either receptor, 6-bromoaplysinopsin displayed slight selectivity toward the 5-HT2C receptor, while 6-bromo-2′-de-N-methylaplysinopsin exhibited strong selectivity for 5-HT2C, highlighting its potential therapeutic value (Munir et al. 2020).

Additionally, two indole alkaloids, 5-bromo-DMT and 5,6-dibromo-DMT, isolated from marine sponges, demonstrate both antimicrobial and neurological activity. 5-bromo-DMT has shown sedative effects, whereas 5,6-dibromo-DMT exhibits antidepressant-like activity in animal models, as well as anticancer properties against colorectal cancer cells (Tasdemir et al. 2002). When administered intraperitoneally at 20 mg/kg, 5,6-dibromo-DMT reduced depressive-like behaviors in forced swim and tail suspension tests in mice (Diers et al. 2008). The forced swim test (FST) and tail suspension test (TST) are commonly used behavioral assays to evaluate antidepressant-like effects in mice. In the FST, mice are placed in a container of water, and their immobility time is measured, with reduced immobility interpreted as a sign of decreased depressive-like behavior, suggesting an antidepressant effect (Armario 2021). Similarly, in the TST, mice are suspended by their tails, and the amount of time they spend immobile is recorded (Steru et al. 1985). Reduced immobility in this test is also associated with a reduction in depressive-like symptoms. Both tests are widely used to assess the efficacy of potential antidepressants, with a decrease in immobility time generally indicating improvement in mood or motivation. While its exact mechanism of action remains unclear, similar compounds' affinity for 5-HT2 receptors suggests that its antidepressant effects may be related to serotonin reuptake inhibition (Hu et al. 2002).

Furthermore, the marine-derived fungus *Aspergillus amstelodami* produces Neoechinulin A, an isoprenyl indole alkaloid with antioxidant, antitumor, and anti-apoptotic activities (Wang et al. 2024b). In mouse models, intracerebroventricular administration of Neoechinulin A at 300 ng/kg significantly improved LPS-induced memory deficits and decreased immobility time in the FST (Sasaki-Hamada et al. 2016). These findings suggest its effects may be mediated through modulation of the serotonin system.

In addition to their affinity for serotonin receptors, indole alkaloids (IAs) may exert antidepressant effects through multiple complementary mechanisms. Beyond 5-HT2A and 5-HT2C receptor modulation, some IAs show activity at other serotonin receptor subtypes such as 5-HT1A and 5-HT7, both of which are closely associated with mood regulation and antianxiety effects (Ikarashi et al. 2018). Also, many IAs demonstrate affinity for neurotransmitter systems beyond serotonin, including dopamine, norepinephrine, and even glutamate, which are involved in mood regulation and cognitive processes. The interaction with these receptors suggests a broader therapeutic potential for treating depression and related disorders. This dual mechanism, receptor modulation and neurotransmitter preservation through inhibition of MAO, could offer advantages over traditional single-target antidepressants. Compared to synthetic drugs, natural IAs may also exhibit higher receptor selectivity or partial agonism, potentially reducing common side effects.

Pharmacokinetic properties of IAs are another area of interest. While their abilities to cross the blood-brain barrier and their metabolic stability are still under investigation, preliminary studies suggest that structural modifications or natural halogenation may enhance their CNS availability (Pauletti et al. 2010). Halogens are incorporated in the design of many pharmaceuticals, mainly to enhance membrane permeability, occupy spaces in the binding pocket, and reduce metabolic degradation (Ibrahim et al. 2017). They can also improve potency and target selectivity by influencing factors such as pKa, conformation, lipophilicity, and hydrophobic interactions, particularly in hydrophobic binding pockets like those of serotonin receptors 5-HT1A and 5-HT7 (Ibrahim et al. 2017). The anti-inflammatory and antioxidant properties of certain IAs, such as Neoechinulin A, further support their potential in addressing neuroinflammation, a process recognized in the pathophysiology of depression. By targeting both neurotransmitter systems and inflammatory pathways, IAs may provide a multifaceted approach to treatment.

However, significant research gaps remain. Most findings are currently limited to preclinical models, with no human clinical trials yet conducted. Toxicological profiles and long-term safety data are sparse, and pharmacokinetic properties such as bioavailability, half-life, and metabolic pathways require further analysis. Future studies should focus on optimizing these compounds for human use and investigating the potential for combination therapies with existing antidepressants. Overall, marine-derived indole alkaloids represent a promising class of compounds with the potential to contribute to depression treatment (Table 8.2).

8.3.3 Carotenoids

Carotenoids are naturally occurring pigments present in plants and microorganisms. In photosynthetic organisms, including plants and microalgae, carotenoids serve crucial roles as accessory pigments in light harvesting and as protectors of the photosynthetic machinery by scavenging reactive oxygen species (ROS) generated by excess light (Vilchez et al. 2011). Their powerful antioxidant properties, stemming from their molecular structure, enable them to neutralize ROS and combat oxidative

Table 8.2 ADME parameters of methylaplysinopsin, Neoechinulin A, and 5,6-dibromo-DMT by SwissADME

SwissADME		Methylaplysinopsin	Neoechinulin A	5,6-dibromo-DMT
Physicochemical properties	Molecular formula	$C_{15}H_{16}N_4O$	$C_{19}H_{21}N_3O_2$	$C_{12}H_{14}Br_2N_2$
	Molecular weight (g/mol)	268.31	323.39	346.06
	# heavy atoms	20	24	16
	# aromatic heavy atoms	11	9	9
	Fraction Csp3	0.2	0.26	0.33
	Num. rotatable bonds	1	3	3
	Num. H-bond acceptors	3	2	1
	Num. H-bond donors	1	3	1
	Molar refractivity	84.19	102.94	75.98
	TPSA (Å^2)	54.81	73.99	19.03
Lipophilicity	Log Po/w (iLOGP)	2.43	2.71	2.99
	Log Po/w (XLOGP3)	0.79	3.33	3.73
	Log Po/w (WLOGP)	1.37	1.74	3.8
	Log Po/W(Mlogp)	1.68	1.53	3.16
	Log Po/w (SILICOS-IT)	2.62	3.87	4.11
	Consensus log Po/w	1.78	2.63	3.56
Water solubility	Log S(ESOL)	−2.34	−4.02	−4.55
	Solubility	1.22e+00 mg/ml; 4.55e-03 mol/l	3.07e-02 mg/ml; 9.50e-05 mol/l	9.67e-03 mg/ml; 2.79e-05 mol/l
	Class	Soluble	Moderately soluble	Moderately soluble
	Log S (Ali)	−1.52	−4.56	−3.82
	Solubility	8.07e+00 mg/ml; 3.01e-02 mol/l	8.90e-03 mg/ml; 2.75e-05 mol/l	5.22e-02 mg/ml; 1.51e-04 mol/l
	Class	Very soluble	Moderately soluble	Soluble

(continued)

Table 8.2 (continued)

SwissADME		Methylaplysinopsin	Neoechinulin A	5,6-dibromo-DMT
	LogS (SILICOS-IT)	−3.14	−5.71	−5.87
	Solubility	1.96e-01 mg/ml; 7.29e-04 mol/l	6.35e-04 mg/ml; 1.96e-06 mol/l	4.66e-04 mg/ml; 1.35e-06 mol/l
	Class	Soluble	Moderately soluble	Moderately soluble
Pharmacokinetics	GI absorption	High	High	High
	BBB permeant	Yes	Yes	Yes
	P-gp substrate	No	No	No
	CYP1A2 inhibitor	No	Yes	Yes
	CYP2C19 inhibitor	No	Yes	Yes
	CYP2C9 inhibitor	No	Yes	No
	CYP2D6 inhibitor	No	No	Yes
	CYP3A4 inhibitor	No	Yes	No
	log Kp (cm/s)	−7.38	−5.91	−5.76
Drug-likeness	Lipinski	Yes; 0 violation	Yes; 0 violation	Yes; 0 violation
	Ghose	Yes	Yes	Yes
	Veber	Yes	Yes	Yes
	Egan	Yes	Yes	Yes
	Muegge	Yes	Yes	Yes
	Bioavailability score	0.55	0.55	0.55
Medicinal chemistry	PAINS	0 alert	0 alert	0 alert
	Brenk	0 alert	2 alerts: isolated_alkene, michael_acceptor_1	0 alert
	Lead-likeness	Yes	Yes	No; 1 violation: XLOGP3 > 3.5
	Synthetic accessibility	3.6	3.77	1.93

stress, a key factor in the development of neurodegenerative diseases such as Alzheimer's disease and Parkinson's disease (Vilchez et al. 2011). Beyond antioxidant activity, carotenoids also influence cell signaling, modulate the immune response, and contribute to anti-inflammatory processes. Carotenoids exhibit unique biological activities, making them important in therapeutic research for both physical and mental health.

Astaxanthin is one such carotenoid, and it is found in salmon, trout, microalgae, and marine invertebrates, with multiple pharmacological properties, such as anti-inflammatory and antioxidant activities (Wang et al. 2024b). Preclinical studies have demonstrated that trans-astaxanthin in an oral dose of 20–80 mg/kg administered over 7 days prevented depressive-like symptoms in mice exposed to lipopolysaccharide (LPS), a model for inflammation-induced depression (Jiang et al. 2017). Neurochemical analysis from the same study revealed that trans-astaxanthin could reverse LPS-induced overexpression of inflammatory cytokines (e.g., IL-1β, IL-6, and TNF-α), and it could reduce the expression of pro-inflammatory enzymes such as inducible nitric oxide synthase, neuronal nitric oxide synthase, and cyclooxygenase-2 (COX-2) (Jiang et al. 2017). These effects are thought to be mediated through modulation of the NF-κB pathway, suggesting that astaxanthin's antidepressant potential is linked to its anti-inflammatory properties (Wang et al. 2024b). Beyond preclinical models, clinical evidence also supports astaxanthin's potential antidepressant effects. One clinical trial investigated astaxanthin supplementation on 28 adults diagnosed with depression and fatigue, as well as on a group of healthy, nondepressed individuals (Talbott et al. 2019). Subjects who received 12 mg of astaxanthin daily for 8 weeks demonstrated significant reductions in both depression and fatigue compared to the placebo group (Talbott et al. 2019).

While conventional antidepressant medications like SSRIs are widely used to treat major depressive disorder, they are often accompanied by multiple side effects. In contrast, astaxanthin has a high safety profile, having been extensively utilized in nutraceuticals, cosmetics, and food products without significant adverse effects (Peng et al. 2023). Evidence also suggests that astaxanthin may have beneficial effects on sleep, which is relevant given the established connection between insomnia and genes regulating circadian rhythms, such as *Per* and *Bmal1 (*Peng et al. 2023*)*. Although the molecular mechanisms underlying these effects are not yet fully understood, preliminary research indicates that astaxanthin may influence the expression of *Per2*, a gene involved in circadian regulation (Peng et al. 2023).

An important consideration in developing astaxanthin as a therapeutic agent is its limited solubility and bioavailability, which may hinder its clinical applications. To overcome these challenges, ongoing research has focused on enhancing astaxanthin bioavailability through advanced delivery systems such as polymeric, lipid-based, inorganic, and hybrid nanocarriers (Liu et al. 2019). As research continues, there is potential for astaxanthin to become significant in the pharmaceutical and biomedical fields. Future studies should aim to determine the exact mechanism of astaxanthin's antidepressant activity and its impact on the regulation of melatonin expression and sleep quality in depressed patients (Peng et al. 2023).

Another carotenoid to note is β-carotene. In marine environments, β-carotene is primarily derived from the microalgae *Dunaliella salina* (Han et al. 2019). Similar to astaxanthin, β-carotene exhibits notable antioxidant and anti-inflammatory properties (Wang et al. 2024b). Experimental studies have further demonstrated its potential antidepressant effects. For instance, oral administration of β-carotene once daily for 28 days significantly reduced immobility time in FST in mice (Kim et al. 2016). Additionally, compared to the control group, β-carotene treatment resulted in a marked decrease in pro-inflammatory cytokines TNF-α and IL-6 while increasing the levels of BDNF and phosphorylated extracellular signal-regulated kinase (pERK) (Kim et al. 2016). Both of these factors are associated with neuroplasticity and mood regulation.

Multiple studies have shown that higher serum levels of β-carotene are associated with a reduced likelihood of experiencing depression. For example, a cross-sectional study involving 1798 US adults found that elevated serum β-carotene concentrations correlated with lower depressive symptom scores (Beydoun et al. 2013). Similarly, some long-term studies confirmed that higher plasma β-carotene levels significantly reduced the risk of depression over extended follow-up periods (Bardinet et al. 2023). These findings suggest that a diet rich in carotenoids, such as from abundant fruit and vegetable consumption, may protect against depression.

Fucoxanthin is a marine-derived carotenoid found abundantly in brown seaweed (Wang et al. 2024b). Like other carotenoids, it exhibits antioxidant and anti-inflammatory properties. In animal studies, fucoxanthin has shown promising antidepressant potential. Mice subjected to LPS-induced depressive-like behaviors displayed improvement in the FST and TST after receiving fucoxanthin of a dose of 200 mg/kg intravenously (Jiang et al. 2017). Further biochemical analysis revealed that fucoxanthin suppressed the LPS-induced overexpression of IL-1β, IL-6, TNF-α, iNOS, and COX-2 in brain regions associated with mood regulation, such as the hippocampus, frontal cortex, and hypothalamus (Jiang et al. 2017). These effects were mediated by modulation of the AMPK-NF-κB signaling pathway.

Other carotenoids with the same proven antioxidant and anti-inflammatory capabilities include zeaxanthin and lutein. Lutein is primarily found in microalgae such as *Chlorella vulgaris* and has demonstrated neuroprotective properties (Wang et al. 2024b). In animal studies, oral administration of lutein at a dose of 10 mg/kg for 7 or 21 days significantly reversed depressive-like behaviors in the TST, OFT, and ST induced by corticosterone (Wang et al. 2024b). These findings suggest that lutein may modulate the HPA axis, reducing glucocorticoid levels to exert its protective effects on the brain (Zeni et al. 2019). Additionally, zeaxanthin, a carotenoid in the xanthophyll family, can be extracted from cyanobacteria such as *Synechocystis* sp., *Microcystis aeruginosa*, and the microalga *Nannochloropsis oculata* (Wang et al. 2024b). Zeaxanthin has also been linked to mood regulation. In diabetic rat models, daily oral zeaxanthin administration from weeks 6–19 reduced depressive-like behaviors in both the OFT and FST (Zhou et al. 2018). This improvement was accompanied by a decrease in the overproduction of IL-6, IL-1, and TNF-α (Zhou et al. 2018) (Table 8.3).

Table 8.3 ADME parameters of astaxanthin, β-carotene, and fucoxanthin by SwissADME

SwissADME		Astaxanthin	β-carotene	Fucoxanthin
Physicochemical properties	Molecular formula	$C_{40}H_{52}O_4$	$C_{40}H_{56}$	$C_{42}H_{58}O_6$
	Molecular weight (g/mol)	596.84	536.87	658.91
	# heavy atoms	44	40	48
	# aromatic heavy atoms	0	0	0
	Fraction Csp3	0.4	0.45	0.55
	Num. rotatable bonds	10	10	12
	Num. H-bond acceptors	4	0	6
	Num. H-bond donors	2	0	2
	Molar refractivity	187.16	184.43	197.07
	TPSA ($Å^2$)	74.6	0	96.36
Lipophilicity	Log Po/w (WLOGP)	8.91	12.61	8.69

There was no data for the other parameters

8.4 Future Prospects

Marine-derived natural bioactives, including omega-3 fatty acids, indole alkaloids, and carotenoids, have demonstrated therapeutic potential for depression treatment. Their efficacy has been shown through multiple mechanisms, such as anti-inflammatory, antioxidant, neuroprotective, and neurotrophic activities. These compounds modulate key pathways involved in depression pathology, including neuroinflammation, oxidative stress, neurotransmitter regulation, and the gut-brain axis. Despite their promising effects in preclinical and clinical studies, several challenges and limitations must be addressed before their widespread therapeutic application.

One major limitation in the clinical translation of marine-derived bioactives is their bioavailability. While EPA and DHA have shown antidepressant effects, their ability to cross the blood-brain barrier efficiently remains a challenge. The development of LPC-based delivery systems has enhanced brain uptake of DHA and EPA, but LPC itself has been associated with potential pro-inflammatory effects (Ahmmed et al. 2023). Similarly, carotenoids like astaxanthin, fucoxanthin, lutein, and zeaxanthin have demonstrated neuroprotective effects, yet their hydrophobic nature and limited absorption hinder their therapeutic potential. Strategies such as lipid-based formulations and nanoparticle delivery systems could improve their stability and bioavailability, making them more viable for clinical use.

Another challenge is the variability in study designs and dosing regimens, which complicates the translation of preclinical findings to human applications. Many animal studies use high doses of marine bioactives that may not be feasible or safe for long-term human consumption. Additionally, differences in metabolism and absorption between animal models and humans can lead to discrepancies in efficacy. Clinical studies have also produced mixed results; for example, while EPA has shown greater efficacy than DHA in reducing depressive symptoms, some trials have reported inconsistent outcomes, especially in specific populations like pregnant women with postpartum depression (Wang et al. 2024b). Standardized dosing protocols and large-scale, well-controlled human trials are needed to establish optimal intake levels and treatment durations for different depressive disorders.

The gut-brain axis represents an exciting area for future research on marine-derived bioactives in depression. Emerging evidence suggests that omega-3s may influence gut microbiota composition, thereby modulating neuroinflammation and neurotransmitter production. However, more research is required to fully understand the interactions between these compounds and the gut microbiome, as well as their long-term effects on mental health. Integrating microbiome analysis in future clinical trials could provide deeper insights into their role in depression treatment.

Additionally, while marine-derived compounds have shown neuroprotective effects, their mechanisms of action remain incompletely understood. Many studies focus on individual pathways, such as BDNF signaling or NF-κB inhibition, but depression is a complex disorder involving multiple interconnected pathways. Future research should take a systems biology approach, examining how these bioactives influence broader neurobiological networks. Advanced techniques, such as metabolomics and transcriptomics, could help explain molecular interactions and identify targets for intervention (Li et al. 2024).

From a practical standpoint, sustainability and accessibility pose additional challenges. Marine resources, like fish and algae, are subject to environmental changes and overexploitation. The growing demand for omega-3 supplements has raised concerns about ecological impact, leading to increased interest in alternative sources such as microalgae (Liu et al. 2022). Similarly, carotenoids from marine organisms may need special biotechnological approaches, such as microbial fermentation or synthetic biology, to ensure a stable and eco-friendly supply (Wang et al. 2024a).

8.5 Conclusion

Despite these challenges, marine-derived bioactives offer a promising, natural approach to depression treatment with multifaceted mechanisms that target key aspects of the disorder. As research advances, overcoming bioavailability issues, optimizing dosing strategies, and exploring new therapeutic targets will be important to assess their potential. By integrating these compounds into personalized treatment approaches, future therapies may provide safer and more effective alternatives to traditional antidepressants, improving mental health outcomes with fewer side effects.

References

Ahmmed MK, Hachem M, Ahmmed F, Rashidinejad A, Oz F, Bekhit AA, Carne A, Bekhit AEA (2023) Marine fish-derived lysophosphatidylcholine: properties, extraction, quantification, and brain health application. Molecules 28(7):3088. https://doi.org/10.3390/molecules28073088

American Psychiatric Association (2013) DSM-5 task force. In: Diagnostic and statistical manual of mental disorders: DSM-5, 5th edn. American Psychiatric Association

Armario A (2021) The forced swim test: historical, conceptual and methodological considerations and its relationship with individual behavioral traits. Neurosci Biobehav Rev 128:74–86. https://doi.org/10.1016/j.neubiorev.2021.06.014

Bakunina N, Pariante CM, Zunszain PA (2015) Immune mechanisms linked to depression via oxidative stress and neuroprogression. Immunology 144(3):365–373. https://doi.org/10.1111/imm.12443

Bardinet J, Pouchieu C, Chuy V, Helmer C, Etheve S, Gaudout D, Samieri C, Berr C, Delcourt C, Cougnard-Gregoire A, Feart C (2023) Plasma carotenoids and risk of depressive symptomatology in a population-based cohort of older adults. J Affect Disord 339:615–623. https://doi.org/10.1016/j.jad.2023.07.076

Bergmann W, Feeney RJ (1951) Contributions to the Study of Marine Products. XXXII. The nucleosides of sponges. I. J Org Chem 16(6):981–987. https://doi.org/10.1021/jo01146a023

Beydoun MA, Beydoun HA, Boueiz A, Shroff MR, Zonderman AB (2013) Antioxidant status and its association with elevated depressive symptoms among US adults: National Health and Nutrition Examination Surveys 2005-6. Br J Nutr 109(9):1714–1729. https://doi.org/10.1017/S0007114512003467

Correia AS, Cardoso A, Vale N (2023) Oxidative stress in depression: the link with the stress response, neuroinflammation, serotonin, neurogenesis and synaptic plasticity. Antioxidants (Basel) 12(2):470. https://doi.org/10.3390/antiox12020470

Costantini L, Molinari R, Farinon B, Merendino N (2017) Impact of omega-3 fatty acids on the gut microbiota. Int J Mol Sci 18(12):2645. https://doi.org/10.3390/ijms18122645

de Jesus Raposo MF, de Morais AM, de Morais RM (2015) Marine polysaccharides from algae with potential biomedical applications. Mar Drugs 13(5):2967–3028. https://doi.org/10.3390/md13052967

Diers JA, Ivey KD, El-Alfy A, Shaikh J, Wang J, Kochanowska AJ, Stoker JF, Hamann MT, Matsumoto RR (2008) Identification of antidepressant drug leads through the evaluation of marine natural products with neuropsychiatric pharmacophores. Pharmacol Biochem Behav 89(1):46–53. https://doi.org/10.1016/j.pbb.2007.10.021

Echave J, Otero P, Garcia-Oliveira P, Munekata PES, Pateiro M, Lorenzo JM, Simal-Gandara J, Prieto MA (2022) Seaweed-derived proteins and peptides: promising marine bioactives. Antioxidants (Basel) 11(1):176. https://doi.org/10.3390/antiox11010176

Haenel T (1986) Historical notes on the therapy of depression. Schweiz Med Wochenschr 116(47):1652–1659. https://www.ncbi.nlm.nih.gov/pubmed/3541174 (Zur Geschichte der Depressionsbehandlung)

Han SI, Kim S, Lee C, Choi YE (2019) Blue-Red LED wavelength shifting strategy for enhancing beta-carotene production from halotolerant microalga, Dunaliella salina. J Microbiol 57(2):101–106. https://doi.org/10.1007/s12275-019-8420-4

Hillhouse TM, Porter JH (2015) A brief history of the development of antidepressant drugs: from monoamines to glutamate. Exp Clin Psychopharmacol 23(1):1–21. https://doi.org/10.1037/a0038550

Hu JF, Schetz JA, Kelly M, Peng JN, Ang KK, Flotow H, Leong CY, Ng SB, Buss AD, Wilkins SP, Hamann MT (2002) New antiinfective and human 5-HT2 receptor binding natural and semisynthetic compounds from the Jamaican sponge Smenospongia aurea. J Nat Prod 65(4):476–480. https://doi.org/10.1021/np010471e

Ibrahim MA, El-Alfy AT, Ezel K, Radwan MO, Shilabin AG, Kochanowska-Karamyan AJ, Abd-Alla HI, Otsuka M, Hamann MT (2017) Marine inspired 2-(5-Halo-1H-indol-3-yl)-N,N-dimethylethanamines as modulators of Serotonin receptors: an example illustrating the power of bromine as part of the uniquely marine chemical space. Mar Drugs 15(8):248. https://doi.org/10.3390/md15080248

Ikarashi Y, Sekiguchi K, Mizoguchi K (2018) Serotonin receptor binding characteristics of geissoschizine methyl ether, an indole alkaloid in Uncaria hook. Curr Med Chem 25(9):1036–1045. https://doi.org/10.2174/0929867324666170320114713

Jiang X, Zhu K, Xu Q, Wang G, Zhang J, Cao R, Ye J, Yu X (2017) The antidepressant-like effect of trans-astaxanthin involves the serotonergic system. Oncotarget 8(15):25552–25563. https://doi.org/10.18632/oncotarget.16069

Kendler KS (2020) The origin of our modern concept of depression-the history of melancholia from 1780-1880: a review. JAMA Psychiatry 77(8):863–868. https://doi.org/10.1001/jamapsychiatry.2019.4709

Kim NR, Kim HY, Kim MH, Kim HM, Jeong HJ (2016) Improvement of depressive behavior by Sweetme sweet pumpkin and its active compound, beta-carotene. Life Sci 147:39–45. https://doi.org/10.1016/j.lfs.2016.01.036

Li Y, Li L, Wei S, Yao J, Liang B, Chu X, Wang L, Liu H, Liao D, Liu D, Jiang P (2024) Integrating transcriptomics and metabolomics to elucidate the mechanism by which taurine protects against DOX-induced depression. Sci Rep 14(1):2686. https://doi.org/10.1038/s41598-023-51138-5

Liu C, Zhang S, McClements DJ, Wang D, Xu Y (2019) Design of astaxanthin-loaded core-shell nanoparticles consisting of chitosan oligosaccharides and poly(lactic- co-glycolic acid): enhancement of water solubility, stability, and bioavailability. J Agric Food Chem 67(18):5113–5121. https://doi.org/10.1021/acs.jafc.8b06963

Liu Y, Ren X, Fan C, Wu W, Zhang W, Wang Y (2022) Health benefits, food applications, and sustainability of microalgae-derived N-3 PUFA. Foods 11(13):1883. https://doi.org/10.3390/foods11131883

Maurer DM, Raymond TJ, Davis BN (2018) Depression: screening and diagnosis. Am Fam Physician 98(8):508–515. https://www.ncbi.nlm.nih.gov/pubmed/30277728

Menaa F, Wijesinghe U, Thiripuranathar G, Althobaiti NA, Albalawi AE, Khan BA, Menaa B (2021) Marine algae-derived bioactive compounds: a new wave of nanodrugs? Mar Drugs 19(9):484. https://doi.org/10.3390/md19090484

Menke A (2024) The HPA axis as target for depression. Curr Neuropharmacol 22(5):904–915. https://doi.org/10.2174/1570159X21666230811141557

Munir S, Shahid A, Aslam B, Ashfaq UA, Akash MSH, Ali MA, Almatroudi A, Allemailem KS, Rajoka MSR, Khurshid M (2020) The therapeutic prospects of naturally occurring and synthetic indole alkaloids for depression and anxiety disorders. Evid Based Complement Alternat Med 2020:8836983. https://doi.org/10.1155/2020/8836983

Noori T, Sureda A, Sobarzo-Sanchez E, Shirooie S (2022) The role of natural products in treatment of depressive disorder. Curr Neuropharmacol 20(5):929–949. https://doi.org/10.2174/1570159X20666220103140834

Osher Y, Belmaker RH (2009) Omega-3 fatty acids in depression: a review of three studies. CNS Neurosci Ther 15(2):128–133. https://doi.org/10.1111/j.1755-5949.2008.00061.x

Pangestuti R, Siahaan EA, Kim SK (2018) Photoprotective substances derived from marine algae. Mar Drugs 16(11):399. https://doi.org/10.3390/md16110399

Pauletti PM, Cintra LS, Braguine CG, da Silva Filho AA, Silva ML, Cunha WR, Januario AH (2010) Halogenated indole alkaloids from marine invertebrates. Mar Drugs 8(5):1526–1549. https://doi.org/10.3390/md8051526

Peng YF, Wang LL, Gu JH, Zeng YQ (2023) Effects of astaxanthin on depressive and sleep symptoms: a narrative mini-review. Heliyon 9(7):e18288. https://doi.org/10.1016/j.heliyon.2023.e18288

Robinson M, Turnbull S, Lee BY, Leonenko Z (2020) The effects of melatonin, serotonin, tryptophan and NAS on the biophysical properties of DPPC monolayers. Biochim Biophys Acta Biomembr 1862(9):183363. https://doi.org/10.1016/j.bbamem.2020.183363

Santhiravel S, Dave D, Shahidi F (2025) Bioactives from marine resources as natural health products: a review. Pharmacol Rev 77(1):100006. https://doi.org/10.1124/pharmrev.123.001227

Sasaki-Hamada S, Hoshi M, Niwa Y, Ueda Y, Kokaji A, Kamisuki S, Kuramochi K, Sugawara F, Oka J (2016) Neoechinulin a induced memory improvements and antidepressant-like effects in mice. Prog Neuro-Psychopharmacol Biol Psychiatry 71:155–161. https://doi.org/10.1016/j.pnpbp.2016.08.002

Schepers M, Martens N, Tiane A, Vanbrabant K, Liu HB, Lutjohann D, Mulder M, Vanmierlo T (2020) Edible seaweed-derived constituents: an undisclosed source of neuroprotective compounds. Neural Regen Res 15(5):790–795. https://doi.org/10.4103/1673-5374.268894

Steru L, Chermat R, Thierry B, Simon P (1985) The tail suspension test: a new method for screening antidepressants in mice. Psychopharmacology 85(3):367–370. https://doi.org/10.1007/BF00428203

Talbott S, Hantla D, Capelli B, Ding L, Li Y, Artaria C (2019) Effect of Astaxanthin supplementation psychophysiological heart-brain axis dynamics in healthy subjects. Funct Foods Health Dis 9(8):521–531. https://doi.org/10.31989/ffhd.v9i8.636

Tasdemir D, Bugni TS, Mangalindan GC, Concepcion GP, Harper MK, Ireland CM (2002) Cytotoxic bromoindole derivatives and terpenes from the Philippine marine sponge Smenospongia sp. Z Naturforsch C J Biosci 57(9–10):914–922. https://doi.org/10.1515/znc-2002-9-1027

Vilchez C, Forjan E, Cuaresma M, Bedmar F, Garbayo I, Vega JM (2011) Marine carotenoids: biological functions and commercial applications. Mar Drugs 9(3):319–333. https://doi.org/10.3390/md9030319

Wang J, Ma W, Ma W, Fang Z, Jiang Y, Jiang W, Kong X, Xin F, Zhang W, Jiang M (2024a) Strategies for the efficient biosynthesis of beta-carotene through microbial fermentation. World J Microbiol Biotechnol 40(5):160. https://doi.org/10.1007/s11274-024-03955-7

Wang X, Yang C, Zhang X, Ye C, Liu W, Wang C (2024b) Marine natural products: potential agents for depression treatment. Acta Biochim Pol 71:12569. https://doi.org/10.3389/abp.2024.12569

Wise T, Cleare AJ, Herane A, Young AH, Arnone D (2014) Diagnostic and therapeutic utility of neuroimaging in depression: an overview. Neuropsychiatr Dis Treat 10:1509–1522. https://doi.org/10.2147/NDT.S50156

Zeni ALB, Camargo A, Dalmagro AP (2019) Lutein prevents corticosterone-induced depressive-like behavior in mice with the involvement of antioxidant and neuroprotective activities. Pharmacol Biochem Behav 179:63–72. https://doi.org/10.1016/j.pbb.2019.02.004

Zhao T, Liu T, Wang L, Xie K, Tang H, Tang M (2024) Dysfunction of neurotransmitter metabolism is associated with the severity of depression in first-diagnosed, drug-naive depressed patients. J Affect Disord 349:332–341. https://doi.org/10.1016/j.jad.2024.01.023

Zhou X, Gan T, Fang G, Wang S, Mao Y, Ying C (2018) Zeaxanthin improved diabetes-induced anxiety and depression through inhibiting inflammation in hippocampus. Metab Brain Dis 33(3):705–711. https://doi.org/10.1007/s11011-017-0179-x

Zhu F, Tu H, Chen T (2022) The microbiota-gut-brain axis in depression: the potential pathophysiological mechanisms and microbiota combined antidepression effect. Nutrients 14(10):2081. https://doi.org/10.3390/nu14102081

Zirpoli H, Chang CL, Carpentier YA, Michael-Titus AT, Ten VS, Deckelbaum RJ (2020) Novel approaches for omega-3 fatty acid therapeutics: chronic versus acute administration to protect heart, brain, and spinal cord. Annu Rev Nutr 40:161–187. https://doi.org/10.1146/annurev-nutr-082018-124539

Marine-Derived Natural Bioactives for Neuronal Cancer and Tumor Remedy

9

Tyler King, Suhrud Pathak, Muralikrishnan Dhanasekaran, Jaufar Palekkodan, Myiesha S. Ameen, and Hemalatha Selvaraj

Abstract

A vast array of naturally occurring bioactive chemicals with substantial therapeutic potential is produced by the remarkable diversity of species found in marine habitats. The utilization of bioactives generated from marine sources to treat tumors and malignancies of the neurons is the main topic of this chapter. Marine chemicals provide new methods of action and focused therapeutic benefits as traditional cancer treatments encounter more difficulties, such as medication resistance and adverse effects. In preclinical research, bioactive peptides, polysaccharides, alkaloids, and other metabolites obtained from marine organisms such as algae, sponges, and microbes have shown encouraging anticancer qualities. Species like red algae, which have thousands of undiscovered variants, are noteworthy because they offer a significant reservoir of potentially useful

T. King · S. Pathak
Drug Discovery and Development, Harrison College of Pharmacy, Auburn University, Auburn, AL, USA
e-mail: tpk0010@auburn.edu

M. Dhanasekaran (✉)
Drug Discovery and Development, Harrison College of Pharmacy, Auburn University, Auburn, AL, USA

Graduate School, Auburn University, Auburn, AL, USA
e-mail: dhanamu@auburn.edu

J. Palekkodan
Illham House, Changuvetti Kund, Kotakkal, Kerala, India

M. S. Ameen
University of Maryland, School of Public Health, College Park, MD, USA

H. Selvaraj
Karpagam Academy of Higher Education, Coimbatore, Tamilnadu, India

M. Bhupathyraaj et al. (eds.), *Neuroprotective Marine Bioactives*, Nutritional Neurosciences, https://doi.org/10.1007/978-981-95-5791-2_9

chemicals. This chapter examines recent research results, mechanisms of action, and possible avenues for marine bioactives to support novel brain cancer treatment strategies.

9.1 Introduction

Natural products have been used to treat injuries, diseases, and many other things since the beginning of humankind. As time passes, humans keep looking for different remedies to treat diseases in our world. Marine bioactive compounds are very promising as Earth is around ¾ water and so much of this is unexplored. The marine environment has enormous potential for compounds to help treat diseases. Marine bioactive compounds hold significant promise for treating various diseases (Karthikeyan et al. 2022). The marine environment has many different species adapting and evolving to live in excruciatingly harsh conditions. Many of these conditions have caused these species to create specific and useful compounds. These compounds can be used as a resistance to extreme temperatures, abnormal pH, saltiness, and contaminants. Marine bioactive compounds from the marine environment are very useful in developing remedies. One disease for which marine bioactive compounds have a future is neuronal cancer.

9.2 History

It appears that brain and central nervous system cancer in Western civilizations has been increasing, but it has been linked to improving diagnostic technology. There is about a five- to-six-fold difference between rates in Europe, the highest, and lowest in Asia (Miranda-Filho et al. 2017). Brain and CNS cancer make up about 3% of cancer cases worldwide. In this study, 78,304 cancer cases were studied between 2003 and 2007; and the results showed 54% in men and 46% in women. There was also an almost twofold difference between white populations and black populations in the United States.

9.3 Pathophysiology

The most common type of neuronal cancer or tumor is a glioma. Though they only make up 1–2% of all cancers, they are considered very disabling and lethal (Jung et al. 2020). These tumors are believed to take advantage of neurodevelopmental mechanisms involved in proliferation, migration, and network formation.

Regardless of all the advances in malignancies, some tumors are still deemed incurable. Among these are gliomas, which barely respond to chemotherapy. Gliomas are tumors of the brain that stem from glial cells in the central nervous system (Khotimchenko et al. 2021). Many consider malignant tumor chemotherapies' biggest problems to be multidrug resistance and their ability to evade apoptosis. Furthermore, chemotherapy used to treat glioblastomas can have side effects like hepatotoxicity, myelosuppression, hepatotoxicity, cerebral edema, blood clots, and many more. The discovery of new compounds that could fight glioblastomas without these side effects while being effective would be revolutionary.

Traditional antitumor compounds can be divided into two groups: cytostatic agents and cytotoxic agents. Cytostatics are essentially used to stop tumor growth, while cytotoxicity is used to kill the tumor cells. A more targeted approach has emerged. These "targeted drugs" specifically act on molecular targets in cancer cells. These drugs can then influence pathways that control whether the cell dies. Glioblastomas often have genetic changes in TERT, EGFR, PTEN, and TP53. Blocking multiple pathways at once could be very effective. Glioblastoma cells also often have irregular cell cycle regulation. This is usually due to the overactivation of cyclin-dependent kinases (CDKs), so a way to target CDKs specifically could be very beneficial.

Glial tumors are divided into two groups based on invasiveness. These two groups are called diffuse gliomas and circumscribed gliomas (Aldape et al. 2015). Diffuse gliomas infiltrate surrounding brain tissue and tend to come back even after extensive surgical resection. Circumscribed gliomas exhibit more localized growth behavior. Many of these tumors don't respond well to chemotherapy and radiation, and tumor control is not very reliable.

Tumors can be noncancerous or cancerous. Some examples of noncancerous tumors are meningiomas, pituitary adenomas, craniopharyngiomas, schwannomas, nasopharyngeal angiofibromas, choroid plexus tumors, dysembryoplastic neuroepithelial tumors, neurofibromas, hemangioblastomas, and chondromas (Johns Hopkins Medicine n.d.-a). Meningiomas form in the meninges. Craniopharyngiomas may press on adjacent brain structures, leading to vision and hormonal problems. Schwannomas can cause hearing loss. Dysembryoplastic neuroepithelial tumors can cause seizures. Other brain tumors can be more malignant, like gliomas, ependymal tumors, hemangiopericytomas, and pineal tumors (Johns Hopkins Medicine n.d.-a). Gliomas make up about 33% of brain tumors. Ependymal tumors may exert pressure on the brain. Typically, malignant brain tumors include chordomas, chondrosarcomas, medulloblastomas, olfactory neuroblastomas, lymphomas, gliosarcomas, and rhabdomyosarcomas. Chordomas are a rare bone cancer found at the base of the skull or lower back. Medulloblastomas most commonly affect children, particularly from ages five to nine. Olfactory neuroblastomas can lead to nosebleeds, loss of smell, and breathing difficulties.

Neuronal brain tumors arise from cells with neuronal differentiation, and the tumor's location significantly influences the effects of these tumors. The temporal lobe is the most common tumor location (Mezzacappa and Thorell 2025). Tumors in the temporal lobe often lead to seizures. These neuronal and glioneuronal tumors usually arise from mutations in the mitogen-activated protein kinase pathway (MAPK). MAPK is a pathway that influences cellular growth, proliferation, and differentiation. Mutations to the BRAF gene are also commonly found in these tumors. A common mutation is the fusion of one BRAF gene with another gene, resulting in the BRAF gene being activated more than usual. This enhances the growth of the tumor, allowing it to survive. Another frequent mutation is in the fibroblast growth factor. Specifically, the FGFR-1 gene is mutated upstream from the mitogen-activated protein kinase pathway (Mezzacappa and Thorell 2025). Fibroblast growth factors are receptor tyrosine kinases, and when these are mutated, it often leads to uncontrolled tumor growth.

Intrinsic oncogenic programs and extrinsic microenvironmental factors like neurons and glial cells influence brain tumors (Pan and Monje 2022). Glial cells include astrocytes, oligodendrocytes, and microglia. Glial cells comprise about half of the brain's cells. They help regulate neurodevelopment and plasticity through interactions with neurons. Many brain tumor cells exhibit similar gene expression to macroglial cells. This could mean that tumor cells hijack typical neuron-glial interactions to promote tumor growth. Neuronal activity promotes the growth and proliferation of gliomas. Glioma cells establish synaptic connections with neurons similar to OPC-neuron interactions, and these synapses exhibit excitatory currents mediated by calcium-permeable AMPA receptors. Blocking these neuron-to-glioma synapses inhibits tumor growth. Neuronal activity also supports glioma growth through paracrine signaling by releasing factors like BDNF, IGF-1, and a shed form of neuroligin-3. Nlgn3 promotes oncogenic signaling and further synapse formation in glioma cells. Reactive astrocytes can enhance tumor cell proliferation, invasion, and therapy resistance and can influence the tumor's immune environment. BDNF promotes glioma growth and reinforces neuron-to-glioma synapses by boosting glutamate responses. IGF-1 is critical for fostering growth in olfactory bulb high-grade gliomas (Pan and Monje 2022).

9.3.1 Symptoms

A brain tumor is an abnormal, uncontrolled growth of cells located within or next to the brain. They can develop inside the brain or even in surrounding structures like nerves, the pineal and pituitary gland, and membranes covering the brain's surface. There are primary and secondary tumors. Primary tumors originate directly in the brain, while secondary tumors develop when cancer cells from other parts of the body metastasize to the brain (Mayo Clinic n.d.). Benign or noncancerous tumors grow slowly and may put pressure on the brain without invading it. Malignant or cancerous tumors grow quickly and can invade and destroy adjacent brain tissue. Tumors can range from being very small to being very large. Small tumors can be detected early before they grow if they cause noticeable symptoms. Sometimes, tumors in less active brain areas don't cause noticeable symptoms and grow large before being detected. Brain tumors can cause many different symptoms (Mayo Clinic n.d.). Some of these symptoms include headaches, pressure in the head, nausea, vomiting, eye problems, loss of feeling and movement in arms or legs, balance problems, speech problems, excessive tiredness, confusion, memory problems, personality changes, seizures, hearing problems, dizziness, and changes in hunger and weight. The most common symptom is headaches. Headaches are very common for people with brain tumors; about half experience headaches. This occurs because a growing tumor can push on nearby cells, leading to uncomfortable pressure on them. These headaches are often worse in the morning because when one sleeps, the pressure builds up in one's skull from cerebrospinal fluid accumulating around the brain. Tumors in the front of the brain, also called the frontal lobe, may cause disruption in balance and, therefore, difficulty walking. They can also lead to personality changes and forgetfulness. Tumors in the middle of the brain, also called the parietal lobe, can cause sensory problems such as vision and hearing issues. Tumors in the back of the brain, also called the occipital lobe, can lead to problems with vision. Tumors on the sides and lower part of the brain, also called the temporal lobe, can result in memory problems and abnormal sensory experiences (Mayo Clinic n.d.).

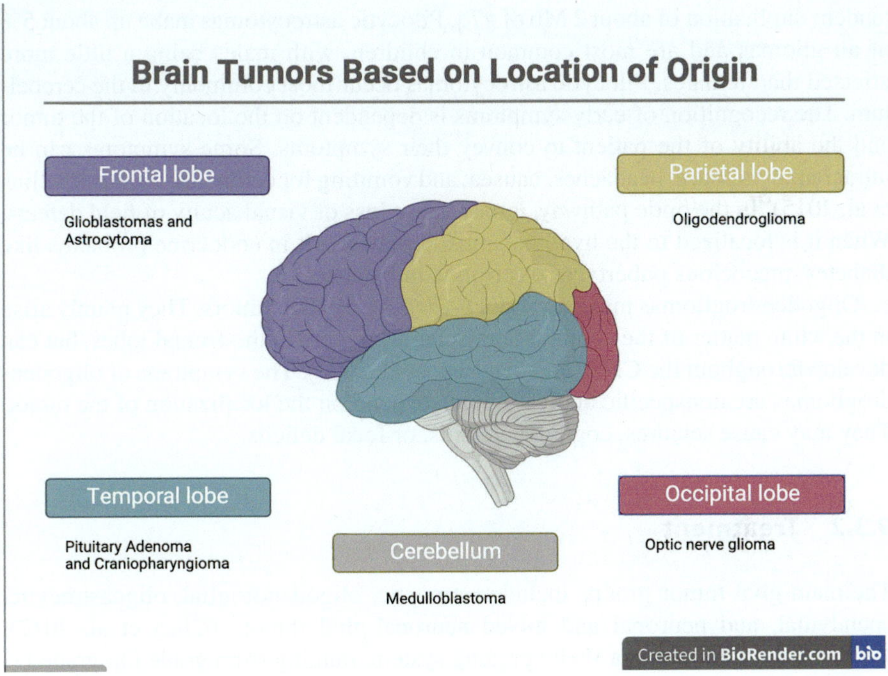

Cancer is the leader among everything in the world, responsible for morbidity and mortality. Central nervous system cancer is the third most common type of cancer even though it only accounts for 0.2% of registered cases of cancer (Anand et al. 2023).

The blood-brain barrier prevents the central nervous system from directly contacting blood. This can make treatment harder as drugs and other compounds have a hard time coming into contact with the cancer in the central nervous system.

Central neurocytomas are uncommon tumors of the central nervous system. Usually, they affect the lateral ventricles of young adults. Maximal resection is the ideal option, but radiotherapy is helpful for patients with incomplete resection

(Sharma et al. 2006). Central neurocytoma patients have a slow clinical course. They present raised intracranial pressure due to obstructive hydrocephalus caused by impaired cerebrospinal fluid flow, also paired with seizures, headache, nausea, vomiting, and memory and visual disturbances. In imaging, it generally shows discrete, solitary masses near the foramen of Monro. On computed tomography scans, they are usually isodense or mildly hyperdense. Calcifications and cystic changes have been seen. Magnetic resonance imaging shows high signals on T1- and T2-weighted images (Sharma et al. 2006).

Pilocytic astrocytomas are relatively benign, with a 10-year survival rate of over 90% (Collins et al. 2015). Most require surgical removal, but rarely, they progress to a more malignant glioma. Researching the genome has shown single irregularities of the MAPK pathway in almost all cases. The most common abnormality is a tandem duplication of about 2 Mb of #7q. Pilocytic astrocytomas make up about 5% of all gliomas and are most common in children, with males being a little more affected than females. Pilocytic astrocytomas occur most commonly in the cerebellum. The recognition of early symptoms is dependent on the location of the tumor and the ability of the patient to convey their symptoms. Some symptoms can be intracranial pressure, headaches, nausea, and vomiting for cerebellar tumors (Collins et al. 2015). In the optic pathway, it may cause loss of visual acuity of field defects. When it is localized to the hypothalamus, it may result in endocrine problems like diabetes, precocious puberty, or electrolyte imbalance.

Oligodendrogliomas make up about 5–20% of all glial tumors. They mainly arise in the white matter of the cerebral hemispheres, mostly in the frontal lobes, but can develop throughout the CNS (Van den Bent et al. 2008). The symptoms of oligodendrogliomas are nonspecific and very much depend on the localization of the tumor. They may cause seizures, cognitive deficits, or focal deficits.

9.3.2 Treatment

The main glial tumor groups include astrocytic, oligodendroglial, oligoastrocytic, ependymal, and neuronal and mixed neuronal-glial tumors (Chen et al. 2017). Tumors are graded using a WHO grading system, ranging from grade I to grade IV, generally based on an escalation in malignancy. Biomarkers have helped make diagnostic conditions more precise. One example of a molecular biomarker of glioma subtypes is MGMT (O6-methylguanine-DNA methyltransferase) promoter methylation. TMZ methylates purine bases, including O6 and N7 sites on guanine, and N3 sites on adenine. O6-methylguanine induction is believed to be the main cytotoxic event that leads to the addition of thymine rather than cytosine during replication. This can cause double-stranded breaks, leading to incorrect DNA replication and, eventually, cell death. MGMT removes the O6 methylguanine adducts, removing their cytotoxic effects (Friedman et al. 2000). Repair of DNA removes the MGMT protein, so higher levels of MGMT are predicted to lead to temozolomide resistance.

Treatment for brain cancer depends on many things, like the size of the tumor, the type of the tumor, and the location of the tumor. It also depends on the patient's health as to what would be the best course of treatment. One treatment for brain cancer would be surgery, and there are many different types. Some of these examples include craniotomy, awake brain tumor surgery, neuroendoscopy, MRI laser-guided ablation, laser interstitial thermal therapy, biopsy, and neuroplastic surgery (Johns Hopkins Medicine n.d.-b). Craniotomy is the removal of a piece of the skull to assess the brain. Awake brain tumor surgery is surgery that is done while the patient is awake so that one can monitor for normal brain function. Neuroendoscopy is a minimally invasive surgery that goes through a natural opening, like the nose, to look at the brain. MRI laser-guided ablation and laser interstitial thermal therapy use computer technology to locate a tumor and destroy cancerous cells with a laser or heat. There is also radiation therapy. Some popular examples of radiation therapy include external beam radiation, stereotactic radiosurgery, and proton therapy. External beam radiation is the standard radiation therapy for brain tumors. It can be directed specifically at the tumor, or it can be directed at the whole brain for metastatic brain tumors. Stereotactic radiosurgery uses small, precise beams of X-rays to destroy the tumor while saving the healthy tissue. Proton therapy uses protons to deliver radiation. It is used sometimes because it emits less radiation to surrounding healthy tissue (Johns Hopkins Medicine n.d.-b). Chemotherapy is another form of treatment, and it is often used alongside surgery or radiation. Chemotherapy uses specific drugs to kill cancer cells, but these drugs often harm normal cells. Targeted drug therapy are medications that selectively attack specific traits of cancer cells. They attack by stopping the tumor's growth and spreading to the surrounding tissue. Tumor-treating fields use electrical impulses to interrupt cancer cell division in the brain.

Glioblastomas are the most common malignant brain tumor in adults and are very aggressive, with a median survival of only 8 months. The standard treatment is surgical resection in combination with radiotherapy and chemotherapy (Kotecha et al. 2023). Radiation therapy has shown an improvement in survival without a big decline in the quality of life. There are also tumor-treating fields that can be used. These have no survival benefit and are used to improve the quality of life of patients with glioblastomas. Tumor treating fields are low-intensity, intermediate-frequency alternating electric fields from a device that inhibit cell division. Despite these alternative treatments, chemotherapy is the only systemic therapy with survival benefits for glioblastomas. The standard compound used for chemotherapy is temozolomide, often called TMZ. It extends median survival from 12.1 to 14.6 months for a newly diagnosed glioblastoma. It was shown to be the most effective against patients with glioblastoma who had O6-methylguanine-DNA-DNA methyltransferase (MGMT) promoter methylation (Kotecha et al. 2023).

9.3.3 Marine Natural Products

As humans have started living longer, neurological and neurodegenerative diseases have become more prevalent. The extreme biodiversity of the marine environment produces unique metabolites that could be very beneficial for therapeutic, preventive, and diagnostic purposes.

δ-Conotoxins are derived from marine cone snails, or *Conus striatus* L. δ-Conotoxins have been shown to disrupt the rapid inactivation process of voltage-gated sodium channels (Grosso et al. 2014). Cancer cells often have abnormal ion channel activity, and δ-conotoxins will be able to control neuronal excitability and, therefore, affect downstream pathways. μ-Conotoxins also block voltage-gated sodium channels but do this by blocking the channels directly. ω-Conotoxins are similar to the other conotoxins, but they are able to modulate calcium channels. Even though they modulate different channels, they can all alter downstream signaling by changing the channels, potentially stopping neuronal cancer growth and proliferation.

Sea cucumbers contain triterpene glycosides. One example is Ds-echinoside A, which was shown to reduce the expression of MMP9, a regulator of treatment resistance and invasiveness, by almost 90% in a hepatocellular carcinoma cell line model (Karbhari and Khagi 2024). It was also shown to reduce NfkB levels, which is useful because when NfkB is turned on, it helps cells survive and avoid apoptosis (Dolcet et al. 2005).

Another triterpene glycoside known as holothurin A has been shown to decrease the expression of MMP9 and vascular endothelial growth factor (Karbhari and Khagi 2024), a regulator of physiological and pathological angiogenesis (Ferrara 2009).

Another triterpene glycoside known as Terminoside A has been shown to hinder the production of nitric oxide through inhibition of nitric oxide synthetase (Karbhari and Khagi 2024). Nitric oxide plays a role in multiple tumorigenic pathways, and inhibiting this could slow tumorigenesis.

Alkaloids are nitrogen-containing cyclic compounds. They are usually found in plants, but an increasing number of alkaloids are being found in marine species. Some of these marine species include algae and sponges. This could be very beneficial because alkaloids show antitumor effects by interfering with cellular pathways due to their unique chemical structures. Zorrimidazolone, a bioactive alkaloid derived from Mediterranean ascidian Polyandrocarpa zorritensis, showed cytotoxic effects against C6 cells, resulting in a 60% decreased in viability when the concentration was 250 μM over 48 h (Khotimchenko et al. 2021). This significant reduction in cell viability by zorrimidazolone can cause cancer cell death. Meridianins are isolated from the ascidian Aplidium meridianum, and variolin B is isolated from an Antarctic sponge. These marine bioactives both inhibit cdk, which are very important in controlling cell division and survival in cancer cells (Khotimchenko et al. 2021). By targeting and inhibiting cdk, they can disrupt the abnormal and uncontrolled proliferation of tumor cells by cancer cells. Fascaplysin is isolated from a Fijian sponge and is a cdk4 inhibitor, has antiangiogenic effects, and leads to apoptosis of the cancer cell (Khotimchenko et al. 2021).

Another type is carboline alkaloids. α-Carbolines are naturally found in marine organisms, and the natural and synthetic versions of α-carbolines have shown anticancer activity. In glioma xenograft models, TJY-16, an α-carboline analogue, significantly inhibited tumor growth, even outperforming conventional treatments like temozolomide. Tetrahydroisoquinoline alkaloids are commonly categorized into ecteinascidins and renieramycins. Ecteinascidin-743 is isolated from Caribbean ascidians and is used as a treatment option for inoperable soft tissue sarcomas that are highly resistant to traditional chemotherapies. This works by binding to the minor groove of the DNA double helix and alkylating N2 guanine (Khotimchenko et al. 2021). It disrupts cell proliferation, impedes DNA repair, and hinders transcription activation, ultimately leading to apoptosis of the target cells. This not only stops cancer cell replication but also triggers cell death, which is very useful for controlling the cancer. Several pyrrospirone alkaloids (C, D, E, F, G, H, and J) and penicipyrroether A are derived from a marine fungus called Penicillium, the species, ZZ380 strain. These can be found in the sea crab Pachygrapsus crassipes. They belong to a family of fungal secondary metabolites known as hirsutellones, which have a 13-membered ether ring structure. Pyrrospirone G shows strong inhibition of glioma cell proliferation, with effective concentrations between 1 and 8 μM. Penicipyrroether A demonstrates strong antiglioma effects, comparable or even superior to doxorubicin, a standard anticancer drug, all while being less toxic to normal cells. The potent antiglioma activity and strong toxicity of these compounds suggest their potential as a safer alternative to other chemotherapies (Khotimchenko et al. 2021). Trichobamide A is a pyrrocidine alkaloid isolated from an ascidian-derived fungus called Trichobotrys effusus 4729. This marine bioactive contains a tetrahydro-5H-furo[2,3-b]pyrrol-5-one ring. This marine bioactive inhibits glioma cell growth and induces apoptosis. Its specific and unique structure shows potential as a targeted cancer therapy. Pseurotin A, isolated from a marine bacterium, Bacillus sp., and a barnacle called Lepas anatifera, has been shown to reduce glioma cell proliferation and alter tumor metabolism by lowering the glycolytic enzymes PKM2 and LDH5. Furthermore, it was shown to boost the TCA cycle/oxidative phosphorylation enzymes PDHB, ATPB, and cytochrome C. Overall, this marine bioactive has been shown to shift tumor energy production toward normal cellular processes. This reprogramming from a glycolytic to an oxidative state by Pseurotin A disrupts the energy necessary for a cancer cell to proliferate (Khotimchenko et al. 2021) uncontrollably. Polycyclic diamine alkaloids are marine bioactives found in sponges, built from molecules like ammonia and aldehydes that contain two key compounds: papuamine and haliclonadiamine. Papuamine and haliclonadiamine are both isolated from Neopetrosia cf. exigua. Papuamine was shown to inhibit glioblastoma cell growth strongly and is about 8 times more potent than haliclonadiamine. This huge difference in potency shows the potential that some compounds have to work much more effectively than others while being similar. Granulatimide and isogranulatimide are alkaloids isolated from the ascidian Didemnum granulatum and inhibit checkpoint kinase 1, an enzyme involved in cell cycle regulation. Disrupting the checkpoint in the cell cycle can help control the uncontrolled proliferation of cancer cells. Some antibiotics can inhibit tumor cell growth and are valuable in the chemotherapy of malignant tumors. Actinomycins are chromopeptide compounds with both anticancer and antibiotic properties, produced by various Streptomyces strains (Khotimchenko et al. 2021). Actinomycins have an unmodified phenoxazinone core linked to two pentapeptide lactone rings. About 50 different actinomycins have been identified, including actinomycin D, C, F, Z, G, and Y. In tumor cells, the phenoxazinone core intercalates between DNA base pairs at guanine/cytosine sites, while the peptide rings bind to the minor groove. This blocks DNA replication and transcription and inhibits the proliferation of glioma cells. Over 40 natural antimycins A have been discovered and have shown anticancer properties. They have a nine-membered dilactone core linked to a 3-formyl aminosalicylic acid moiety and are isolated from the culture broth of the marine actinomycete Streptomyces antibiotics H12–15. Neoantimycin A and neoantimycin B, along with known antimycins A1ab, A2a, and A9, were obtained from this marine source. Antimycins A1ab, A2a, and A9 were 20–26 times more potent with an IC50 below 1.6 μg/mL than neoantimycins against human glioblastoma SF-268 cells. This high potency against glioblastoma cells shows the potential of antimycins as strong anticancer treatments that may outperform today's standard treatments. Bagremycins are isolated from a mangrove-derived actinomycete called Streptomyces sp. Q22. Bagremycin C is active against four different glioma cell lines with IC50 values in the range between 2.2 and 6.4 μM at 72 h, compared to the activity of doxorubicin at 0.4–3.3 μM (Khotimchenko et al. 2021). Bagremycin C induces apoptosis in U87MG cells in a dose- and time-dependent manner and causes cell cycle arrest at the G0/G1 phase, with significant increases in apoptotic cell numbers compared to controls (Khotimchenko et al. 2021). By promoting apoptosis and stopping cell cycle progression at the same time, it disrupts multiple pathways that cancer cells use to survive.

Chemotherapy is very difficult to use to treat brain cancer because of the blood-brain barrier. The blood-brain barrier can block many drugs from coming into contact with the cancer in the brain. Researchers have started to look for compounds with unique structures and diverse mechanisms of action to be effective against neuronal tumors. Four structures from sea anemone-derived actinomycete Streptomyces sp. ZZ406 has been shown to reduce proliferation and lactate in glioma cells (Alves et al. 2021). A sulforhodamine B assay was used to test these compounds to inhibit the proliferation of human glioma U87MG, U251, and SHG44 cells. Doxorubicin was used as a positive control. Compound 1,1-hydroxymethyl-8-hydroxy-anthraquinone-3-carboxylic acid showed strong activity against different glioma cells with IC50 values in a range of 4.7–8.1 μM. Compound 2, a chromone, showed IC50 values of 21.6–25.8 μM. Compounds 3 and 4 were inactive. The positive control, doxorubicin, showed antiproliferative activity with IC50 values of 1.9–9.6 μM. Compound 1 data shows it was the most effective because the lower the IC50 value, the more potent the drug is, and therefore, less of the drug is required to produce the desired effect. Compound 1 was more effective than the positive control, showing that marine compounds can be effective (Alves et al. 2021).

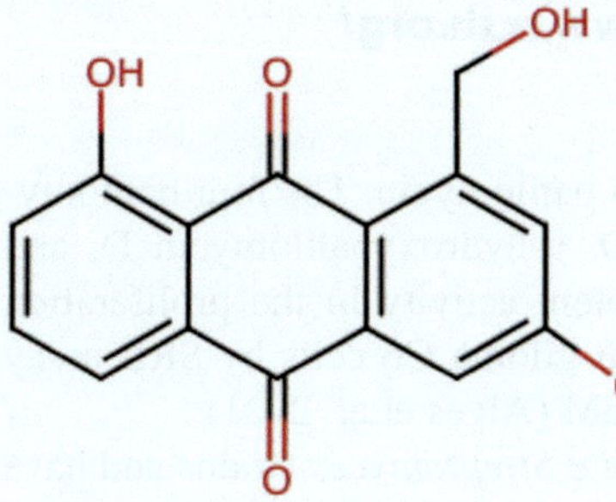

9.3.4 Compound 1- (https://www.rcsb.org/chemical-sketch)

Furthermore, the different kinds of seaweeds make up a large number of species of marine algae. The different kinds of seaweeds have different compounds. Rhodophyta, or red seaweed, has aplysin, which is a bromo sesquiterpene isolated from *Laurencia tristicha.* It was found to lower the number of U87MG, U251MG, and primary invasive glioma cells in a dose-dependent manner. It also deactivated and reactivated the Akt pathway, causing inhibitory effects on U87Mg cells. Phaeophyta, or brown seaweed, contains eckol, which is a phlorotannin component. Eckol was shown to protect cells from oxidative stress caused by H2O2. It protects cells from oxidative stress by radical quenching and catalase activation (Alves et al. 2021). Treatment with eckol in sphere-forming glioma stemlike cells lowered PI3K-Akt activities and completely halted Ras-Raf-1 interaction and Raf-1 and Erk activations. This was thought to enhance the sensitivity of glioma stemlike cells to anticancer treatments. Another compound from brown seaweed is fucoidan, a sulfated saccharide. The effects of fucoidan were tested on T98G-induced or THP1-induced angiogenesis. They were evaluated by tube formation assay using a conditioned medium of fucoidan-treated cells. After 24 h, when using 100 mg/ml of fucoidan, T98G and THP1 showed inhibition in cell-induced tube formation (Alves et al. 2021).

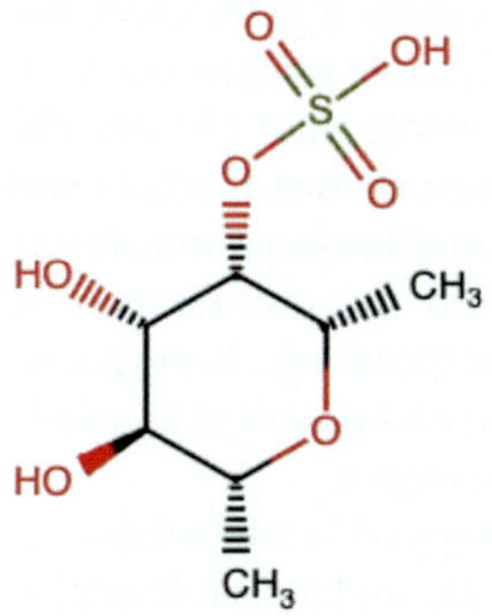

9.3.5 Fucoidan Compound 1 (https://www.rcsb.org/chemical-sketch)

Another compound from green seaweeds is called bafilomycin. The four bafilomycins tested were bafilomycin A1, bafilomycin D, 9-hydroxybafilomycin D, and 23-*O*-butyrylbafilomycin D. They all showed potent activity in the proliferation against U87-MG, U251, and SHG44 cells and rat glioma C6 cells by SRB assay with IC50 values ranging between 0.35 and 2.95 μM (Alves et al. 2021).

Grincamycins are isolated from the actinomycete *Streptomyces* strains and have antitumor properties with unknown mechanisms. In this study, isocitrate dehydrogenase 1(IDH1) was found to be a target of grincamycin b (GCN B). Isocitrate dehydrogenase 1 converts isocitrate to 2-oxoglutarate (Wang et al. 2021). IDH1 is one of the antitumor targets of GCNs, so an effective target to study and test more for intervention could be IDH1. It was predicted by molecular docking that GCN B binds at the dimer interface of IDH1 and forms hydrogen bonds with residues N271 and S280 there. By targeting IDH1, GCN B may disrupt the cancer's metabolism and redox balance, which can impair its growth. Grincamycin was tested with both wild-type and mutant IDH1 R132C and mutant IDH1 R132H. GCN B had an inhibitory effect with an IC50 value of 1.67 for wild-type IDH1, 0.78 for the R132C mutant, and 3.44 μM for the R132H mutant (Wang et al. 2021). GCN B was shown to induce endoplasmic reticulum stress, reactive oxygen species accumulation, and apoptotic cell death. Further studies of marine bioactives that could inhibit IDH1 in different ways could be very useful for finding a way to control neuronal cancer through targeting IDH1.

Largazole is a histone deacetylase inhibitor. Glioblastoma multiforme RNA sequencing showed altered histone deacetylases coding genes. The Cancer Genome Atlas RNA sequencing showed an increase in the expression levels of HDACs 1–3 and HDAC 7 in high-grade gliomas (Al-Awadhi et al. 2020). HDAC enzymes remove acetyl groups from histone proteins. This leads to the inhibition of transcription and an increase in chromosome condensation. Cancers can utilize this because it can silence tumor suppressor genes. Another challenging aspect of cancer in the CNS is that it is very difficult for drug concentrations to reach an appropriate level within the tumor. This is because the brain uptake of drugs causes it to be lower, but largazole is effective at crossing the blood-brain barrier. Largazole is a prodrug with a thioester moiety that undergoes hydrolysis to make largazole thiol (Al-Awadhi et al. 2020). This is important because the thiol form is the one that binds to and inhibits HDACs. Largazole was tested in vitro with two glioblastoma multiforme cancer cell lines, SF-268 and SF-295. Largazole inhibited the proliferation of SF-268 and SF-295 with IC50 values of 62 and 68 nM, respectively. Largazoles' bioavailability and mass-producible synthesis further enhance its potential as a therapeutic agent for glioblastoma multiforme and other CNS cancers.

The manzamine derivative, PCTC, specifically 1-(9′-propyl-3′-carbazole)-1, 2, 3, 4-tetrahydro-β-carboline, has been shown to have cytotoxic and antiproliferative effects in human and murine glioma cell lines. PCTC did this through increasing the expression of apoptotic enzymes, increasing intracellular reactive oxygen species, and blocking the G1/S phase of the cell cycle (Lin et al. 2022). PCTC was tested on A172, U87MG, and GL261 glioma cell lines. These cell lines were then tested for cytotoxicity using a CCK-8 assay after treatment with PCTC. All three cell lines showed a significant inhibitory effect with a concentration greater than or equal to 2 μM. At doses greater than or equal to 4 μM, microscopic analysis revealed cell shrinkage and apoptotic-like morphological destruction. PCTC also inhibited colony formation in GL261 cells after exposure for 9 days. Furthermore, after exposure to PCTC or DMSO for 6 h, increased populations of early and late Annexin V apoptotic cells were observed in all three cell lines. Levels of caspase-3/7 were then tested to confirm the mechanism of apoptosis, and increased levels of caspase-3/7 were found. This assay directly suggests that PCTC initiates programmed cell death. Next, cell cycle stages were tracked in the GL261 cells using flow cytometry. After a treatment with four μM PCTC, there was a significant increase in the subG1 phase and decreases in the S and G2 phases. When cells are unable to progress normally through the cell cycle, they can trigger apoptosis. This data showed cytotoxicity of PCTC in the GL261 cells. Lastly, the study was tested to determine if there was a synergistic effect of TMZ and PCTC on GL261 cells. TMZ, alone with concentrations of 0.1 and 0.2 mM, and PCTC, alone with concentrations of 2 and 4 μM, showed insignificant cytotoxic effects, but treatment with both TMZ and PCTC showed significant effects (Lin et al. 2022). PCTC with a concentration of 4 μM and TMZ with a concentration of 0.2 mM showed the strongest antiproliferative effect in GL261 glioma cells. PCTC may be a very beneficial therapeutic option for glioblastoma, especially when combined with current therapies.

Medulloblastoma is the most common malignant brain tumor in children. High MYC levels drive group 3 medulloblastomas. MYC is an oncogene that encodes a transcription factor responsible for regulating cell growth, proliferation, metabolism, and apoptosis (Xue et al. 2021). To test a marine bioactive called frondoside A, human-induced pluripotent stem cells were used to generate neuronal precursors. These neuronal precursors were then transformed by introducing the MYC oncogene and turning off p53. P53 normally prevents "healthy" cells from turning into cancer cells by stopping the cell cycle, starting DNA repair, or triggering apoptosis. This MYC expression and non-expression of P53 mimicked group 3 medulloblastomas. Frondoside A was found to suppress MYC expression and its gene targets in medulloblastoma cells. Through testing the signal intensity of nuclear MYC immunostaining in iMB and two patient-derived MB cells, it was found that frondoside A was able to decrease the signal, showing that it may inhibit MYC expression (Xue et al. 2021). Furthermore, frondoside A was shown to suppress MYC's downstream target genes, including cyclin-dependent kinases and other oncogenic regulators. These genes were downregulated, further supporting the inhibitory effect of frondoside A on MYC transcriptional activity. Frondoside A also showed selective cytotoxicity to iMB tumors but not the normal brain cells. In the normal brain regions surrounding frondoside A-treated tumor xenografts, immunostaining of caspase-3 did not detect cytotoxicity in normal brain cells.

Fucoxanthin is a marine bioactive product derived from algae. Fucoxanthin was tested against U251-human glioma cells. Fucoxanthin inhibits the growth of

U251-human glioma cells in a dose- and time-dependent manner (Wu et al. 2019). For dose-dependent testing, when U251 cells were treated with 10, 20, and 40 μM fucoxanthin for 72 h, cell viability significantly decreased from 100% in the control to 70.2%, 55.6%, and 28.4%, respectively. For time-dependent testing, when cells were exposed to 20 μM fucoxanthin for 24, 48, and 72 h, cell viability was significantly reduced from 100% in the control to 80.3%, 69.2%, and 56.4%, respectively. Flow cytometry analysis was then used to test fucoxanthin-induced cell growth. Fucoxanthin treatment showed significant U251 cell apoptosis through the increase in the sub-G1 peak. A sub-G1 peak means that they have a reduced amount of DNA from apoptotic fragmentation (Wu et al. 2019). After exposing cells to 10, 20, and 40 μM of fucoxanthin, they all showed a significant increase in apoptosis, with percentages increasing from 1.4% in the control to 11.6%, 33.8%, and 62.4%, respectively.

Fucoxanthin also caused overproduction of reactive oxygen species in a time-dependent manner (Fig. 9.1).

When cells were treated with 20 μM fucoxanthin, the phosphorylation of DNA damage markers, such as Ser1981-ATM, Ser428-ATR, Ser15-p53, and Ser139-histone, increased over time. The increase in DNA damage markers suggests that fucoxanthin induced DNA damage. Next, it was shown that fucoxanthin inhibited U251 cell growth in a MAPK- and PI3K-AKT-dependent manner. Fucoxanthin increased the phosphorylation of Thr183-JNK, Thr180-p38, and Thr202-ERK and inhibited the PI3K-AKT pathway by reducing phosphorylation of AKT at Ser473 from 12 to 72 h. Activating stress signals through MAPKs like JNK, p38, and ERK encouraged cell death of the U251 human glioma cells. Furthermore, the inactivation of survival signals like PI3K-AKT through reduced phosphorylation stops cell proliferation and leads to apoptosis (Wu et al. 2019).

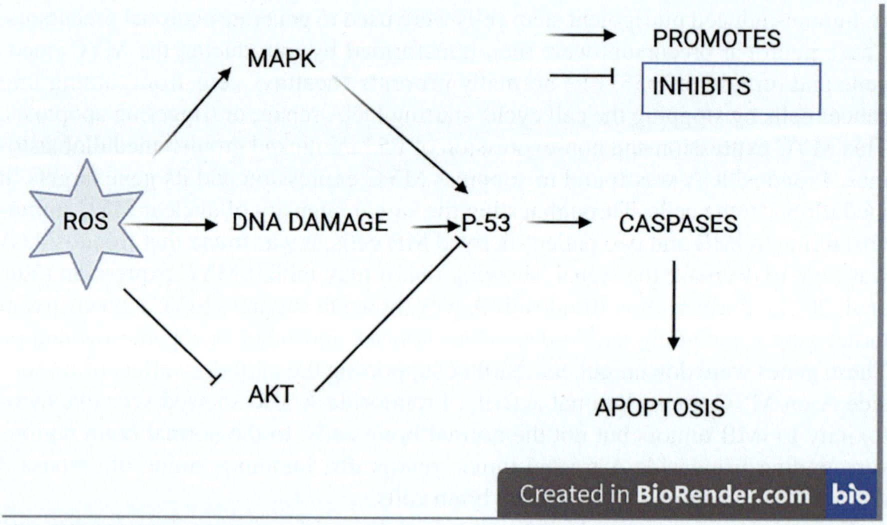

Fig. 9.1 Fucoxanthin (Wu et al. 2019)

Physiochemical properties	Formula	C42H5806
	Molecular weight	658.91 g/mol
	No. of heavy atoms	48
	No. of aromatic heavy atoms	0
	Fraction Csp3	0.55
	No. of rotatable bonds	12
	No. of H-bond acceptors	6
	No. of H-bond donors	2
	Molar refractivity	197.07
	TPSA	96.36 Å^2
Lipophilicity	Log Po/w (iLOGP)	7.17
	Log Po/w (XLOGP3)	8.05
	Log Po/w (WLOGP)	8.69
	Log Po/w (MLOGP)	4.74
	Log Po/w (SILICOS-IT)	10.32
	Consensus Log Po/w	7.8
Water solubility	ESOL Log S	−8.2
	ESOL solubility	4.11e–06 mg/ml; 6.24e–09 mol/L
	ESOL solubility Class	Poorly soluble
	Ali Log S	−9.93
	Ali solubility	7.77e-08 mg/ml; 1.18e-10 mol/L
	Ali solubility class	Poorly soluble
	SILICOS-IT Log S	−5.9
	SILICOS-IT solubility	8.31e-04 mg/ml; 1.26e-06 mol/L
	SILICOS-IT solubility class	Moderately soluble
Pharmacokinetics	GI absorption	Low
	BBB permeant	No
	P-gp substrate	Yes
	CYP1A2 inhibitor	No
	CYP2C19 inhibitor	No
	CYP2C9 inhibitor	No
	CYP2D6 inhibitor	No
	CYP3A4 inhibitor	No
	Log Kp (Skin permeation)	−4.60 cm/s
Drug-likeness	Lipinski	No; 2 violations: MW > 500, MLOGP>4.15

(continued)

Physiochemical properties	Formula	C42H5806
	Ghose	No; 4 violations: MW > 480, WLOGP>5.6, MR > 130, #atoms>70
	Veber	No, 1 violation: Rotors>10
	Egan	No, 1 violation: WLOGP>5.88
	Muegge	No; 2 violations: MW > 600, XLOGP3 > 5
	Bioavailability score	0.17
Medicinal chemistry	PAINS	0 alert
	Brenk	3 alerts Three-membered_heterocycle, michael_acceptor_1, polyene
	Leadlikeness	No; 3 violations: MW > 350, Rotors>7, XLOGP3 > 3.5
	Synthetic accessibility	7.56

In-silico evaluation of ADME parameters of fucoxanthin (http://www.swissadme.ch/).

Two similar marine bioactives have shown antitumor effects on glioblastoma. Astaxanthin is a red pigment found in crustaceans and fish. Adonixanthin is an intermediate product of astaxanthin. Cell viability assays were used on the murine glioblastoma cell line GL261 and the human glioblastoma cell line U251MG for testing. In GL261 cells, astaxanthin inhibited cell viability at concentrations above 5 μM, while adonixanthin did so at levels over 0.1 μM (Tsuji et al. 2020). In U251MG cells, astaxanthin reduced viability at concentrations greater than 1 μM, and adonixanthin reduced viability at concentrations over 0.1 μM. In many tumors, ERK1/2 and Akt phosphorylation are increased, which can lead to cell growth and survival. ERK has been shown to lead to proliferation, survival, and motility/invasion (Montagut and Settleman 2009). Six hours posttreatment, both compounds had reduced phosphorylation of ERK1/2 and Akt. These reduced phosphorylations could potentially slow tumor growth. Furthermore, after 48 h, both astaxanthin and adonixanthin increased the activation of p38 MAPK. P38 MAPK is a signal often linked to stress responses and cell cycle arrest. Activation of this signal could inhibit tumor progression and growth. Treatment of astaxanthin and adonixanthin for 48 h further evidenced the inhibition of tumor growth through the decreased expression of cyclin D1 and increased expression of p27 (Tsuji et al. 2020). Cyclin D1 helps drive cell cycle progression, and p27 helps stop cell division. Next, the levels of reactive oxygen species were tested. High levels of reactive oxygen species can promote tumor progression by activating survival signals like ERK1/2 and Akt. Both compounds reduced the levels of intracellular reactive oxygen species in glioblastoma cells, but adonixanthin was more effective than astaxanthin. To verify, the NOX4 enzyme that produces reactive oxygen species was tested. Adonixanthin significantly decreased Nox4 expression, while astaxanthin did not affect it.

Radiotherapy and temozolomide chemotherapy are used to treat glioblastoma. Both radiotherapy and temozolomide chemotherapy are DNA damage-inducing therapies that do not always give the desired result, and have a high risk of resistance. Stellettin B (STELB) was tested and showed an increase in the sensitivity of GBM to ionizing radiation and TMZ in vitro and in vivo (Peng et al. 2023). GBM cells efficiently repair DNA double-strand breaks using the homologous recombination pathway, using proteins like BRCA1, BRCA2, and RAD51. STELB was found to inhibit BRCA1, BRCA2, and RAD5, therefore inhibiting glioblastoma cells from repairing double-strand breaks. STELB was able to inhibit BRCA1/2 and RAD5 by ubiquitinating PI3Kα. Furthermore, STELB was found to be able to penetrate the blood-brain barrier quickly in the study. STELB alone reduced cell viability in a dose-dependent manner, but when combined with IR, STELB significantly increased cytotoxicity and apoptosis. This was further evidenced by increased cleavage of caspase-3 and PARP. STELB also increased both single- and double-strand DNA breaks in GBM cells, but the combination of STELB with IR resulted in even more DNA damage than either treatment by itself. It was also found that after 24 h of treatment, STELB did not cause significant apoptosis in GBM cells. This suggests that STELB's early effect is to impair DNA repair, leading to damage accumulation, and eventually, enough buildup of damaged DNA leads to apoptosis (Peng et al. 2023). Next, STELB was found to enhance the sensitivity of GBM cells to TMZ. This was tested through the use of TMZ alone and in combination. TMZ adds methyl groups to DNA, leading to double-strand breaks. These double-stranded breaks are then fixed by an enzyme called MGMT. It was found that the combined treatment of TMZ and STELB led to much lower cell viability and greater DNA damage and apoptosis compared to treatment with TMZ alone, especially in MGMT-negative GBM cells. Amazingly, it was found that STELB specifically inhibited HR repair in GBM cell lines SF295 and U87. It inhibited HR repair by reducing the expression of proteins such as BRCA1, BRCA2, and RAD51 and blocking RAD51 foci formation. It did this while leaving NHEJ largely unaffected. By specifically blocking HR, STELB prevents tumor cells from repairing damage, making them more sensitive to the treatment of TMZ, which causes double-stranded breaks. To test this further, RNA sequencing of SF295 GBM cells treated with STELB was tested. Gene set enrichment analysis confirmed that genes involved in homologous recombination repair, like BRCA1/2 and RAD51, were significantly downregulated. STELB was found not to directly inhibit PI3K's kinase activity through an Adapta kinase assay at concentrations below 10 μM, but it was found to downregulate all four PI3K isoforms' expression through a Western blot analysis. PI3Kα levels dropped by about 50% at 6 h after STELB treatment [27]. Overall, the effectiveness of STELB as an effective GBM therapeutic is very promising because of its ability to cross the blood-brain barrier and work together with TMZ, creating a synergistic effect of increasing sensitivity to TMZ double-stranded breaks.

Coibamide A is isolated from a marine cyanobacterium. Coibamide A induced cytotoxicity in a concentration and time-dependent manner, with an EC50 below 100 nM, in both human U87-MG and SF-295 glioblastoma cells as well as in mouse embryonic fibroblasts (Hau et al. 2013). It did this by inducing autophagosome accumulation via an mTOR-independent mechanism. The cyclic structure of coibamide A is critical for its activity; when it is linearized, it removes the cytotoxic effects. Furthermore, different cell types had different responses to coibamide A. SF-295 cells exhibited caspase-3 activation, and U87-MG cells displayed cytoplasmic vacuolization and did not display the typical indicators of apoptosis. In U87-MG and SF-295 glioblastoma, within the first 24–36 h of exposure, no cytotoxic effects were observed, but at 72 h with 20 nM coibamide A, both cell lines exhibited reduced proliferation. Next, trypan blue exclusion was used to look at cell viability (Peng et al. 2023). U87-MG cells with a concentration of 20 nM of coibamide A maintained integrity until between 24 and 36 h, but after that, the number of viable cells began to decrease. After 96 h, very few cells remained viable compared to the control. The release of LDH was then tested to determine membrane damage. LDH release did not show significant differences when treated with 20 nM coibamide A compared to the control, but in cultures exposed to 100 nM coibamide A, there was a significant difference compared to the control. SF-295 glioblastoma cells showed similar results in both the cell viability and LDH release results. Next, the activity of caspase-3/7 was tested. Caspase-3/7 plays a very important role in apoptosis. Even though U87-MG cells were overall more sensitive to coibamide A-induced cell death, coibamide A triggered a significantly stronger and more efficient activation of caspase-3/7 in the SF-295 glioma cell line (Hau et al. 2013).

Bryostatin-1 is known to sensitize cells to chemotherapeutic drugs. Bryo-1's effect on radiation response was tested in Chinese hamster AA8 fibroblasts and human malignant glioma BMG-1 cells (Dagur et al. 2015). Phorbol esters and non-phorbol esters alter the way cells respond to various stressors, including radiation. Phorbol esters are highly tumorigenic; luckily, bryostatin-1 is a non-phorbol ester and is a weak carcinogen. Bryostatin-1 is derived from the marine bryozoan Bugula neritina. Bryostatin-1 has a dual effect on PKCδ. Short-term exposure activates PKCδ, and prolonged exposure can lead to the degradation of PKCδ. The activation of PKCδ can lead to the phosphorylation of the MEK/MAP kinase pathway. Furthermore, bryostatin-1 temporarily elevates p21 levels, deactivates cdk2, and lowers cyclin-B expression, which collectively can cause cell cycle arrest in the G2 phase and promote cell death. Bryo-1's effect was tested with radiation, and it was shown that in BMG-1 cells pretreated with bryostatin-1, growth inhibition was significantly increased at a dose of 2 Gy (Dagur et al. 2015). Bryostatin-1 pretreatment did not change radiation-induced growth inhibition on AA8 Chinese hamster fibroblasts. Overall, this study shows bryostatin-1's potential as a selective radiosensitizer in malignant glioma cells and how it can selectively target.

Lamellarins are isolated from Lamellaria mollusks. Lamellarin D is a marine alkaloid with cytotoxic effects against many different cancer cell lines. Lamellarin D has been shown to inhibit topoisomerase I, but it can also target cancer cell mitochondria to trigger apoptosis (Kluza et al. 2006). It does this by disrupting the inner mitochondrial transmembrane potential in the P388 leukemia cell line. This can lead to mitochondrial swelling and cytochrome c leakage. After 24 h of exposure to 5 μmol/L of lamellarin D or camptothecin, chromosome condensation and nuclear fragmentation were observed in both. These are key markers of late-stage apoptosis.

Lamellarin D also induced $\Delta\psi_m$ disruption in P388 cells within 4 h. Immunoblot showed a significant increase in the amount of cytochrome c in the cytosolic fraction only 15 min after treatment with lamellarin D. A 1-h treatment with lamellarin D at 5 μmol/L caused both the cytoplasm and nucleus of P388 cells to shrink. Furthermore, in lamellarin D-treated cells, the mitochondria appeared enlarged and showed a decrease in electron density compared to untreated cells. These two features indicate swelling of the mitochondrial matrix. Lamellarin D was unable to promote apoptosis of isolated nuclei, but when mitochondria were added, lamellarin D was able to induce apoptosis of the isolated nuclei (Kluza et al. 2006). This points to the fact that lamellarin D is dependent on mitochondria to induce apoptosis. Lamellarin D is a promising anticancer agent, a marine bioactive that could be very useful in inducing mitochondrial-dependent apoptosis in gliomas that often resist other forms of cancer therapies.

9.4 Conclusion

"Let food be the medicine and medicine be the food" was said by Hippocrates (Lobine et al. 2021). Cancer research is becoming more and more extensive with new chemicals and different uses of radiation, but the answer might be right in front of our eyes. Some cures my come from sources we already think of as nourishment: marine bioactives. Drug resistance is becoming more and more of an issue as diseases and bacteria evolve, but it seems that people think this is the future and the best way to fight diseases. This is why marine bioactives are vital to our future and need to be studied more. There is a vast amount of marine bioactives that still need to be tested, and one could be the exact one we need to fight neuronal cancer. For example, just in red algae, it is estimated that there are around 8000 species all loaded with marine bioactive peptides (Ahmed et al. 2022). In conclusion, while finding sure answers in drug research can help fight diseases, maybe the natural bioactives of the ocean that have been evolving and learning to fight diseases naturally may be the key to enhancing human life.

References

Ahmed I, Asgher M, Sher F, Hussain SM, Nazish N, Joshi N, Sharma A, Parra-Saldívar R, Bilal M, Iqbal HMN (2022) Exploring marine as a rich source of bioactive peptides: challenges and opportunities from marine pharmacology. Mar Drugs 20(3):208. https://doi.org/10.3390/md20030208

Al-Awadhi FH, Salvador-Reyes LA, Elsadek LA, Ratnayake R, Chen QY, Luesch H (2020) Largazole is a brain-penetrant class I HDAC inhibitor with extended applicability to glioblastoma and CNS diseases. ACS Chem Neurosci 11(13):1937–1943. https://doi.org/10.1021/acschemneuro.0c00093

Aldape K, Zadeh G, Mansouri S, Reifenberger G, von Deimling A (2015) Glioblastoma: pathology, molecular mechanisms and markers. Acta Neuropathol 129(6):829–848. https://doi.org/10.1007/s00401-015-1432-1

Alves A, Costa P, Pinto M, Ferreira D, Correia-da-Silva M (2021) Small molecules of marine origin as potential anti-glioma agents. Molecules 26(9)

Anand R, Kumar L, Mohan L, Bharadvaja N (2023) Nano-inspired smart medicines targeting brain cancer: diagnosis and treatment. JBIC J Biol Inorg Chem 28(1):1–15. https://doi.org/10.1007/s00775-022-01981-0

Chen R, Smith-Cohn M, Cohen AL, Colman H (2017) Glioma subclassifications and their clinical significance. Neurotherapeutics 14(2):284–297. https://doi.org/10.1007/s13311-017-0519-x

Collins VP, Jones DTW, Giannini C (2015) Pilocytic astrocytoma: pathology, molecular mechanisms and markers. Acta Neuropathol 129(6):775–788. https://doi.org/10.1007/s00401-015-1410-7

Dagur RS, Hambarde S, Chandna S (2015) Bryostatin-1 causes radiosensitization of BMG-1 malignant glioma cells through differential activation of protein kinase-Cδ not evident in the non-malignant AA8 fibroblasts. Mol Cell Biochem 401(1):49–59. https://doi.org/10.1007/s11010-014-2291-0

Dolcet X, Llobet D, Pallares J, Matias-Guiu X (2005) NF-kB in development and progression of human cancer. Virchows Arch 446(5):475–482. https://doi.org/10.1007/s00428-005-1264-9

Ferrara N (2009) Vascular endothelial growth factor. Arterioscler Thromb Vasc Biol 29(6):789–791. https://doi.org/10.1161/ATVBAHA.108.179663

Friedman HS, Kerby T, Calvert H (2000) Temozolomide and treatment of malignant Glioma1. Clin Cancer Res 6(7):2585–2597

Grosso C, Valentão P, Ferreres F, Andrade PB (2014) Bioactive marine drugs and marine biomaterials for brain diseases. Mar Drugs 12(5):2539–2589

Hau AM, Greenwood JA, Löhr CV, Serrill JD, Proteau PJ, Ganley IG, McPhail KL, Ishmael JE (2013) Coibamide A induces mTOR-independent autophagy and cell death in human glioblastoma cells. PLoS One 8(6):e65250. https://doi.org/10.1371/journal.pone.0065250

http://www.swissadme.ch/

https://www.rcsb.org/chemical-sketch

Johns Hopkins Medicine (n.d.-a) Brain tumor types. Retrieved April 12, 2025, from: https://www.hopkinsmedicine.org/health/conditions-and-diseases/brain-tumor/brain-tumor-types

Johns Hopkins Medicine (n.d.-b) Brain tumor treatment. Retrieved April 12, 2025, from: https://www.hopkinsmedicine.org/health/conditions-and-diseases/brain-tumor/brain-tumor-treatment

Jung E, Alfonso J, Monyer H, Wick W, Winkler F (2020) Neuronal signatures in cancer. Int J Cancer 147(12):3281–3291. https://doi.org/10.1002/ijc.33138

Karbhari N, Khagi S (2024) Marine-derived therapeutics for the management of glioblastoma: a case series and comprehensive review of the literature. Onco 4(4):369–380

Karthikeyan A, Joseph A, Nair BG (2022) Promising bioactive compounds from the marine environment and their potential effects on various diseases. J Genet Eng Biotechnol 20(1):14. https://doi.org/10.1186/s43141-021-00290-4

Khotimchenko R, Bryukhovetskiy I, Khotimchenko M, Khotimchenko Y (2021) Bioactive compounds with antiglioma activity from marine species. Biomedicine 9(8)

Kluza JRM, Gallego M-A, Loyens A, Beauvillain J-C, Sousa-Faro J-MF, Cuevas C, Marchetti P, Bailly C (2006) Cancer cell mitochondria are direct proapoptotic targets for the marine antitumor drug Lamellarin D. Cancer Res 66(6):3177–3187. https://doi.org/10.1158/0008-5472.CAN-05-1929

Kotecha R, Odia Y, Khosla AA, Ahluwalia MS (2023) Key clinical principles in the management of glioblastoma. JCO Oncol Pract 19(4):180–189. https://doi.org/10.1200/OP.22.00476

Lin YJ, Huang CY, Shen YC, Wei KC, Chuang CC, Hsu PW, Huang YC, Hwang TL, Chen PY (2022) A manzamine-derived compound as a potential therapeutic agent for glioma by inducing apoptosis and cell cycle arrest. Am J Cancer Res 12(4):1740–1751

Lobine D, Rengasamy KRR, Mahomoodally MF (2021) Functional foods and bioactive ingredients harnessed from the ocean: current status and future perspectives. Crit Rev Food Sci Nutr 62(21):5794–5823. https://doi.org/10.1080/10408398.2021.1893643

Mayo Clinic (n.d.) Brain tumor—symptoms & causes. Retrieved April 12, 2025, from: https://www.mayoclinic.org/diseases-conditions/brain-tumor/symptoms-causes/syc-20350084

Mezzacappa FM, Thorell W (2025) Neuronal brain tumors. [Updated 2022 Nov 21]. In: StatPearls [Internet]. StatPearls Publishing, Treasure Island (FL). Available from: https://www.ncbi.nlm.nih.gov/books/NBK576406/

Miranda-Filho A, Piñeros M, Soerjomataram I, Deltour I, Bray F (2017) Cancers of the brain and CNS: global patterns and trends in incidence. Neuro-Oncology 19(2):270–280. https://doi.org/10.1093/neuonc/now166

Montagut C, Settleman J (2009) Targeting the RAF–MEK–ERK pathway in cancer therapy. Cancer Lett 283(2):125–134. https://doi.org/10.1016/j.canlet.2009.01.022

Pan Y, Monje M (2022) Neuron-glial interactions in health and brain cancer. Adv Biol 6(9):e2200122. https://doi.org/10.1002/adbi.202200122

Peng X, Zhang S, Wang Y, Zhou Z, Yu Z, Zhong Z, Zhang L, Chen ZS, Claret FX, Elkabets M, Wang F, Sun F, Wang R, Liang H, Lin HW, Kong D (2023) Stellettin B sensitizes glioblastoma to DNA-damaging treatments by suppressing PI3K-mediated homologous recombination repair. Adv Sci (Weinh) 10(3):e2205529. https://doi.org/10.1002/advs.202205529

Sharma MC, Deb P, Sharma S, Sarkar C (2006) Neurocytoma: a comprehensive review. Neurosurg Rev 29(4):270–285. https://doi.org/10.1007/s10143-006-0030-z

Tsuji S, Nakamura S, Maoka T, Yamada T, Imai T, Ohba T, Yako T, Hayashi M, Endo K, Saio M, Hara H, Shimazawa M (2020) Antitumour effects of astaxanthin and adonixanthin on glioblastoma. Mar Drugs 18(9). https://doi.org/10.3390/md18090474

Van den Bent MJ, Reni M, Gatta G, Vecht C (2008) Oligodendroglioma. Crit Rev Oncol Hematol 66(3):262–272. https://doi.org/10.1016/j.critrevonc.2007.11.007

Wang Z, Li ZX, Zhao WC, Huang HB, Wang JQ, Zhang H, Lu JY, Wang RN, Li W, Cheng Z, Xu WL, Di Z, Zhou LS, Jiang W, Yu L, Liu JY, Luo C, Zhu H, Dan, Y.,…Dang, Y. J. (2021) Identification and characterization of isocitrate dehydrogenase 1 (IDH1) as a functional target of marine natural product grincamycin B. Acta Pharmacol Sin 42(5):801–813. https://doi.org/10.1038/s41401-020-0491-6

Wu H-I, Fu X-Y, Cao W-Q, Xiang W-Z, Hou Y-J, Mau J-K, Wang Y, Fan C-D (2019) Induction of apoptosis in human glioma cells by fucoxanthin via triggering of ROS-mediated oxidative damage and regulation of MAPKs and PI3K–AKT pathways. J Agric Food Chem 67(8):2212–2219. https://doi.org/10.1021/acs.jafc.8b07126

Xue Y, Fu Y, Zhao F, Gui G, Li Y, Rivero-Hinojosa S, Liu G, Li Y, Xia S, Eberhart CG, Ying M (2021) Frondoside a inhibits an MYC-driven medulloblastoma model derived from human-induced pluripotent stem cells. Mol Cancer Ther 20(6):1199–1209. https://doi.org/10.1158/1535-7163.MCT-20-0603

Marine-Derived Bioactive Compounds: Promising Agents for the Prophylaxis and Treatment of Neuronal Infections

10

Hailey Lushington, Suhrud Pathak, Rishi Nadar, Muralikrishnan Dhanasekaran, and Jaufar Palekkodan

Abstract

With the rise in antimicrobial resistance and the advent of new neurotropic pathogens, neuronal infections—which are frequently accompanied by inflammation, neurodegeneration, and a lack of effective treatment options—represent a significant global health problem. Because of their distinct structural diversity and multifunctional bioactivities, marine-derived natural bioactives have become attractive options for the prevention and treatment of CNS illnesses. These substances have strong antibacterial, anti-inflammatory, and neuroprotective qualities that allow them to disrupt several phases of the infection process, such as immune evasion, pathogen reproduction, and host neuronal injury. The discovery pipeline has been further extended by developments in marine biotechnology and bioprospecting, which have made it possible to identify and optimize new bioactives with potential therapeutic uses. Furthermore, marine-derived compounds are being investigated for their potential involvement in vaccine production and immunomodulation, providing a more comprehensive approach to infection prevention. This study focuses on the existing environment, problems, and future possibilities of marine natural products as new weapons for fighting

H. Lushington · S. Pathak · R. Nadar
Drug Discovery and Development, Harrison College of Pharmacy, Auburn University, Auburn, AL, USA

M. Dhanasekaran (✉)
Drug Discovery and Development, Harrison College of Pharmacy, Auburn University, Auburn, AL, USA

Graduate School, Auburn University, Auburn, AL, USA
e-mail: dhanamu@auburn.edu

J. Palekkodan
Illham House, Changuvetti kund, Kottakkal, Kerala, India

M. Bhupathyraaj et al. (eds.), *Neuroprotective Marine Bioactives*, Nutritional Neurosciences, https://doi.org/10.1007/978-981-95-5791-2_10

brain infections, emphasizing their revolutionary potential in neuroinfectious disease treatment.

10.1 Introduction

Neuronal infections involve the invasion of pathogens such as bacteria, fungi, parasites, and viruses into the central nervous system (CNS). These can lead to cellular damage and inflammation and occasionally cause irreversible neurological damage. The CNS, which consists of the brain and spinal cord, is protected by a selective barrier called the blood-brain barrier (BBB); it's meant to block the entry of most microorganisms and toxins. However, certain pathogens can bypass or even breach the BBB and can result in life-threatening infections such as meningitis, encephalitis, and brain abscesses. Untreated infections can trigger a cascade of immune responses that may lead to the negative effects listed above, such as permanent neurological impairment. Neuronal infections occur through multiple pathways, such as hematogenous and direct inoculation, and through the PNS retrograde transport. Hematogenous spread is where pathogens enter the bloodstream and cross the BBB. Direct inoculation can occur through trauma or surgical procedures. Peripheral nervous system (PNS) retrograde transport is where viruses like herpes simplex virus (HSV) travel via axonal transport to reach the CNS. Common pathogenic organisms include bacteria such as *Streptococcus pneumoniae* and *Neisseria meningitidis*, viruses like herpesviruses and enteroviruses, fungi such as *Cryptococcus neoformans*, and parasites like *Toxoplasma gondii*. All of these pathogens can cause a variety of adverse consequences not only on the nervous system but also on the body as a whole.

Neuronal infections have been documented throughout history, with multiple ancient texts describing symptoms consistent with neurological infection, such as bacterial and viral meningitis. Hippocrates, the Greek philosopher, described symptoms that are comparable to those of meningitis and encephalitis. Ancient Egyptian texts from as early as the seventeenth century documented fevers and neurological disturbances related to infections. The connection between microorganisms and neuronal infections became more distinct once microbiology was developed in the nineteenth century. During this time, key bacterial pathogens such as *Neisseria meningitidis* and *Streptococcus pneumoniae* were identified as major causes of bacterial meningitis. Once vaccines and antibiotics began to evolve in the twentieth century, there was a significant reduction in the severity and occurrence of bacterial infections in the CNS. Moreover, in the 1980s, the viral pathogens that cause encephalitis were discovered, and the rise of HIV displayed that finding effective antiviral therapies was necessary.

Emerging viruses that spread from animals to humans, like West Nile virus and Zika virus, highlight the growing danger of viral infections that affect the nervous system. According to Stefano (2021), advancements in molecular diagnostics and genomic techniques have enhanced the identification and characterization of emerging pathogens, enabling better management and treatment of neuronal infections.

The introduction of metagenomic next-generation sequencing (mNGS) has transformed pathogen identification, facilitating the rapid detection of rare and novel pathogens that may contribute to CNS infections.

The growing prevalence of antibiotic-resistant pathogens is adding another level of complexity in managing and treating CNS diseases. For example, treatment for meningitis and brain abscesses is thwarted by multidrug-resistant bacteria strains such as *Klebsiella pneumoniae* and *Acinetobacter baumannii*, requiring the investigation of novel treatments (Kim et al. 2009). Treatment of CNS infections like *Candida auris* is made more challenging due to the proliferation of this drug-resistant fungus, requiring the creation of innovative therapeutic approaches. Furthermore, the presence of dormant viruses like HIV and HSV in the CNS emphasizes the need for advanced antiviral treatments that can target these latent infections and effectively cross the BBB.

The pathophysiology of neuronal infections involves a set of complex interactions among pathogens, the immune system, and neuronal tissues. When pathogens breach the BBB, an inflammatory response is triggered that can result in neural damage. One of the key mechanisms for inflammatory response includes BBB disruption, during which pathogens release toxins and inflammatory mediators that compromise the integrity of the BBB, allowing more immune cells and pathogens to enter the CNS (Chen et al. 2024).

Another way neuroinflammation can occur is when microglia and astrocytes are activated. This activation will lead to the production of cytokines, chemokines, and reactive oxygen species (ROS), all of which can contribute to neuronal damage (DiSabato et al. 2016). Activation of glial cells and neuronal apoptosis can lead to prolonged inflammation and excessive cell death. Additionally, excitotoxicity, caused by an overabundance of glutamate release, can further contribute to neuronal cell death (Mark et al. 2001). Persistent infections, which will cause prolonged inflammation, could result in chronic neurological conditions. Neurological conditions can include cognitive decline, epilepsy, and different neurodegenerative diseases.

Neuronal infections can vary depending on the type of pathogen exposed to, the severity of the infection, and the part of the CNS affected (Shahan et al. 2021). However, most neuronal infections have common symptoms that include fever and headache, which are notable features of systemic infection and neuroinflammation. For meningitis specifically, neck stiffness and photophobia are classical signs, while altered mental status, such as disorientation and lethargy, often are markers of severe infections. Additionally, seizures and neurological deficits could indicate encephalitis or abscess formation, and focal neurological signs such as weakness, cranial nerve palsies (in any of the 12 cranial nerves), and sensory disturbances may occur in severe cases. Accurate and early diagnosis is crucial to effectively treating a neurological disease. The principal analysis for diagnosing all forms of meningitis is cerebrospinal fluid (CSF) analysis through lumbar puncture. Polymerase chain reaction (PCR) assays, serological tests, and advanced imaging techniques like magnetic resonance imaging (MRI) also contribute to the process of detecting pathogens and the evaluation of the extent of damage done to the CNS (Syed Hussin et al. 2025).

Once the diagnosis has been made, the next step for managing a neuronal infection is deciding on effective treatment. This includes finding the appropriate antimicrobial therapy and supportive care. Some available treatment options include synthetic FDA-approved drugs and natural bioactive compounds derived from marine organisms. FDA-approved treatments include antibiotics. Antibiotics for neuronal infections include ceftriaxone and vancomycin for bacterial meningitis (Saez-Llorens et al. 2002), rifampin and isoniazid for tuberculosis meningitis, and antiviral agents like acyclovir for herpes simplex encephalitis (HSE) and ganciclovir for cytomegalovirus infections. Furthermore, antifungal agents like amphotericin B and fluconazole treat cryptococcal meningitis, whereas antiparasitic drugs like pyrimethamine and sulfadiazine are effective against toxoplasmosis. Recent studies, such as those conducted by Smaga et al. (2020) and Ruiz-Bedoya et al. (2022), highlight the increasing role of combination therapies and the importance of antimicrobial stewardship in improving clinical outcomes for CNS infections. The increasing prevalence of drug-resistant pathogens emphasizes the growing need for alternative treatment methods. Recently, there has been a rise in treatments involving natural compounds with characteristic bioactive properties.

Bioactive compounds that are found in marine ecosystems are becoming an increasingly popular alternative to treating neuronal infections. Marine ecosystems have bioactives that are abundant with antimicrobial, antiviral, and anti-inflammatory properties that could suggest some novel treatment for these infections. Notable marine bioactives include bryostatin, which is derived from the *Bugula neritina*, which modulates protein kinase C (PKC) activity. Bryostatin appears to have both neuroprotective and anti-inflammatory effects. Manoalide, a marine sponge-derived compound, also exhibits anti-inflammatory response as well as antibacterial properties. Another example of a marine product includes ziconotide, which is derived from the venom of the cone snail *Conus magus*, and has received approval for the treatment of chronic pain. Plitidepsin, a cyclic depsipeptide from *Aplidium albicans*, exhibits antiviral activity against certain neurotropic viruses, while squalamine, a natural steroid derived from dogfish sharks, includes broad-spectrum antibacterial and antiviral effects. According to a study by Mehla et al. (2012), marine natural products have shown immense potential in modulating neuroinflammation and neurodegenerative diseases.

The mechanisms of marine bioactives include anti-inflammatory effects by inhibiting pro-inflammatory cytokines and modulating immune responses. Anti-inflammatory effects include antiviral and antimicrobial actions. These actions include disrupting biofilm formation, direct inhibition of pathogen replication, prevention of neuronal apoptosis, and reduction of oxidative stress. Research by Hannan et al. (2020) highlights the antimicrobial and neuroprotective potential of marine-derived peptides and polysaccharides. Similarly, Valia-Vera et al. (2007) demonstrated the therapeutic efficacy of marine bioactives in targeting CNS pathogens and the regulation of immune responses. A recent review by Elbandy (2022) further emphasized the antiviral potential of various marine compounds in preventing neuronal infections caused by neurotropic viruses (any virus that can cause a CNS disease).

Despite the growing recognition of marine bioactives in treating neuronal infections, there are many challenges that must be addressed before these compounds can be effectively integrated into any clinical practice (Rizvi and Saleh 2018). One critical area for further research involves understanding the pharmacokinetics and bioavailability of marine-derived compounds in the CNS. Since the BBB is highly selective and limits the entry of many therapeutic agents, it is vital to understand how these marine bioactives will cross into the BBB in high enough concentrations to have the desired effect on the CNS. There are some advanced drug delivery systems to the CNS that could be used for marine bioactives to cross. These include systems like nanoparticles, liposomes, and polymeric carriers. For instance, nanoparticle delivery or marine bioactives could prolong their half-life, enhance stability, and assist in targeted delivery of the neuronal infection in the tissue. A study by Dando et al. (2014) discussed that nanoparticle-encapsulated marine bioactives showed that penetration across the BBB was significantly improved. Not only was penetration improved, but the clinical effectiveness in models of neuronal infections were improved as well. Using nanotechnology in drug delivery would revolutionize a drug's bioavailability and also decreases the potential for the entire system to become toxic.

Marine-derived natural bioactives offer promising alternatives for the prevention and treatment of neuronal infections by targeting factors such as inflammation and microbial replication and also provide neuroprotection. As antimicrobial resistance continues to increase and synthetic drug efficacy declines, using the potential of marine bioactives may provide novel therapeutic strategies to combat neuronal infections in an effective manner. According to research by Perry and Schentag (2001), natural marine compounds exhibit unique structures and potent biological activities that make them excellent alternatives to previously known treatments meant to target CNS infections. Additionally, investigations by Tajerian et al. (2024) emphasize the symbiotic behavior between novel marine-derived therapeutic compounds and conventional therapies. This cooperation should allow for a highly effective treatment option for neuronal infections.

In future research, scientists should attempt to isolate these unique marine bioactives, then characterize them, attempt to understand their various mechanisms of action, and implement trials with other therapeutic treatments to see if they can integrate for a useful treatment. Before treatment can be undergone, these bioactives must undergo various clinical trials. Clinical trials should test not only if the marine bioactives work for treatment but also if they are safe for human use. If these clinical trials are successful, marine bioactives are vast and will assist in the revolution of future neuronal treatments for various infections. These marine bioactive treatments will be used so that infections that are becoming increasingly more complex and diverse will be able to be treated to keep patients not only alive but make a full recovery once treatment ends.

Marine ecosystems cover about 70% of the surface of the Earth, allowing for a rich source of biologically active compounds which are known as marine natural products (MNP). These MNPs are mainly unused and provide a novel source for potential therapeutic use (Yang et al. 2025). These bioactive molecules are derived from a large variety of marine organisms such as algae, bacteria, coral, fungi, mollusks, sponges, and tunicates. Marine environments usually undergo extreme conditions marked by high pressure, low temperatures, and limited light penetration. However, the harsh conditions assist in the evolution of these unique biochemical pathways in the various marine organisms. This leads to the production of advanced secondary metabolites with unique molecular structure and potent biological activities (Karthikeyan et al. 2022).

Marine-derived alkaloids have demonstrated possible uses that include protection against various neuronal infections and immunomodulation. According to Rehman et al. (2019), marine alkaloids, like manzamine A that is isolated from marine sponges, exhibit potent anti-inflammatory, antimicrobial, and neuroprotective effects. Bioactive compounds, like manzamine A, work by inhibiting the pro-inflammatory cytokines and reduce oxidative stress. These inhibitions will assist in preventing any neuronal damage that could have been caused by bacterial and viral infections. Additionally, Grosso et al. (2014) state that marine alkaloids also demonstrate possible antiviral activity by inhibiting viral replication and limiting pathogen spread in the CNS. Both of which make them extremely promising candidates for the development of new therapies against neuronal infections.

Over the past five decades, there has been a headway in marine bioactive research that has led to scientists identifying over 300,000 marine compounds. There have been multiple MNPs that have undergone clinical trials in an effort to gain FDA approval for clinical use. Some examples of MNPs that have been FDA approved include ziconotide, which provides pain relief and reduces neuroinflammation by blocking N-type calcium channels, and trabectedin (Chaigneau et al. 2022) from the tunicate *Ecteinascidia turbinata* is used as a chemotherapy treatment for more advanced cancers. Both these examples and other marine bioactives have extremely diverse chemical structures, like alkaloids, peptides, terpenoids, and polyketides. These structures give these bioactives a pharmacological advantage that sets them apart from other various natural products used in therapeutic settings.

Antimicrobial resistance is becoming more and more problematic with the emerging infectious diseases. Many are turning to marine-derived products as potential sources for antibiotic and antiviral agents. Compounds such as marinopyrrole A from marine *Streptomyces* and plumbagin from *Plumbago europaea* have demonstrated potent antibacterial and antifungal activities (Haste et al. 2011). Marine-derived peptides such as polybrominated diphenyl ethers (PBDEs) found in marine sponges exhibit antibacterial activity, making them candidates for combating drug-resistant bacterial infections that are rising (Tan and Salleh 2024). PBDEs and other newly discovered bioactives will help in the expansion of new potential therapeutic agents for these antimicrobial-resistant pathogens that induce CNS infections. Any further research into these potential marine-derived bioactives should focus on minimizing side effects, bioavailability, and optimization of pharmacokinetics.

Marine natural products can also exhibit antiviral properties, particularly against viruses that negatively impact the CNS and respiratory tract. For example, griffithsin, a lectin derived from the red algae *Griffithsia*, seems to have potent activity against viral infections such as HIV, SARS-CoV, and other viruses that are in an envelope by binding to the glycoproteins and preventing viral entry into host cells (Ferir et al. 2012). These antiviral agents provide a new approach for the development of therapies against viral infections.

Some research is beginning to suggest that there is a possible connection between neuronal infections and developing certain types of cancers. For example, there are certain chronic CNS infections that occur due to neurotropic viruses like human papillomavirus (HPV), Epstein-Barr virus (EBV), and human herpesvirus 6 (HHV-6) that have been associated with the etiology of brain tumors and other types of malignant cancers. Persistent inflammation and immune dysregulation induced by these pathogens contribute to genetic mutations, cellular transformation, and tumorigenesis. This could be explained because these infections can cause inflammation; constant inflammation and immune system dysfunction could contribute to factors such as cellular transformation, tumorigenesis, and genetic mutations. All of these factors can make one more at risk for developing various types of cancer. According to Shin et al. (2002), chronic infection with EBV could be linked to CNS lymphomas, particularly in those who are immunocompromised; extended exposure to virus and various immune mechanisms could lead to the transformation of B cells that are infected into malignant tumors. Chronic inflammation brought on by viral infections can create a tumor inducing microenvironment, enhancing angiogenesis, and evading immune detection. If scientists can find the molecular mechanisms that connect neuronal infections and cancer, there may be the ability for new treatments to be found to target these mechanisms. Marine bioactives are the best place to look for these novel treatments.

As previously stated, marine-derived compounds have become for pivotal oncological research, with several of these compounds demonstrating potent anticancer properties through various molecular mechanisms. Some of those include angiogenesis inhibition, apoptosis induction, and immunomodulation. Notable examples include Trabectedin, brand name Yondelis (2012), which interferes with DNA binding and inhibits transcription factors, leading to apoptosis in cancer cells. Similarly, plinabulin, derived from Dendrobaena veneta, disrupts microtubule dynamics, inhibiting cancer cell proliferation and enhancing immune responses.

A significant breakthrough in marine-derived cancer therapy is the discovery of salinosporamide A (marizomib) from Salinispora tropica, a potent proteasome inhibitor that has shown efficacy against multiple myeloma and glioblastoma (Potts et al. 2011). Additionally, didemnin B, a cyclic depsipeptide derived from *Trididemnum solidum*, has demonstrated anticancer activity through its inhibition of protein synthesis and induction of apoptosis. According to Dalisay et al. (2024) , marine-derived anticancer agents are increasingly being integrated into combination therapies to enhance treatment efficacy and reduce resistance. They go on to cite that there are at least 21 marine-derived compounds that could be potentially used for cancers, including those rare few that are considered neuronal infections.

Chronic inflammation and oxidative stress are key contributors to neurodegenerative diseases such as Alzheimer's disease, Parkinson's disease, and multiple sclerosis (MS). Marine-derived bioactives such as bryostatin, derived from the bryozoan Bugula neritina, modulate PKC activity and reduce neuroinflammation, providing neuroprotective benefits in models of AD and stroke. According to Sun and Alkon (2006), bryostatin has demonstrated the ability to reduce neuroinflammation, protect against neuronal damage, and enhance synaptic plasticity, which are essential for mitigating the harmful effects of neurotropic infections. Additionally, bryostatin's ability to stimulate immune responses and modulate cytokine production may help control infection-induced inflammation and limit neuronal damage, offering a potential therapeutic strategy for combating CNS infections. Similarly, manoalide, a marine sponge-derived compound, targets inflammation, oxidative stress, and microbial proliferation. According to Yeom et al. (2021), marine natural products, particularly sulfated polysaccharides and bioactive peptides, exhibit potent immunomodulatory and antimicrobial effects that may prevent and mitigate CNS infections. These compounds enhance host immune responses by modulating cytokine production and inhibiting pathogen replication, thereby reducing the neuroinflammatory damage caused by bacterial and viral infections. The study emphasizes the therapeutic potential of marine bioactives in protecting neuronal tissues from infection-induced damage and promoting CNS homeostasis, paving the way for novel treatment strategies.

Marine-derived omega-3 polyunsaturated fatty acids (PUFAs), particularly docosahexaenoic acid (DHA), eicosapentaenoic acid (EPA), and docosapentaenoic acid (DPA), have demonstrated neuroprotective effects by reducing neuroinflammation, promoting synaptic plasticity, and preventing neuronal apoptosis. According to Dyall (2015), DHA, in particular, accumulates in the brain and exerts neuroprotective effects by modulating the inflammatory response, thereby limiting the damage caused by infections and preventing neuronal apoptosis. EPA and DPA further contribute by enhancing immune function and supporting tissue repair, making omega-3 fatty acids valuable adjuncts in the prevention and management of CNS infections.

Marine natural products have shown promise as adjuvants and immunomodulatory agents, enhancing the efficacy of vaccines and promoting immune responses, which is particularly relevant in the context of preventing and treating neuronal infections. Monophosphoryl lipid A (MPL), derived from marine bacteria, according to Ko et al. (2017) is a potent adjuvant used in vaccines to enhance adaptive immunity, offering protection against pathogens that can cause CNS infections. Similarly, zymosan, a polysaccharide derived from marine yeast, activates toll-like receptor 2 (TLR2) and induces innate immune responses, making it an effective immunomodulatory agent that can strengthen the body's defense against neurotropic pathogens. As stated in an experimental study by Volman et al. (2005), marine-derived natural products have demonstrated significant anti-inflammatory, antioxidant, and immunomodulatory properties, making them valuable candidates for therapeutic applications in various diseases, including neuroinflammatory conditions. Marine compounds, such as polyphenols, sterols, and alkaloids, modulate inflammatory pathways by inhibiting pro-inflammatory cytokines and reducing oxidative stress, which are critical factors in the progression of neuronal infections. Inflammatory responses triggered by bacterial, viral, and fungal pathogens in the CNS often lead to excessive cytokine release and oxidative damage, contributing to neuronal apoptosis and tissue destruction. By mitigating these harmful effects, marine-derived anti-inflammatory agents can protect against neuronal damage and improve clinical outcomes in CNS infections. Additionally, these compounds may complement standard antimicrobial therapies by reducing inflammation and enhancing immune responses, making them promising adjunctive treatments for managing neuronal infections. By enhancing immune surveillance and reducing inflammation, these marine-derived compounds may help prevent the establishment and progression of neuronal infections, providing an additional layer of protection in vaccine strategies and immunotherapy.

Marine-derived polysaccharides, such as fucoidan from brown algae and laminarin from marine fungi, exhibit immunomodulatory properties by stimulating macrophages, enhancing phagocytosis, and promoting the production of cytokines. These compounds play a critical role in enhancing host defense mechanisms, making them valuable candidates for vaccine adjuvants and immunotherapeutic agents. Oxidative stress and inflammation play key roles in the pathogenesis of cardiovascular diseases, diabetes, and metabolic disorders. Marine-derived antioxidants such as astaxanthin, a carotenoid derived from *Haematococcus pluvialis*, exhibit potent antioxidant and anti-inflammatory properties, protecting against oxidative stress-induced cellular damage. Astaxanthin, as stated by Ambati et al. (2014), has been shown to improve lipid metabolism, reduce atherosclerosis, and prevent endothelial dysfunction, making it a valuable preventive agent against cardiovascular diseases.

Similarly, squalene, a natural lipid derived from deep-sea sharks and algae, exhibits cardioprotective effects by reducing cholesterol levels, preventing oxidative stress, and enhancing mitochondrial function. Though squalene is mainly known for cardiac function, Bate et al. (2004) suggest that marine natural products exhibit diverse bioactivities, including antimicrobial and anti-inflammatory effects, which are crucial in combating neuronal infections. These compounds can inhibit the growth of neurotropic pathogens and modulate immune responses, reducing inflammation and preventing further neuronal damage.

Marine natural products have emerged as valuable tools in diagnostic imaging and biomarker discovery. Conotoxins, derived from cone snails, have been utilized in positron emission tomography (PET) imaging to target voltage-gated ion channels and assess neuronal activity. Similarly, brevetoxins, derived from Karenia brevis, are used in bioassays to detect harmful algal blooms and marine biotoxins. Marine-derived nanoparticles, such as chitosan-based nanoparticles, have been developed as carriers for targeted drug delivery and molecular imaging. These nanoparticles enhance the bioavailability of therapeutic agents and enable real-time monitoring of disease progression. Gao and Wu (2022 discuss molecular imaging techniques, such as PET and MRI, that are crucial for detecting neuroinflammation, monitoring pathogen dynamics, and evaluating treatment responses in neuronal infections, enabling more effective therapeutic interventions. Marine-derived enzymes and peptides have been employed in the development of biosensors and bioassays for detecting pathogens, toxins, and environmental pollutants. Luminase, a marine-derived luciferase, has been integrated into biosensors for real-time detection of bacterial pathogens and biotoxins. Additionally, lectins derived from marine algae have been utilized in bioassays for detecting glycoprotein biomarkers associated with cancer and infectious diseases.

Marine natural products represent a vast and largely unexplored reservoir of bioactive compounds with immense therapeutic, preventive, and diagnostic potential. As antimicrobial resistance rises and chronic diseases become increasingly prevalent, marine-derived bioactives offer innovative solutions for addressing unmet medical needs. Continued research into the isolation, characterization, and clinical evaluation of marine compounds is essential for unlocking their full therapeutic potential. By integrating marine natural products into modern medical practice, researchers can develop novel therapies, improve patient outcomes, and revolutionize the future of medicine.

Neuronal infections, caused by bacterial, viral, fungal, and parasitic pathogens, pose a significant threat to the CNS, leading to conditions such as meningitis, encephalitis, brain abscesses, and neurodegenerative complications. The BBB serves as a critical defense mechanism, restricting the entry of pathogens and toxins into the CNS. However, certain pathogens have evolved mechanisms to breach the BBB, resulting in neuroinflammation, neuronal damage, and long-term neurological dysfunction. Traditional antimicrobial therapies often fail to penetrate the BBB effectively or address the inflammatory cascade that contributes to neuronal damage. As antimicrobial resistance continues to rise, the search for novel therapeutic and preventive agents becomes imperative. Marine-derived bioactive compounds, sourced from marine sponges, algae, bacteria, and invertebrates, have emerged as promising candidates for addressing these challenges by modulating immune responses, inhibiting pathogen replication, and preventing neuroinflammation.

Neuronal infections arise when pathogens evade host immune defenses and invade the CNS, triggering inflammatory responses that can lead to irreversible tissue damage. Pathogens can enter the CNS through several mechanisms. Hematogenous spread occurs when pathogens enter the bloodstream and cross the BBB through endothelial cells or by inducing inflammatory responses that disrupt the BBB. One of the more common bacterial pathogens, such as *Streptococcus pneumoniae* and *Neisseria meningitidis,* exploits this mechanism and eventually leads to the contraction of bacterial meningitis. Retrograde axonal transport is another pathway where neurotropic viruses such as HSV and rabies virus utilize peripheral nerves to reach the CNS, leading to encephalitis and neuronal apoptosis. Additionally, direct inoculation through trauma or surgical procedures may introduce pathogens directly into the CNS, causing abscess formation and subsequent inflammation. Once inside the CNS, pathogens trigger a cascade of inflammatory responses involving the activation of microglia and astrocytes, the release of pro-inflammatory cytokines (including IL-1β, TNF-α, and IL-6), and the production of ROS. Prolonged neuroinflammation contributes to neuronal apoptosis, excitotoxicity, and synaptic dysfunction, exacerbating neuronal damage and increasing the risk of chronic neurological conditions (Adamu et al. 2024).

Marine natural products have demonstrated potential in mitigating the pathogenesis of neuronal infections by inhibiting microbial replication, modulating immune responses, and protecting neurons from inflammation-induced damage. Several marine-derived compounds exhibit potent antimicrobial and antiviral activities by targeting key microbial processes and preventing pathogen replication. Bryostatin, derived from the marine bryozoan Bugula neritina, is a modulator of PKC that exhibits both antimicrobial and neuroprotective properties. As discussed previously, bryostatin has been shown to inhibit the replication of HIV and herpesviruses, which are known to cause neuronal infections. By modulating PKC activity, bryostatin enhances immune responses and reduces inflammation, preventing neuronal damage associated with viral infections. Squalamine, a natural steroid isolated from the dogfish shark (Mammari et al. 2022), disrupts microbial membranes and inhibits biofilm formation, making it effective against bacterial and fungal pathogens implicated in meningitis and brain abscesses. Squalamine's ability to enhance antimicrobial efficacy while reducing inflammation makes it a promising candidate for treating drug-resistant CNS infections. Similarly, plitidepsin, derived from the marine tunicate Aplidium albicans, inhibits elongation factor-1α (eEF1A), a key factor in viral replication. Plitidepsin has demonstrated potent antiviral activity against neurotropic viruses such as Zika and enteroviruses, preventing viral spread and limiting inflammation in the CNS.

Neuroinflammation plays a pivotal role in the pathogenesis of neuronal infections, contributing to neuronal apoptosis and tissue damage. Marine-derived bioactives offer anti-inflammatory and immunomodulatory properties that mitigate neuroinflammation and promote neuronal survival. Manoalide, isolated from marine sponges, inhibits phospholipase A2 (PLA2), an enzyme involved in the production of pro-inflammatory mediators. By inhibiting PLA2, manoalide reduces the release of cytokines such as IL-1β and TNF-α, mitigating inflammation and preventing neuronal apoptosis. Manoalide's ability to modulate neuroinflammatory responses makes it a promising candidate for treating bacterial and viral encephalitis. Fucoidan, a sulfated polysaccharide derived from brown algae (Luthuli et al. 2019), activates toll-like receptors (TLRs) and enhances phagocytic activity, leading to the clearance of pathogens and resolution of inflammation. Fucoidan modulates immune responses by promoting the production of anti-inflammatory cytokines, reducing ROS production, and protecting against neuronal damage caused by chronic inflammation. Monophosphoryl lipid A (MPL), derived from marine bacteria, is a potent vaccine adjuvant that activates TLR4 and enhances adaptive immunity. MPL has been incorporated into vaccines to protect against bacterial and viral infections that can lead to neuronal damage. By enhancing adaptive immune responses, MPL promotes long-term immunity and reduces the risk of recurrent CNS infections.

Neurodegeneration and apoptosis resulting from neuronal infections contribute to long-term cognitive impairment and neurological dysfunction. Marine bioactives have demonstrated neuroprotective properties by modulating oxidative stress, reducing excitotoxicity, and enhancing neuronal survival. Ziconotide, as discussed earlier, blocks N-type calcium channels, reducing neuronal excitotoxicity and providing neuroprotection. Ziconotide has been approved for the treatment of chronic pain associated with neurodegenerative conditions, highlighting its potential in preventing neuronal damage following infections. Phlorotannins, polyphenolic compounds derived from brown algae, exhibit potent antioxidant and anti-inflammatory properties that protect against oxidative stress-induced neuronal damage (McGivern 2007). By neutralizing ROS and inhibiting pro-inflammatory signaling, phlorotannins prevent neuronal apoptosis and promote synaptic integrity. Similarly, plakortide E, isolated from marine sponges, modulates mitochondrial function and reduces oxidative stress, protecting neurons from apoptosis triggered by neuroinflammation (Sun et al. 2011). Plakortide E has shown potential in preventing neurodegenerative damage associated with chronic viral infections such as HIV and HSV.

This table illustrates that ziconotide administered through continuous intrathecal infusion or epidural delivery demonstrated significant efficacy in relieving differing types of severe pain. This specific table includes pain that was cancer related, nonmalignant, chronic, neuropathic, and postoperative. Both the double-blind and the placebo studies showed that ziconotide reduced pain scores anywhere from 15% to 53% with consistent effectiveness throughout the maintenance phases.

A. Double-blind, placebo-controlled studies		
Study report	**Ziconotide dosing**	**Clinical observations**
Intractable severe pain due to cancer or AIDS (McGivern 2007)	68 patients received ziconotide and 40 received placebo	Ziconotide provided moderate to complete pain relief. Pain scores were reduced by an average of 53% during the initial phase, with no loss of efficacy during the maintenance phase
	Ziconotide was administered by continuous intrathecal infusion at 0.1–2.4 μg/h or 0.4–7.0 μg/h for up to 6 days, followed by a 5-day maintenance phase for responders	Adverse events included dizziness, nystagmus, nausea, postural hypotension, somnolence, and confusion
Intractable non-malignant severe chronic pain (McGivern 2007)	169 patients received ziconotide and 86 received placebo	Ziconotide provided moderate to complete pain relief. Pain scores were reduced by an average of 31% during the initial phase, with no loss of efficacy during the maintenance phase
	Ziconotide was administered by continuous intrathecal infusion at 0.4–7.0 μg/h or 0.1–2.4 μg/h for 6 days, followed by a 5-day maintenance phase for responders	Adverse events included dizziness, nausea, nystagmus, and a bnormal gait

(continued)

Intractable severe chronic pain (McGivern 2007)	112 patients received ziconotide and 108 received placebo	Ziconotide provided significant pain relief. Pain scores were reduced by an average of 15% and there was decreased consumption of opioids Adverse events included dizziness, confusion, ataxia, abnormal gait and memory impairment
	Ziconotide was administered by continuous intrathecal infusion at 0.1–0.9 μg/h for up to 3 weeks	
Post-operative pain (McGivern 2007)	18 patients received ziconotide and 12 received placebo	Ziconotide provided significant pain relief at both low and high doses. Pain scores were reduced by an average of 50% and there was decreased consumption of morphine
	Ziconotide was administered by continuous intrathecal infusion at 0.7 or 7.0 μg/h for 48–72 h	Adverse events included dizziness, blurred vision, nystagmus, and sedation
B. Open-label studies		
Study report	**Ziconotide dosing, efficacy and adverse events**	
Intractable de-afferentation pain (McGivern 2007)	One patient received ziconotide	
	Ziconotide was administered by continuous intrathecal infusion at 0.3–3 ng/kg/h for 8 days	
	Complete pain relief was experienced with 2–3 ng/kg/h – the patient's pain score was reduced from 85 mm to 0 mm	
	Adverse events included dizziness, blurred vision, and lateral gaze nystagmus	
Neuropathic pain (McGivern 2007)	22 patients received ziconotide	
	Ziconotide was administered by intrathecal infusion of 1, 5, 7.5, or 10 μg over 1 h	
	Analgesia was dose-dependent, being greatest in the 7.5 μg group and lowest in the 1 μg group	
	Adverse events included myasthenia, dizziness, headache, nausea, abnormal gait, and somnolence	

(continued)

Neuropathic pain (McGivern 2007)	Patient 1 received a single epidural dose of ziconotide at 5 μg. Significant pain relief was achieved, with the patient's pain score reduced from 85 mm to <10 mm. Adverse events included light-headedness, headache, and somnolence	
	Patient 2 received a single epidural dose of ziconotide at 10 μg. Complete pain relief was experienced, with the patient's pain score reduced from 68 mm to 0 mm. Adverse events included decreased blood pressure, somnolence, itching, nausea, and headache	
	Patient 3 received continuous intrathecal infusion of ziconotide at 0.3–100 ng/kg/h for up to 34 days. Ziconotide delivered significant pain relief at 5–12 ng/kg/h, with the patient's pain score reduced from 67 mm to 4 mm. Adverse events included sedation, confusion, memory impairment, slurred speech, and double vision	

Vaccination remains a cornerstone in preventing neuronal infections caused by bacterial and viral pathogens. Marine bioactives have shown promise as adjuvants that enhance vaccine efficacy by promoting robust immune responses. Zymosan, a polysaccharide derived from marine yeast, activates TLR2 and induces innate immune responses, making it an effective immunomodulatory agent in vaccine formulations. According to Dillon et al. (2006), zymosan enhances antigen presentation and promotes the production of memory T cells, improving long-term immunity against neurotropic pathogens. Carrageenan, derived from red algae, inhibits viral attachment and entry into host cells, preventing viral replication. Carrageenan has been explored as a potential preventive agent in nasal sprays and topical formulations to protect against viral infections that can lead to neuronal damage.

Chronic neurological sequelae, including cognitive decline, epilepsy, and motor dysfunction, often follow unresolved neuronal infections. Marine bioactives play a crucial role in preventing these long-term complications by modulating inflammation, preventing neuronal apoptosis, and enhancing synaptic plasticity. Astaxanthin, a carotenoid derived from marine microalgae, exhibits potent antioxidant activity that protects against oxidative stress and mitochondrial dysfunction. By preventing oxidative damage, astaxanthin reduces the risk of neurodegenerative complications following CNS infections. Eicosapentaenoic acid (EPA) and docosahexaenoic acid (DHA), long-chain omega-3 fatty acids derived from marine fish oils, exhibit anti-inflammatory and neuroprotective properties that enhance neuronal repair and reduce inflammation following neuronal infections. Clinical studies have shown that EPA and DHA improve outcomes in patients with neuroinflammatory conditions, making them promising candidates for preventing postinfection complications.

This figure illustrates that the relationship between inflammation and oxidative stress is bidirectional (Kousparou et al. 2023). ROS can trigger inflammation by acting as an inflammatory agent. Conversely, inflammation can induce oxidative stress. Key components in this process include IkB, which is the inhibitory subunit of NFkB; MAPK, which stands for mitogen-activated protein kinase; and NFkB, or nuclear factor kappa-light-chain enhancer of activated B cells, along with phosphate and ROS.

The exploration of the synergistic effects of marine bioactives with conventional antimicrobial therapies is another promising area of investigation. Combination therapies involving marine bioactives and FDA-approved antimicrobial agents may enhance therapeutic efficacy by targeting multiple pathways simultaneously. For instance, combining bryostatin with antiviral agents such as acyclovir or ganciclovir may improve outcomes in viral encephalitis by reducing viral replication while modulating inflammation. Similarly, marine-derived alkaloids with anti-inflammatory properties, such as manzamine A, may complement antibiotic therapy in bacterial meningitis by mitigating the harmful effects of excessive neuroinflammation. Studies by Farghaly et al. (2021) have shown that marine bioactives can potentiate the activity of existing antibiotics, reducing the required dosage and minimizing the risk of antibiotic resistance. Investigating the precise mechanisms of these synergistic interactions could pave the way for more effective combination therapies for neuronal infections.

Moreover, future research should also focus on identifying new marine-derived compounds with potent antimicrobial, antiviral, and neuroprotective properties. Marine ecosystems, particularly deep-sea environments, harbor a wealth of unexplored biodiversity that may contain novel bioactives capable of combating drug-resistant pathogens and modulating immune responses in the CNS. Advanced techniques such as metagenomics, metabolomics, and bioinformatics can accelerate the discovery of novel marine compounds and predict their potential therapeutic applications. High-throughput screening of marine microbial metabolites can help identify candidates with potent activity against neurotropic pathogens. Additionally, investigating the ecological roles of these bioactives in their natural environment may provide insights into their mechanisms of action and potential therapeutic targets.

Rigorous preclinical and clinical evaluation is necessary to establish the safety, efficacy, and long-term effects of marine bioactives in treating neuronal infections. While many marine compounds have demonstrated promising results in vitro and in animal models, translating these findings into effective human therapies requires well-designed clinical trials. Phase I and II clinical trials should assess the pharmacokinetics, tolerability, and potential side effects of these compounds, while Phase III trials can evaluate their efficacy in larger patient populations. Ethical considerations, including sustainable sourcing of marine bioactives and minimizing ecological impact, should also be prioritized to ensure that the exploitation of marine resources does not harm marine ecosystems. By addressing these challenges and capitalizing on the untapped potential of marine biodiversity, future research can unlock new avenues for developing innovative therapies to combat neuronal infections. Marine bioactives offer a promising and sustainable solution to the growing burden of antimicrobial resistance and the limitations of current treatment modalities. Through continued exploration and integration of marine-derived compounds into therapeutic regimens, researchers can significantly improve outcomes for patients suffering from neuronal infections, ultimately contributing to the advancement of global health.

Marine-derived bioactives offer unparalleled potential in addressing the pathology of neuronal infections by modulating immune responses, preventing microbial replication, and protecting against neuronal damage. Their diverse pharmacological activities, including antimicrobial, anti-inflammatory, and neuroprotective properties, make them ideal candidates for developing novel therapeutic and preventive strategies for CNS infections. As antimicrobial resistance rises and the burden of neurodegenerative complications associated with infections increases, harnessing the potential of marine bioactives could revolutionize the management of neuronal infections. Future research should focus on optimizing the isolation, characterization, and clinical validation of these compounds, paving the way for their integration into mainstream medical practice. Through continued exploration of marine biodiversity, researchers can unlock new therapeutic avenues to combat neuronal infections and improve long-term neurological outcomes.

Neuronal infections, resulting from bacterial, viral, fungal, and parasitic pathogens, continue to pose significant challenges in healthcare due to their potential to cause severe neurological damage and long-term cognitive and motor dysfunction. Conventional antimicrobial therapies, while effective in managing many infections, are often limited by their inability to cross the BBB efficiently, leading to incomplete pathogen clearance and prolonged neuroinflammation. The alarming rise of antimicrobial resistance further complicates the treatment landscape, necessitating the search for novel and more effective therapeutic and preventive agents. Marine natural products, derived from diverse marine organisms such as sponges, algae, bryozoans, fungi, and invertebrates, have shown immense potential in filling this therapeutic gap. These bioactive compounds offer antimicrobial, anti-inflammatory, immunomodulatory, and neuroprotective properties that target multiple pathways involved in the pathogenesis of neuronal infections. Moving forward, marine natural products have the potential to revolutionize the prevention, diagnosis, and treatment of neuronal infections.

Marine-derived bioactives provide a unique advantage over conventional antimicrobials by targeting both the pathogens and the host immune responses that contribute to neuronal damage. Future research should focus on optimizing these compounds to enhance their therapeutic efficacy and bioavailability. Bryostatin, derived from the *Bugula neritina*, has demonstrated remarkable potential in modulating PKC activity. PKC is vital to neuronal survival, immune responses, and synaptic plasticity. Bryostatin's ability to inhibit viral replication and prevent the spread of neurotropic viruses such as HIV and HSV highlights its potential as a future antiviral therapeutic. Further research should explore the development of bryostatin analogs with improved stability, reduced toxicity, and enhanced BBB permeability to maximize their efficacy in treating viral encephalitis and other CNS infections.

Similarly, squalamine, a natural steroid derived from the dogfish shark, offers broad-spectrum antimicrobial properties by disrupting bacterial membranes and inhibiting biofilm formation. Squalamine's ability to prevent biofilm formation is particularly significant in the context of bacterial meningitis, where biofilms often shield bacteria from immune responses and antibiotics, contributing to persistent infections. Future studies should focus on the synergistic effects of squalamine with existing antibiotics, potentially reducing the risk of antimicrobial resistance and improving clinical outcomes in bacterial CNS infections. Plitidepsin, derived from the marine tunicate *Aplidium albicans*, targets elongation factor-1α (eEF1A) and inhibits viral replication, making it a promising candidate for the treatment of neurotropic viral infections such as Zika and enteroviruses. By enhancing the stability and bioavailability of plitidepsin through nanoparticle-based delivery systems, future research can improve its therapeutic efficacy in treating viral infections that target the CNS.

The future of marine natural products extends beyond antimicrobial activity to include immunomodulation, which plays a critical role in managing neuronal infections. Excessive or prolonged neuroinflammation is a hallmark of CNS infections, contributing to neuronal apoptosis, excitotoxicity, and chronic neurological dysfunction. Marine-derived immunomodulators, such as manoalide and fucoidan, offer promising avenues for controlling inflammation and preventing long-term neuronal damage. Manoalide, isolated from marine sponges, inhibits phospholipase A2 (PLA2) and reduces the production of pro-inflammatory cytokines such as IL-1β and TNF-α. This mechanism holds great potential for mitigating neuroinflammation in bacterial meningitis and viral encephalitis, where excessive cytokine production contributes to neuronal apoptosis. Future research should focus on the clinical evaluation of manoalide in animal models of CNS infections and its potential as an adjunct therapy to reduce inflammation and prevent long-term sequelae.

Fucoidan, a sulfated polysaccharide derived from brown algae, activates toll-like receptors (TLRs) and enhances phagocytic activity, promoting pathogen clearance and immunomodulation. Fucoidan's ability to balance pro-inflammatory and anti-inflammatory responses makes it an ideal candidate for preventing chronic inflammation in the CNS. Clinical studies investigating the immunomodulatory effects of fucoidan in models of bacterial and viral meningitis can pave the way for its inclusion in combination therapies to optimize immune responses while preventing excessive neuroinflammation.

Neuronal infections often result in long-term neurological sequelae, including cognitive decline, epilepsy, and motor dysfunction, primarily due to persistent neuroinflammation and neuronal apoptosis. Marine-derived bioactives such as ziconotide, phlorotannins, and plakortide E have demonstrated potent neuroprotective properties that can mitigate these long-term complications. Ziconotide, a peptide derived from the venom of the cone snail *Conus magus*, blocks N-type calcium channels, preventing excitotoxicity and reducing neuronal apoptosis. Ziconotide's ability to protect neurons from calcium-induced damage makes it a valuable candidate for preventing chronic neurological dysfunction following CNS infections. Future research should explore the repurposing of ziconotide for postinfection neuroprotection and its potential for enhancing synaptic recovery after viral encephalitis and meningitis.

Phlorotannins, polyphenolic compounds derived from brown algae, exhibit potent antioxidant and anti-inflammatory properties that protect neurons from oxidative stress and neuroinflammation. By scavenging ROS and modulating inflammatory signaling, phlorotannins prevent neuronal apoptosis and preserve synaptic integrity. Future clinical trials evaluating the efficacy of phlorotannins in preventing postinfection cognitive decline and neurodegenerative complications can open new avenues for their use in neuroprotection. Similarly, *plakortide E* modulates mitochondrial function and reduces oxidative stress, protecting neurons from apoptosis triggered by chronic inflammation. Investigating the synergistic effects of *plakortide E* with conventional neuroprotective agents can enhance its efficacy in preventing long-term neuronal damage following CNS infections.

Prevention of neuronal infections remains a priority, particularly in high-risk populations and regions where emerging neurotropic pathogens pose a threat. Marine bioactives have demonstrated potential as vaccine adjuvants and prophylactic agents that enhance immune responses and prevent pathogen entry into the CNS. Monophosphoryl lipid A (MPL), derived from marine bacteria, is a potent vaccine adjuvant that activates TLR4 and enhances adaptive immunity. MPL has been incorporated into vaccines targeting bacterial and viral pathogens that can cause neuronal infections, promoting long-term immunity and reducing the risk of recurrent CNS infections. Future research should focus on expanding the use of MPL-based adjuvants in vaccines against emerging neurotropic viruses such as West Nile virus and Zika virus, which pose a growing threat to global health.

Zymosan, a polysaccharide derived from marine yeast, activates TLR2 and induces innate immune responses, making it an effective immunomodulatory agent in vaccine formulations. Zymosan enhances antigen presentation and promotes the production of memory T cells, improving long-term immunity against neurotropic pathogens. Exploring the potential of zymosan in mucosal vaccines that prevent viral entry through the nasal epithelium can further enhance its efficacy in preventing viral encephalitis and meningitis. Carrageenan, derived from red algae, inhibits viral attachment and entry into host cells, preventing viral replication. Carrageenan-based formulations, such as nasal sprays and topical gels, have shown promise in preventing viral infections that can lead to neuronal damage. Future research should focus on optimizing the stability and bioavailability of carrageenan-based prophylactic agents to maximize their efficacy in preventing CNS infections.

The future of marine natural products in treating and preventing neuronal infections hinges on continued exploration of marine biodiversity and the discovery of novel bioactive compounds with therapeutic potential. Advances in marine biotechnology, genomics, and metabolomics have enabled the identification and characterization of novel compounds with diverse pharmacological activities. Future research should focus on high-throughput screening of marine extracts for antimicrobial, anti-inflammatory, and neuroprotective activities, followed by structure-activity relationship (SAR) studies to optimize their efficacy and safety. The development of novel drug delivery systems, such as nanoparticle-based carriers and liposomal formulations, can enhance the bioavailability and BBB penetration of marine bioactives, ensuring their effective delivery to the CNS.

Marine symbionts, including bacteria and fungi associated with marine invertebrates, represent an untapped source of bioactive compounds with unique structures and potent biological activities. Future studies should explore the symbiotic relationships between marine organisms and their associated microbiota to identify novel bioactives with potential applications in neuronal infection management. Metagenomic and metatranscriptomic approaches can further enhance the discovery of gene clusters responsible for the biosynthesis of novel bioactives, paving the way for their large-scale production and commercialization.

While marine bioactives offer immense promise in the treatment and prevention of neuronal infections, several challenges must be addressed to facilitate their clinical translation. Ensuring the safety, stability, and scalability of marine-derived compounds is essential for their successful integration into mainstream medical practice. Future clinical trials should evaluate the pharmacokinetics, pharmacodynamics, and safety profiles of marine bioactives in animal models and human populations to establish their efficacy and tolerability. Regulatory frameworks should also be adapted to accommodate the unique challenges associated with marine-derived compounds, ensuring their expedited approval and commercialization.

To ensure that marine life remains healthy while harvesting bioactive compounds for therapeutic use, a multifaceted approach that combines sustainable practices, conservation efforts, and biotechnological innovation is essential. Implementing strict guidelines on the collection of marine organisms, including quotas and seasonal restrictions, can prevent overexploitation and allow ecosystems to recover. Establishing marine-protected areas where biodiversity can thrive undisturbed ensures the preservation of critical habitats. Additionally, promoting aquaculture and mariculture for the cultivation of bioactive-producing organisms, such as sponges and algae, reduces the need for wild harvesting. Advances in microbial fermentation and synthetic biology can also enable the production of marine-derived compounds in controlled laboratory settings, minimizing environmental impact. Collaborations between scientists, policymakers, and local communities are crucial to developing frameworks that balance the exploration of marine bioactives with the protection of marine ecosystems, ensuring that future generations can continue to benefit from the ocean's rich biodiversity.

Collaboration between academic institutions, pharmaceutical companies, and regulatory agencies is critical to accelerating the clinical translation of marine bioactives. Public-private partnerships can facilitate large-scale production, formulation optimization, and clinical evaluation of marine-derived compounds, ensuring their timely integration into the treatment and prevention of neuronal infections. By harnessing the untapped potential of marine biodiversity, the scientific community can develop innovative therapies and preventive strategies to combat the growing burden of neuronal infections and improve long-term patient outcomes.

10.2 Conclusion

Marine natural products hold tremendous promise in transforming the landscape of neuronal infection management by offering novel therapeutic and preventive agents that target multiple aspects of the infection process. Their unique ability to modulate immune responses, inhibit pathogen replication, and protect neurons from inflammation-induced damage makes them invaluable candidates for addressing the growing challenges associated with CNS infections. As antimicrobial resistance rises and emerging neurotropic pathogens pose new threats, the future of marine bioactives lies in optimizing their therapeutic potential, expanding their applications in vaccine development, and exploring new sources of bioactive compounds through marine biotechnology. By continuing to explore the vast biodiversity of the marine environment and leveraging advances in biotechnology, researchers can unlock the full potential of marine natural products, ushering in a new era of innovation in the treatment and prevention of neuronal infections.

References

Adamu A, Li S, Gao F, Xue G (2024) The role of neuroinflammation in neurodegenerative diseases: current understanding and future therapeutic targets. Front Aging Neurosci 16:1347987. https://doi.org/10.3389/fnagi.2024.1347987

Ambati RR, Phang SM, Ravi S, Aswathanarayana RG (2014) Astaxanthin: sources, extraction, stability, biological activities and its commercial applications--a review. Mar Drugs 12(1):128–152. https://doi.org/10.3390/md12010128

Bate C, Salmona M, Diomede L, Williams A (2004) Squalestatin cures prion-infected neurons and protects against prion neurotoxicity. J Biol Chem 279(15):14983–14990. https://doi.org/10.1074/jbc.M313061200

Chaigneau L, Jary M, Nerich V, Hervieu A, Aubry S, Charon Barra C, Meynard G, Neumann F, Kalbacher E, Isambert N (2022) Real-world experience of efficacy and safety of Trabectedin in patients with soft tissue sarcoma: A Bicentric retrospective analysis. Oncology 100(12):633–644. https://doi.org/10.1159/000527602

Chen T, Dai Y, Hu C, Lin Z, Wang S, Yang J, Zeng L, Li S, Li W (2024) Cellular and molecular mechanisms of the blood-brain barrier dysfunction in neurodegenerative diseases. Fluids Barriers CNS 21(1):60. https://doi.org/10.1186/s12987-024-00557-1

Dalisay DS, Tenebro CP, Sabido EM, Suarez AFL, Paderog MJV, Reyes-Salarda R, Saludes JP (2024) Marine-derived anticancer agents targeting apoptotic pathways: exploring the depths for novel cancer therapies. Mar Drugs 22(3). https://doi.org/10.3390/md22030114

Dando SJ, Mackay-Sim A, Norton R, Currie BJ, St John JA, Ekberg JA, Batzloff M, Ulett GC, Beacham IR (2014) Pathogens penetrating the central nervous system: infection pathways and the cellular and molecular mechanisms of invasion. Clin Microbiol Rev 27(4):691–726. https://doi.org/10.1128/CMR.00118-13

Dillon S, Agrawal S, Banerjee K, Letterio J, Denning TL, Oswald-Richter K, Kasprowicz DJ, Kellar K, Pare J, van Dyke T, Ziegler S, Unutmaz D, Pulendran B (2006) Yeast zymosan, a stimulus for TLR2 and dectin-1, induces regulatory antigen-presenting cells and immunological tolerance. J Clin Invest 116(4):916–928. https://doi.org/10.1172/JCI27203

DiSabato DJ, Quan N, Godbout JP (2016) Neuroinflammation: the devil is in the details. J Neurochem 139 Suppl 2(Suppl 2):136–153. https://doi.org/10.1111/jnc.13607

Dyall SC (2015) Long-chain omega-3 fatty acids and the brain: a review of the independent and shared effects of EPA, DPA and DHA. Front Aging Neurosci 7:52. https://doi.org/10.3389/fnagi.2015.00052

Elbandy M (2022) Anti-inflammatory effects of marine bioactive compounds and their potential as functional food ingredients in the prevention and treatment of neuroinflammatory disorders. Molecules 28(1). https://doi.org/10.3390/molecules28010002

Ferir G, Huskens D, Palmer KE, Boudreaux DM, Swanson MD, Markovitz DM, Balzarini J, Schols D (2012) Combinations of griffithsin with other carbohydrate-binding agents demonstrate superior activity against HIV Type 1, HIV Type 2, and selected carbohydrate-binding agent-resistant HIV Type 1 strains. AIDS Res Hum Retrovir 28(11):1513–1523. https://doi.org/10.1089/AID.2012.0026

Gao Y, Wu Y (2022) Recent advances of chitosan-based nanoparticles for biomedical and biotechnological applications. Int J Biol Macromol 203:379–388. https://doi.org/10.1016/j.ijbiomac.2022.01.162

Grosso C, Valentao P, Ferreres F, Andrade PB (2014) Bioactive marine drugs and marine biomaterials for brain diseases. Mar Drugs 12(5):2539–2589. https://doi.org/10.3390/md12052539

Hannan MA, Dash R, Haque MN, Mohibbullah M, Sohag AAM, Rahman MA, Uddin MJ, Alam M, Moon IS (2020) Neuroprotective potentials of marine algae and their bioactive metabolites: pharmacological insights and therapeutic advances. Mar Drugs 18(7). https://doi.org/10.3390/md18070347

Haste NM, Hughes CC, Tran DN, Fenical W, Jensen PR, Nizet V, Hensler ME (2011) Pharmacological properties of the marine natural product marinopyrrole A against methicillin-resistant Staphylococcus aureus. Antimicrob Agents Chemother 55(7):3305–3312. https://doi.org/10.1128/AAC.01211-10

Karthikeyan A, Joseph A, Nair BG (2022) Promising bioactive compounds from the marine environment and their potential effects on various diseases. J Genet Eng Biotechnol 20(1):14. https://doi.org/10.1186/s43141-021-00290-4

Kim BN, Peleg AY, Lodise TP, Lipman J, Li J, Nation R, Paterson DL (2009) Management of meningitis due to antibiotic-resistant Acinetobacter species. Lancet Infect Dis 9(4):245–255. https://doi.org/10.1016/S1473-3099(09)70055-6

Ko EJ, Lee YT, Kim KH, Lee Y, Jung YJ, Kim MC, Lee YN, Kang T, Kang SM (2017) Roles of aluminum hydroxide and Monophosphoryl lipid A adjuvants in overcoming CD4+ T cell deficiency to induce isotype-switched IgG antibody responses and protection by T-dependent influenza vaccine. J Immunol 198(1):279–291. https://doi.org/10.4049/jimmunol.1600173

Kousparou C, Fyrilla M, Stephanou A, Patrikios I (2023) DHA/EPA (Omega-3) and LA/GLA (Omega-6) as bioactive molecules in neurodegenerative diseases. Int J Mol Sci 24(13). https://doi.org/10.3390/ijms241310717

Luthuli S, Wu S, Cheng Y, Zheng X, Wu M, Tong H (2019) Therapeutic effects of Fucoidan: a review on recent studies. Mar Drugs 17(9). https://doi.org/10.3390/md17090487

Mammari N, Salles E, Beaussart A, El-Kirat-Chatel S, Varbanov M (2022) Squalamine and its Aminosterol derivatives: overview of biological effects and mechanisms of action of compounds with multiple therapeutic applications. Microorganisms 10(6). https://doi.org/10.3390/microorganisms10061205

Mark LP, Prost RW, Ulmer JL, Smith MM, Daniels DL, Strottmann JM, Brown WD, Hacein-Bey L (2001) Pictorial review of glutamate excitotoxicity: fundamental concepts for neuroimaging. AJNR Am J Neuroradiol 22(10):1813–1824. https://www.ncbi.nlm.nih.gov/pubmed/11733308

McGivern JG (2007) Ziconotide: a review of its pharmacology and use in the treatment of pain. Neuropsychiatr Dis Treat 3(1):69–85. https://doi.org/10.2147/nedt.2007.3.1.69

Mehla R, Bivalkar-Mehla S, Nagarkatti M, Chauhan A (2012) Programming of neurotoxic cofactor CXCL-10 in HIV-1-associated dementia: abrogation of CXCL-10-induced neuroglial toxicity in vitro by PKC activator. J Neuroinflammation 9:239. https://doi.org/10.1186/1742-2094-9-239

Perry TR, Schentag JJ (2001) Clinical use of ceftriaxone: a pharmacokinetic-pharmacodynamic perspective on the impact of minimum inhibitory concentration and serum protein binding. Clin Pharmacokinet 40(9):685–694. https://doi.org/10.2165/00003088-200140090-00004

Potts BC, Albitar MX, Anderson KC, Baritaki S, Berkers C, Bonavida B, Chandra J, Chauhan D, Cusack JC Jr, Fenical W, Ghobrial IM, Groll M, Jensen PR, Lam KS, Lloyd GK, McBride W, McConkey DJ, Miller CP, Neuteboom ST et al (2011) Marizomib, a proteasome inhibitor for all seasons: preclinical profile and a framework for clinical trials. Curr Cancer Drug Targets 11(3):254–284. https://doi.org/10.2174/156800911794519716

Rehman MU, Wali AF, Ahmad A, Shakeel S, Rasool S, Ali R, Rashid SM, Madkhali H, Ganaie MA, Khan R (2019) Neuroprotective strategies for neurological disorders by natural products: an update. Curr Neuropharmacol 17(3):247–267. https://doi.org/10.2174/1570159X16666180911124605

Rizvi SAA, Saleh AM (2018) Applications of nanoparticle systems in drug delivery technology. Saudi Pharm J 26(1):64–70. https://doi.org/10.1016/j.jsps.2017.10.012

Ruiz-Bedoya CA, Mota F, Tucker EW, Mahmud FJ, Reyes-Mantilla MI, Erice C, Bahr M, Flavahan K, de Jesus P, Kim J, Foss CA, Peloquin CA, Hammoud DA, Ordonez AA, Pardo CA, Jain SK (2022) High-dose rifampin improves bactericidal activity without increased intracerebral inflammation in animal models of tuberculous meningitis. J Clin Invest 132(6). https://doi.org/10.1172/JCI155851

Saez-Llorens X, McCoig C, Feris JM, Vargas SL, Klugman KP, Hussey GD, Frenck RW, Falleiros-Carvalho LH, Arguedas AG, Bradley J, Arrieta AC, Wald ER, Pancorbo S, McCracken GH Jr, Marques SR, Trovan meningitis Study, G (2002) Quinolone treatment for pediatric bacterial meningitis: a comparative study of trovafloxacin and ceftriaxone with or without vancomycin. Pediatr Infect Dis J 21(1):14–22. https://doi.org/10.1097/00006454-200201000-00004

Shahan B, Choi EY, Nieves G (2021) Cerebrospinal fluid analysis. Am Fam Physician 103(7):422–428. https://www.ncbi.nlm.nih.gov/pubmed/33788511

Shin JH, Lee HK, Khang SK, Kim DW, Jeong AK, Ahn KJ, Choi CG, Suh DC (2002) Neuronal tumors of the central nervous system: radiologic findings and pathologic correlation. Radiographics 22(5):1177–1189. https://doi.org/10.1148/radiographics.22.5.g02se051177

Smaga I, Fierro D, Mesa J, Filip M, Knackstedt LA (2020) Molecular changes evoked by the beta-lactam antibiotic ceftriaxone across rodent models of substance use disorder and neurological disease. Neurosci Biobehav Rev 115:116–130. https://doi.org/10.1016/j.neubiorev.2020.05.016

Stefano GB (2021) Historical insight into infections and disorders associated with neurological and psychiatric sequelae similar to long COVID. Med Sci Monit 27:e931447. https://doi.org/10.12659/MSM.931447

Sun MK, Alkon DL (2006) Bryostatin-1: pharmacology and therapeutic potential as a CNS drug. CNS Drug Rev 12(1):1–8. https://doi.org/10.1111/j.1527-3458.2006.00001.x

Sun XY, Tian XY, Li ZW, Peng XS, Wong HN (2011) Total synthesis of plakortide E and biomimetic synthesis of plakortone B. Chemistry 17(21):5874–5880. https://doi.org/10.1002/chem.201003309

Syed Hussin SAD, Chen XW, Al-Talib H, Chua AL, Azimi Z, Wang SM (2025) Diagnostic accuracy of multiplex real-time PCR for detecting viruses associated with encephalitis: a systematic review and meta-analysis protocol. PLoS One 20(3):e0318805. https://doi.org/10.1371/journal.pone.0318805

Tajerian A, Sofian M, Zarinfar N, Ramezani A (2024) Manifestations, complications, and treatment of neurobrucellosis: a systematic review and meta-analysis. Int J Neurosci 134(3):256–266. https://doi.org/10.1080/00207454.2022.2100776

Tan LT, Salleh NF (2024) Marine cyanobacteria: a rich source of structurally unique anti-Infectives for drug development. Molecules 29(22). https://doi.org/10.3390/molecules29225307

Trabectedin (2012) In LiverTox: clinical and research information on drug-induced liver injury. https://www.ncbi.nlm.nih.gov/pubmed/31643478

Valia-Vera JC, Villanueva VL, Asensio-Samper JM, Lopez-Alarcon MD, de Andres JA (2007) Ziconotide: an innovative alternative for intense chronic neuropathic pain. Rev Neurol 45(11):665–669. https://www.ncbi.nlm.nih.gov/pubmed/18050098 (Ziconotide: una alternativa innovadora en el dolor cronico neuropatico intenso)

Volman TJ, Hendriks T, Goris RJ (2005) Zymosan-induced generalized inflammation: experimental studies into mechanisms leading to multiple organ dysfunction syndrome. Shock 23(4):291–297. https://doi.org/10.1097/01.shk.0000155350.95435.28

Yang X, Kubican SE, Yi Z, Tong S (2025) Advances in magnetic nanoparticles for molecular medicine. Chem Commun (Camb) 61(15):3093–3108. https://doi.org/10.1039/d4cc05167j

Yeom JH, Kim HY, Lim JH, Yoon KW, Kim HM, Jeong HJ (2021) A calcium channel blocker, manoalide exerts an anti-allergic inflammatory effect through attenuating NF-kappaB activity. Immunopharmacol Immunotoxicol 43(6):799–805. https://doi.org/10.1080/08923973.2021.1988101

Modern Scope, Outcomes, and Potential Application of Marine-Derived Natural Bioactives to Advance the Field of Neuroscience

11

Carson Greer, Keyi Liu, Preston Cook, Suhrud Pathak, Rishi Nadar, Muralikrishnan Dhanasekaran, Nishi Ann, Jaufar Palekkodan, Husni Ahmed Abdullah Al-Goshae, and Vijeyrani Vadipatti Baskarpandian

Abstract

Marine-derived natural products (MNPs) have long been undermined as potential resources for neurodegenerative disease intervention, offering preventive and therapeutic implications through versatile bioactivities. In this systematic review, the exploration of MNPs' chemical features and mechanistic actions will aid in addressing the pathological hallmarks of various neurodegenerative diseases, including Alzheimer's, Parkinson's, and amyotrophic lateral sclerosis. Unique compounds, such as 11-dehydrosinulariolide, bryostatin-1, and dieckol, exhibit neuroprotective capabilities through the mitigation of oxidative stress, neuroin-

C. Greer · K. Liu · P. Cook · S. Pathak · R. Nadar
Drug Discovery and Development, Harrison College of Pharmacy, Auburn University, Auburn, AL, USA

M. Dhanasekaran (✉)
Drug Discovery and Development, Harrison College of Pharmacy, Auburn University, Auburn, AL, USA

Graduate School, Auburn University, Auburn, AL, USA
e-mail: dhanamu@auburn.edu

N. Ann
Department of Zoology, Providence Women's College (Autonomous), Calicut, Kerala, India

J. Palekkodan
Illham House, Changuvetti kund, Kottakkal, Kerala, India

H. A. A. Al-Goshae
International Medical School, Management and Science University, Shah Alam, Malaysia

V. V. Baskarpandian
Department of Mathematics, Yadava College, Thiruppalai, Madurai, India

M. Bhupathyraaj et al. (eds.), *Neuroprotective Marine Bioactives*, Nutritional Neurosciences, https://doi.org/10.1007/978-981-95-5791-2_11

flammation, mitochondrial dysfunction, and protein aggregation. Additionally, various marine bioactives demonstrate significant efficacy in both in vitro and in vivo models, engaging essential molecular pathways such as the PI3K/Akt, Nrf2/ARE, MAPK, and gut-brain axis pathways. Emerging research suggests that many of these natural compounds could potentially play roles in epigenetic modulation and synaptic restoration, pioneering the opportunity for precision medicine, specifically tailored interventions. Overall, this review highlights the transformative potential of MNPs while emphasizing the importance of interdisciplinary collaboration to accelerate clinical breakthroughs. As the ocean continues to yield molecular insights, MNPs offer a compelling avenue toward novel neurotherapeutics in an era of aging populations and escalating neurological burdens.

11.1 Marine Bioactives

The sea is one of the most biodiverse environments on the planet, signified by the coral reefs possessing the most diversity out of any other ecosystem, and thereby giving it the potential to provide medically applicable natural products. Natural products (NPs) are defined as chemicals produced by living organisms, including various plants and animals, that are used therapeutically to treat diseases and maintain health (Jiménez 2018). Marine natural products (MNPs) are natural compounds derived specifically from organisms living in aquatic environments. These marine natural products have gained significant attention in the scientific community due to their bioactive potential and diverse chemical structures (Liang et al. 2019). A wide range of organisms have been recognized for this potential, from sponges to microorganisms and beyond. The chemical complexes in some of these organisms have already been found to possess anti-inflammatory, antiviral, and anti-tumor properties (Mayer et al. 2011).

The chemical diversity of MNPs has been found useful in various disease states, including cancer, neurological disorders, and certain autoimmune diseases. For example, NVP-LAQ824, found in several sea sponges, functions as an inhibitor of histone deacetylases (HDACs) (Simmons et al. 2005). Histone deacetylases (HDACs) are enzymes that remove acetyl groups from histone proteins, leading to chromatin condensation and gene silencing. By altering gene expression, HDACs can promote cancer by silencing tumor suppressor genes and activating oncogenes (Khangura et al. 2017). The compound NVP-LAQ824 inhibits this type of enzyme, thereby creating an anticancer effect. Its anti-cancerous properties give it the potential as an anticancer drug.

This meta-analysis aims to specifically target the properties of MNPs in the potential treatment of neurodegenerative pathologies, such as Parkinson's disease (PD), Alzheimer's disease (AD), amyotrophic lateral sclerosis (ALS), and multiple sclerosis (MS), and to categorize them as either preventative or therapeutic approaches. These marine-derived natural products are valuable for potential drug discovery, as recent studies have demonstrated their efficacy in treating neurodegenerative disorders. The mechanisms in the MNPs include the inhibition of beta-amyloid protein (Aβ) and acetylcholinesterase (AChE) (Alghazwi et al. 2016). These are each reagents that are seen in Alzheimer's, and there has been in vitro evidence of MNPs offering relief. MNPs also align with the current literature regarding the gut-brain axis, working as targeted candidates for alleviating dysregulated reagents in the gut (Fakhri et al. 2021).

Despite these promising developments that MNPs have provided, challenges remain in the research of MNPs for the treatment of neurodegeneration. The convoluted pathways and mechanisms of the chemical complexes associated with MNPs make it challenging to translate them into clinical usage. Above all, the issues of sustainability and the potential ecological impacts of the research on MNPs must always be considered. Still, advances in genome mining and other biotechnologies are proving useful for employing MNPs and all that they can offer (Chu et al. 2020).

There are several core neurodegenerative diseases, one of which is Parkinson's disease. There are specific biochemical mechanisms that have been described in the development of Parkinson's disease. In normal circumstances, dopaminergic cells produce dopamine (DA), a neurotransmitter that regulates voluntary movement and a wide array of behavioral functions (Cacabelos 2017). In the PD model, there is a loss of dopaminergic neurons in the substantia nigra pars compacta. This leads to characteristic movement symptoms, such as rigidity and a resting tremor (Alexander 2004). MNPs have been observed to therapeutically offer relief or decrease the severity of several of these PD symptoms.

The most common neurodegenerative disease is Alzheimer's disease (Lamptey et al. 2022). This is the most prevalent form of dementia, a general term for diseases that cause progressive loss of two or more areas of cognition over time, outside the bounds of normal aging, including memory and other behavioral functions. The presence of abnormal neuritic plaques and neurofibrillary tangles pathologically characterizes AD. There is also evidence that MNPs can therapeutically offer relief from these symptoms. Although current studies and literature have varied opinions, β-amyloid (Aβ) peptides are thought to play a role in the obtrusive neuronal. The exact molecular mechanisms pertaining to the plaques are unclear. The review "Amyloid Beta in Aging and Alzheimer's Disease" highlights the need to understand better the molecular mechanisms of β-amyloid peptides and how they contribute to the formation of degenerative plaques (Sehar et al. 2022). Another description of this specific disease state is what is known as a "tauopathy," which signifies a buildup of abnormal tau proteins in the neurons or glial cells. As mentioned before, one of the primary causes of AD is the formation of neurofibrillary tangles, and these tangles are composed of tau proteins (Mandelkow and Mandelkow 2012). These tau proteins are originally a microtubule-associated protein (MAP), specifically microtubule stabilizers (Alonso et al. 2018). However, it has been demonstrated in vitro that tau becomes mutated through hyperphosphorylation, resulting in the characteristic neurofibrillary tangles. Although amyloid plaque buildup can be observed in those of later ages, it alone does not necessarily denote AD. The same cannot be said for tau-induced neurofibrillary tangles. AD and PD constitute the majority of the current literature and are typically used as models for neurodegenerative diseases, especially in the literature reviewed in this analysis.

In addition to these two, amyotrophic lateral sclerosis (ALS) is another prevalent neurodegenerative disease. This disease state is characterized by the chronic death of motor neurons, which ultimately leads to fatal paralysis (Riancho et al. 2021). It is not necessarily an autoimmune disorder, but rather a neurodegenerative state accompanied by neuroinflammation. The mechanistic pathway of this disease is not yet fully understood; however, mitochondrial dysfunction is believed to play a significant role in its progression. Mutations in the gene responsible for encoding superoxide dismutase-1 (SOD1) have been linked to characteristic mitochondrial dysfunction in ALS (Candelise et al. 2022). Another gene that warrants scrutiny is the TARDBP gene, located on chromosome 1 and encoding the protein TDP-43 (Suk and Rousseaux 2020). When this gene is mutated, this can cause aggregations of TDP-43 in the cytoplasm. This aggregation is characteristic of ALS, suggesting a correlation between the two. One additional factor to consider regarding ALS pathways is the FUS (fused in sarcoma) RNA-binding protein, which can exhibit mutations that impair its ability to perform its role in RNA repair (Sukhanova et al. 2020). Like mutations in TARDBP, this can lead to neuronal dysfunction.

Multiple sclerosis is an autoimmune neurodegenerative disease that is described as the immune system attacking the myelin sheath of neurons in the CNS, which leads to demyelination, inflammation, and progressive neurodegeneration that disrupts nerve signal transmission. Even though this disease is not usually fatal, it can still greatly impact quality of life and life expectancy. MS has a specific pathophysiology that has been gradually revealed over the last decade. Immune activation starts in the periphery, where autoreactive T-cells, particularly Th1 and Th17 subsets, are triggered by myelin proteins and migrate into the CNS, where they initiate inflammation (Damsker et al. 2010). B-cells contribute to the disease process by producing autoantibodies and secreting pro-inflammatory cytokines, such as TNF-α, IL-17, and IFN-γ, which disrupt the blood-brain barrier (BBB), allowing immune cells to infiltrate (Wang et al. 2018). Once in the CNS, these immune cells attack the myelin sheath, leading to demyelination and axonal degeneration, severely impairing nerve conduction. The damage is exacerbated by oxidative stress and mitochondrial dysfunction, which further contribute to neuronal injury. In addition, chronic inflammation is driven by microglial activation, persists over time, and prevents effective remyelination, promoting neurodegeneration. This inflammatory environment results in persistent neurodegeneration, causing irreversible damage and progressive disability in those who suffer from MS. Direct literature and data on diseases other than PD and AD with regard to MNPs are limited, but correlations between neuroprotectant qualities in the MNPs covered may be applied to various other neurodegenerative diseases. Some of these general effects include antioxidative, anti-inflammatory, and, in some cases, antimutagenic properties.

In addition to these neuroprotective effects, MNPs also offer possible therapeutic effects against neurodegenerative diseases. Specific components of MNPs, such as peptides, alkaloids, quinones, and terpenes, could offer reduction of oxidative stress, prevention of mitochondrial dysfunction, and, more specifically for neurodegeneration, prevention of α-synuclein aggregation and excessive interleukin-6 (IL-6) release (Silva et al. 2023), both of which are associated with neurodegenerative disease states (Shan et al. 2024), (Kaplan et al. 2003). In this context, the therapeutic use of an MNP equates to its application during an ongoing/present disease state.

11.2 Preventative

MNPs have the potential to offer preventative neurological effects against neurodegenerative disease states. Pereira and Valado review this neuroprotection in "The Seaweed Diet in Prevention and Treatment of the Neurodegenerative Diseases," where it is suggested that microalgae have a basis of neuroprotective effects in vitro and in vivo animal studies (Zhang et al. 2018). The compounds studied offer protection against properties such as antioxidant, anti-inflammatory, and anti-apoptotic effects (Catanesi et al. 2021). One of the neuroprotective compounds reviewed was 11-dehydrosinulariolide (11-de), a derivative of the soft coral *Sinularia flexibilis (*Lin et al. 2018*)*, which exhibits neuroprotection by upregulating the Akt/PI3K pathway (Feng et al. 2016). The upregulation of this specific pathway protects cells from damage resulting from 6-hydroxydopamine (6-OHDA) (Fig. 11.1). This study provided evidence to suggest that 11-de protects by increasing cytosolic or mitochondrial DJ-1 expression. In the study, the zebrafish that exhibited Parkinson's-like symptoms that were treated with the 11-de swam more, suggesting improved

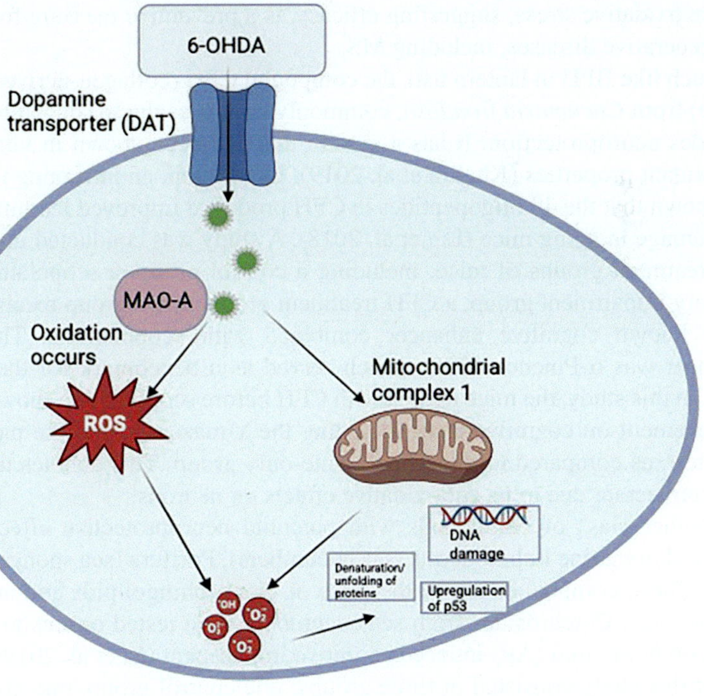

Fig. 11.1 Pathway of 6-OHDA

movement. Additionally, tyrosine hydroxylase (TH), a key biomarker of healthy dopamine-producing cells (Yi et al. 2024), was observed to improve with the application of 11-de.

Bryostatin-1 represents another important marine natural bioactive. This compound is derived from the organism *Bugula neritina,* a sessile bryozoa found to be abundant in the northern Gulf of Mexico (Keough and Chernoff 1987). It has been shown to produce biological activity by binding to protein kinase C (PKC), particularly the PKC-α and PKC-ε isoforms (Tian et al. 2023). This is essential because these involve synaptic plasticity, which is the ability for neurons to form and strengthen connections. This suggests possible therapeutic effects for disease states that impact cognitive ability and memory, such as Alzheimer's.

There have been many targeted compounds with possible neuroprotective effects, including BPH (bioactive protein hydrolysate) from *Benthosema pterotum*, commonly known as the lantern fish (Catanesi et al. 2021). This compound, specifically the dipeptide aspartic acid-tryptophan (Asp-Trp), gains efficacy from its ability to reduce free radicals, which cause oxidative stress and damage to neurons. It does this because the tryptophan(Tyr) residues have electron-rich aromatic rings, which can directly interact with and neutralize free radicals. Also, like many of the other antioxidants, it protects from H_2O_2 (hydrogen peroxide, a common free radical) induced apoptosis. This specific marine bioactive may be applied to ALS, as H_2O_2 is a free radical that has been observed and linked to this disease state (Sanghai and Tranmer 2021). The implementation of the dipeptide Asp-Trp has been shown to reduce oxidative stress, suggesting efficacy as a preventive measure for many neurodegenerative diseases, including MS.

Much like BPH in lantern fish, the compound CFH (collagen-derived fish hydrolysate) from *Cucumaria frondosa,* commonly known as the sea cucumber, similarly provides neuroprotection. It has a system that has been shown in vitro to exhibit antioxidant properties (Khalifa et al. 2019). In addition, an intriguing in vivo study has shown that the 40 oligopeptides in CFH produced improved learning and memory damage in aging mice (Lin et al. 2018). A study was conducted involving various treatment groups of mice, including a control group, a scopolamine-induced memory impairment group, a CFH treatment group, and a group receiving a previously known cognitive enhancer combined with scopolamine. The cognitive enhancer was α-Pinene (APN), which served as a benchmark for the efficacy of CFH. In this study, the mice treated with CFH before scopolamine showed a marked improvement on cognitive tests, including the Y-maze test and the passive avoidance test, as compared to the scopolamine-only group. This denotes the CFH as a neuroprotectant due to its antioxidative effects on neurons.

Another class of compounds with potential neuroprotective effects has been observed in marine Echinoderms (sea cucumbers), Porifera (sea sponges), and mollusks. These compounds are in the class of glycosphingolipids and are known as cerebrosides. Cerebrosides from sea cucumbers were tested on rats to observe the effect on β-amyloid (Aβ)-induced cognitive impairment (Li et al. 2019). The methods of this study consisted of three groups: one control group, one group injected with Aβ1–42 to induce cognitive damage modeling that of AD, and one group injected with Aβ1–42 + sea cucumber cerebrosides. The groups were subjected to two different cognitive tests, one of which was the Morris Water Maze, which assessed spatial memory and recognition by measuring the time the rats took to navigate through the maze. The other test was a general evaluation of locomotor activity. This study concluded that the rats treated with the sea cucumber cerebrosides exhibited improved cognitive ability as compared to the rats not treated with it. They accomplished this by reducing neuron apoptosis and decreasing synaptic dysfunction. This suggests that sea cucumber cerebrosides may possess neuroprotective qualities. Having already been tested in another mammal, this suggests that cerebrosides may also exhibit a neuroprotective effect in humans.

Another compound with in vitro evidence of neuroprotection is the heteropolysaccharide known as SV-2 (Ye et al. 2019). This polysaccharide was isolated from the species *Ommastrephes bartrami,* also known as the neon flying squid. The purpose of this study was to find out if SV-2 provided neuroprotection from the induced neurotoxin 6 6-hydroxydopamine (6-OHDA). The extraction method used to obtain the SV2-1 polysaccharide was autolysis, which involves utilizing the squid's own enzymes to break down the tissue. Both anion-exchange and gel-permeation chromatography were used to analyze the size of the molecule in question, which determined it has a molecular weight of 2.3 kDa (kilodaltons), or relatively small. This study ultimately determined that when tested in PC12 cells (a common model for neuronal cells), the application of 6-OHDA exhibited neuroprotective qualities by numerically improving cell viability. This was accomplished by increasing the activity of superoxide dismutase, which is an antioxidative enzyme, and decreasing lactate dehydrogenase and malondialdehyde, both of which are markers of oxidative stress and damage. These findings suggest that the SV2-1 polysaccharide may serve as a neuroprotective agent, potentially contributing to future therapeutic approaches for neurodegenerative diseases.

Kahalalide F is a cyclic peptide compound taken from the marine algae *Bryopsis pennata* and has shown potential neuroprotective effects, especially with regard to the stabilization of calcium homeostasis. Calcium imbalance is a key feature in many neurodegenerative diseases, including Huntington's disease, where disrupted calcium signaling can lead to neuronal excitotoxicity, cell death, and other forms of damage. The mutation in the huntingtin gene disrupts normal cellular functions, including calcium homeostasis, which is critical for neuron health. In HD, there is an imbalance in intracellular calcium levels, which contributes to neurodegeneration in several ways. First, the extracellular influx of calcium ions through various channels, including NMDA receptors (glutamate receptors), is increased due to dysfunctional glutamate signaling. The excess calcium enters neurons and accumulates in the cytosol and mitochondria. Mitochondria normally act as buffers to regulate calcium, but in HD, their ability to handle high calcium loads is impaired, leading to mitochondrial dysfunction, which exacerbates the damage. High levels of intracellular calcium activate various harmful signaling pathways, including the calcium/calmodulin-dependent protein kinase II (CaMKII) pathway, which can lead to oxidative stress, apoptosis, and the activation of cell death pathways.

11.3 Therapeutic

Several studies have found algae and the bioactive compounds in various algae to exhibit health impacts (Thomas and Kim 2013), (Singh et al. 2017). This marine microalgae can be classified in one of three ways: brown (Phaeophyceae), red (Rhodophyceae), and green algae (Chlorophyceae). Microalgae possess carotenoids, which are natural pigments found in other vegetation, most notably in carrots, which give them their orange color. These carotenoids have been observed by Pangestuti et al. to offer antioxidant effects (Chamorro et al. 2006) by helping to eradicate reactive oxygen species (ROS). These ROS create oxidative stress on neurons, making them less sustainable and leading to diseases such as PD and AD. Another study found the tendency of marine extracts from microalgae to decrease caspase-3 activity (Silva et al. 2018). This is crucial because the function of the enzyme caspase-3 is to incite programmed cell death, or apoptosis (Porter and Jänicke 1999). The decrease in its activity is crucial to prevent the occurrence of apoptosis in neurons, which may ultimately lead to chronic neurodegeneration. Marine algae have also demonstrated potential for protection against neurodegenerative diseases as a whole due to their ability to combat oxidative stress and reduce neuroinflammation and excitotoxicity. This lends credibility to not only improving common disease states such as PD and AD but also in relatively less prevalent ones like ALS and MS (Naik et al. 2024). Many compounds found in various algae have been shown to have neuroprotective effects, including polyunsaturated fatty acids (PUFAs), specific carbohydrates, and phenols. PUFAs have been shown to not only assist in the correction of brain function but also to lower the risk of cardiovascular disease, diabetes, and assist in blood pressure regulation. An example of a marine algae that produces large amounts of PUFAs is *Ulva fasciata,* commonly known as the green algae. Inclusion of this plant in one's diet may help promote neuroprotection against various neurodegenerative diseases, as well as provide other health benefits. A polysaccharide that was found to demonstrate neuroprotective effects is the β-1,3-glucan compound. It has been shown to help stimulate the immune system, as well as scavenge free radicals (Menaa et al. 2021). This is not the only polysaccharide shown to have neuroprotective effects from algae.

Fucoidans, a type of sulfated polysaccharide in the brown algae species *Undaria pinnatifida*, have been found to improve cell viability by inhibition of caspase-3 (Mohd Sairazi and Sirajudeen 2020). This view is further supported by an in vivo study that utilized gerbils to observe whether they would experience neuroprotection from transient global cerebral ischemia (tCGI), a condition characterized by the temporary cessation of blood flow to the hippocampal CA1 pyramidal neurons (Kim et al. 2019). The fucoidan treatment was given at two different doses, 25 mg/kg and 50 mg/kg. The results showed that at the 50 mg/kg dose, the gerbils experienced protection from the induced tCGI. This was determined by the fact that at this dose, there was a decreased activation of astrocytes and microglia in the ischemic area, which indicates less inflammation. This suggests fucoidans from *Undaria pinnatifida* could have future potential as a therapeutic agent in the acute stroke setting or in high-risk stroke patients.

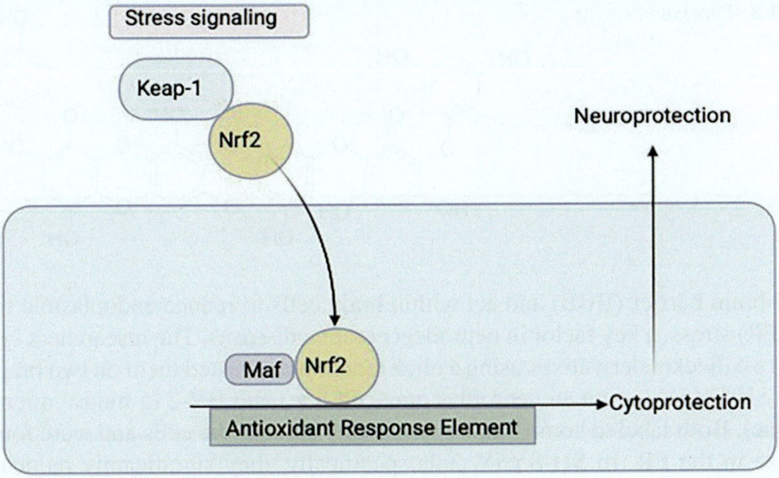

Fig. 11.2 Nrf2 antioxidant pathway

Another brown algae that has provided an intriguing therapeutic neuroprotectant is *Dictyopteris undulata* (Shimizu et al. 2015). The compound presented is a sesquiterpene (a terpene with the formula C15H24) named Zonarol. It exhibits these neuroprotective effects by activating the Nrf2/ARE pathway (Fig. 11.2), which in turn activates phase II enzymes (Johnson et al. 2008), such as HO-1 (Hine 2012), SOD (Wang et al. 2018), and NAD(P)H quinone oxidoreductase-1 (NQO-1) (Ngo and Duennwald 2022). The upregulation of these enzymes produces an exaggerated antioxidative effect, clearing more harmful ROS. The importance of this compound lies in its specification as the first pro-electrophilic neuroprotectant compound extracted from brown algae. The term "pro-electrophilic" translates to that the compound is not originally electrophilic but becomes so through oxidation. It has the structure of para-hydroquinone, meaning that the parent ring is disubstituted with two hydroxyl groups arranged in the para position.

Phlorotannins are a group of natural compounds that can also be found in brown algae, specifically in families such as Sargassaceae and Fucaceae. This nutraceutical is polyphenolic, meaning its organic structure contains multiple phenol groups (OH-Ph) and thereby provides it with inherent antioxidant properties. Structurally, they are very intriguing, as they contain ether bonds, which makes for a better and more stable antioxidant due to a longer delocalization of the radical charge picked up. They also contain dibenzodioxin and dibenzofuran structures. Dibenzodioxin and dibenzofuran structures enhance antioxidant activity by providing a rigid, conjugated ring system that stabilizes unpaired electrons through extensive electron delocalization. Their ability to support multiple phenolic groups also enables efficient radical scavenging at multiple reactive sites. Most notably, the phlorotannin dieckol (Fig. 11.3) has been shown to cross the blood-brain barrier during in vivo studies (Kwak et al. 2015). This specific study investigated whether fluorescein isothiocyanate (FITC)- and rhodamine B-labeled dieckol compounds could cross the

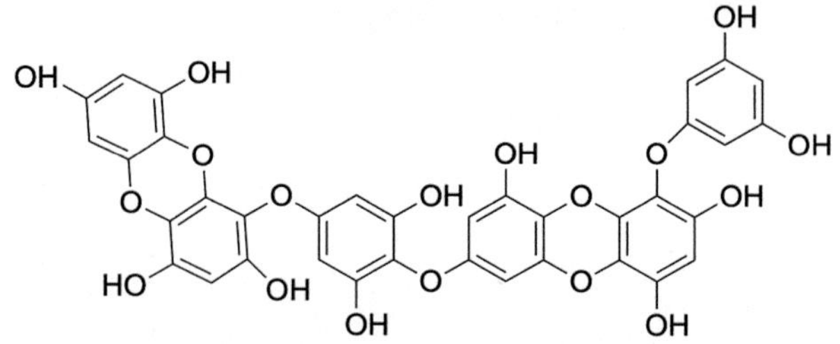

Fig. 11.3 Dieckol structure

blood-brain barrier (BBB) and act within brain cells to reduce endoplasmic reticulum (ER) stress, a key factor in neurodegenerative diseases. The researchers synthesized two dieckol derivatives using a click reaction and tested them on two brain cell lines: SH-SY5Y (a human neuroblastoma cell line) and BV-2 (a mouse microglial cell line). Both labeled compounds successfully entered the cells and were found to localize in the ER. In SH-SY5Y cells specifically, they significantly reduced ER stress. Furthermore, in vivo experiments in rats confirmed that the compounds were capable of crossing the BBB after intravenous administration. This was the first study to show that dieckol derivatives can be delivered systemically, reach the brain, and target ER stress in neural cells, offering a promising therapeutic strategy for neurodegenerative conditions. This specific phlorotannin contains six phloroglucinol units, which can be described as a benzene ring with three hydroxyl groups on each meta position of the molecule. This compound has been observed to inhibit acetylcholinesterase, which supports cholinergic functions, proving valuable in Alzheimer's research. Dieckol has also been found to modulate essential signaling pathways, including the PI3K/Akt and MAPK pathways, suggesting its potential as a multi-target therapeutic agent. Specifically, regarding the PI3K/Akt pathway, it could also prove useful in the prevention of brain cancers.

Cyanobacteria have also been found to have therapeutic effects on the nervous system (Rahman et al. 2021). Cyanobacteria are photosynthetic prokaryotes that tend to live in marine environments, and they represent a notable source of bioactives that play a role in medicine (Khalifa et al. 2019). These are commonly referred to as green-blue algae and belong to the Oscillatoriales family. The species in question is *Spirulina platensis*, a multicellular planktonic cyanobacterium that has anti-inflammatory and antioxidative properties. It has been observed to protect against neuronal loss in the substantia nigra, which was onset by MPTP (1-Methyl-4-phenyl-1,2,3,6-tetrahydropyridine), a neurotoxin associated with PD and the current toxin-induced model thereof. Not only that, but it has also been observed during in vivo studies to protect from neurodegeneration marked by 6-hydroxydopamine (6-OHDA) (Chattopadhyaya et al. 2015). This protection from 6-OHDA may be a generalizable neuroprotective effect, as this is a free radical that has also been linked to Huntington's disease (Chattopadhyaya et al. 2015). Huntington's disease is an autosomal dominant neurodegenerative disorder caused by a mutation in the HTT gene, resulting in the abnormal expansion of CAG repeats and the production of mutant huntingtin protein (Liu et al. 2023).

An intriguing class of compounds, known as Neolamellarin A derivatives, has been demonstrated in vitro to protect against glutamate-induced apoptosis (Zhang et al. 2022). This is key because in ALS, elevated glutamate levels and glutamate receptor activation can lead to excitotoxicity and cell death, a process known as glutamatergic apoptosis (Provenzano et al. 2023). It is thought that this is the case because relative success has been experienced with drugs such as Rizuole, a glutamate-release inhibitor (Blasco et al. 2014). Neolamellarin A derivatives have a polyaromatic structure, which makes them capable of numerous biological activities. The specific pathway by which Neolamellarin A derivatives perform this function is not entirely understood, but it is believed to occur through gene regulation, specifically involving the genes Bcl-2, Bax, and caspase-3 (Zhang and Bhavnani 2006). A compound known as GV-971 has been approved as a drug due to its possession of beneficial therapeutic effects for AD (Hu et al. 2023). This compound is an oligosaccharide derived from brown sea algae, specifically from the species *Ecklonia kurome (*Cao et al. 2022*)*. It became approved in 2020 by the State Drug Administration of China in the form of Sodium Oligomanne Capsules. It achieves its neuroprotection through the gut-brain axis, making it the first of its kind in the world. The development of this potentially novel AD therapeutic was discontinued due to the fact that its manufacturer, Shanghai Green Valley Pharmaceuticals, ran out of funding. However, it still provides an interesting model for therapeutic neuroprotection in MNPs. GV-971 offers neuroprotection and anti-inflammatory properties by inhibiting the amyloid beta protein (Aβ), which tends to form harmful plaques (Zhang et al. 2021). In addition, since the pathway of this drug interacts with the gut-brain axis, it is theorized to have an indirect effect of mitigation of tau protein hyperphosphorylation (Wang et al. 2018). The combination of both of these neuroprotective effects warrants it as a possible future therapeutic for neurodegenerative diseases such as AD.

Moving away from plant-based organisms, a popular dietary supplement is omega-3 fatty acids. These supplements are derived from fish and taken in the effort to improve general cognition. A specific omega-3 fatty acid that plays a role in neuroprotection is docosahexaenoic acid (DHA) (Shimazawa et al. 2009). DHA was studied in vitro in retinal ganglion cells (RGC-5) to observe if there were any neuroprotective qualities present. It was tested among several free radicals, including hydrogen peroxide (H_2O_2), superoxide anion (−O2), and hydroxyl radical (-OH). The method of this experiment was to test how DHA scavenged these radicals, and it was determined that it accomplished this in a concentration-based manner. In this case, DHA scavenged H_2O_2 at 0.1 μM, superoxide at 10 μM, and hydroxyl radicals at 100 μM. In conclusion, DHA exhibited a protective effect on the retinal ganglion cells by scavenging ROS that cause oxidative stress and damage. Additionally, DHA activates the PI3K/Akt pathways, resulting in an inhibitory effect on the GSK-3β enzyme, which is crucial in the hyperphosphorylation of the tau protein (Jia et al. 2025). The inhibition of this enzyme produces another neuroprotective effect. These effects can be applied to neurodegenerative diseases on the whole, as the mentioned pathways are highly applicable to PD, AD, ALS, and MS.

11.4 Future Prospects

There has been a limited amount of literature on specific gene-therapy targeted approaches to neuroprotection regarding MNPs; however, some correlations can be drawn between MNPs and their effects on genes and neuroprotection. For example, *Sargassum integerrimum* has not only been shown to exhibit neuroprotective effects (Yang et al. 2017) but has also demonstrated the ability to activate Nrf2 (nuclear factor erythroid 2–2-related factor 2) (Fig. 11.3), which is a gene that helps in regulating the antioxidant response (Ma et al. 2018) (Wu et al. 2019). Although the correlation between *Sargassum integerrimum* exhibiting neuroprotective qualities through gene therapy has not been directly studied, it would be interesting to explore the relationship in the future.

Another potential future concept to consider is precision medicine. Precision medicine, as defined by the US National Human Genome Research Institute, is the innovative approach that uses information about an individual's genomic, environmental, and lifestyle information to guide decisions related to their medical management (Delpierre and Lefèvre 2023). There is potential to utilize MNPs in precision medicine. Precision medicine has previously been utilized in the context of neurodegenerative diseases, specifically in cases such as genetic profiling and biomarker development (Strafella et al. 2018). Comprehensive studies directly linking MNPs and precision medicine in the context of neurodegenerative diseases are still developing. Some MNPs can pertain to someone's condition in some aspects, yet not specifically genetic, in the context of precision medicine. For example, if someone is located in a tropical environment with high sun exposure and hopes to take an MNP with neuroprotective qualities, Astaxanthin (3,3′-dihydroxy-β,β-carotene-4,4′-dione) might be a great choice.

This compound is a red carotenoid (a class previously discussed) derived from the microalgae *Haematococcus pluvialis*. Astaxanthin offers both UV-protective and neuroprotective effects. In this sense, precision medicine accomplishes a specific aim while accounting for another factor related to the subject's lifestyle. A study among Japanese adults was conducted over a span of 10 weeks to investigate the skin-protective effects of Astaxanthin (Ito et al. 2018). The type of study is known as a double-blind study, or a placebo-controlled study, indicating that neither the groups of subjects nor the researchers knew who was receiving a placebo or the Astaxanthin. The method used in this study was to dose the Astaxanthin group with 4 mg daily, and the placebo group with a capsule containing no active ingredient. Four qualities were measured: minimal erythema dose (MED), which indicates the amount of UV exposure required to cause damage (the higher the MED, the better), skin hydration, transepidermal water loss (TEWL), and the participants' subjective skin condition. The conclusion was that among the Astaxanthin supplementation group, each of these three measured qualities was higher than in the placebo group.

It is essential to note that the hypothetical patient in a tropical environment requires an MNP with neuroprotective properties. Astaxanthin has also been shown to provide neuroprotection through its antioxidative properties. Numerous studies have investigated the neuroprotective effects of Astaxanthin. Not only does it scavenge free radicals to prevent oxidative stress, but it also penetrates the blood-brain barrier, allowing it to protect the brain from oxidative damage directly (Wang and Qi 2022). It has also been found to activate the enzyme superoxide dismutase, which regulates the amount of ROS in cells (Islam et al. 2022). In addition to this effect, Astaxanthin can also function as a neuroprotective agent in precision medicine, particularly in relation to the specific lifestyle factor of increased sun exposure. In the future, there will hopefully be more literature surrounding studies of neuroprotectant MNPs and their roles in precision medicine. With advancing techniques in genomics and personalized diagnostics, it will hopefully soon be possible to prescribe certain MNPs to individuals based on their APOE (apolipoprotein E) genotype, oxidative stress sensitivity, or inflammatory biomarker profiles. Similar to other compounds previously discussed, marine-derived compounds, such as fucoxanthin, a carotenoid found in brown algae, have demonstrated the ability to attenuate oxidative stress and modulate neuroinflammatory pathways, which are particularly relevant for individuals with a genetic predisposition to Alzheimer's disease. Additionally, marine-derived polysaccharides, such as carrageenans and fucoidans, have shown promise in modulating the gut-brain axis, a critical pathway implicated in neurodegenerative conditions (Fakhri et al. 2021). These polysaccharides can influence gut microbiota composition, thereby affecting neuroinflammation and cognitive function. This specific approach is an intriguing combination of nutrigenomics, neurobiology, and marine pharmacology. As high-throughput screening, AI-driven compound analysis, and marine biodiversity research advance, the integration of MNPs into individualized neurotherapeutics may become a cornerstone of precision neurology.

11.5 Conclusion

Unfortunately, there are currently still no highly effective therapeutic options for many of the mentioned neurological conditions. However, the MNPs discussed hold considerable promise for future treatments and therapies. It has been experimentally observed, both in vivo and in vitro, that specific marine natural products offer preventive or therapeutic aid in neurodegeneration. This is significant as neurodegenerative diseases affect an incredibly large portion of the population. In 2019, it was calculated to be ten million deaths and 349.2 million individuals were affected by neurodegenerative disorders. The experimentation and study of these MNPs offer a favorable route toward the development of innovative treatments that target key mechanisms of neurodegeneration, particularly oxidative stress, inflammation, and abnormal protein aggregation. As research on these products remains ongoing, the marine ecosystems continue to offer new compounds with unique chemical and biological properties that are challenging to replicate synthetically. With the high prevalence of neurodegenerative diseases today, it is imperative that the study of these unique and diverse compounds continues and expands.

References

Alexander GE (2004) Biology of Parkinson's disease: pathogenesis and pathophysiology of a multisystem neurodegenerative disorder. Dialogues Clin Neurosci 6(3):259–280
Alghazwi M, Kan YQ, Zhang W, Gai WP, Garson MJ, Smid S (2016) Neuroprotective activities of natural products from marine macroalgae during 1999–2015. J Appl Phycol 28(6):3599–3616
Alonso AD, Cohen LS, Corbo C, Morozova V, ElIdrissi A, Phillips G et al (2018) Hyperphosphorylation of tau associates with changes in its function beyond microtubule stability. Front Cell Neurosci 12:338
Blasco H, Mavel S, Corcia P, Gordon PH (2014) The glutamate hypothesis in ALS: pathophysiology and drug development. Curr Med Chem 21(31):3551–3575
Cacabelos R (2017) Parkinson's disease: from pathogenesis to pharmacogenomics. Int J Mol Sci 18(3):551
Candelise N, Salvatori I, Scaricamazza S, Nesci V, Zenuni H, Ferri A et al (2022) Mechanistic insights of mitochondrial dysfunction in amyotrophic lateral sclerosis: an update on a lasting relationship. Metabolites 12(3):233
Cao Y, Yu F, Lyu Y, Lu X (2022) Promising candidates from drug clinical trials: implications for clinical treatment of Alzheimer's disease in China. Front Neurol 13:1034243
Catanesi M, Caioni G, Castelli V, Benedetti E, d'Angelo M, Cimini A (2021) Benefits under the sea: the role of marine compounds in neurodegenerative disorders. Mar Drugs 19(1):24
Chamorro G, Pérez-Albiter M, Serrano-García N, Mares-Sámano JJ, Rojas P (2006) *Spirulina maxima* pretreatment partially protects against 1-methyl-4-phenyl-1,2,3,6-tetrahydropyridine neurotoxicity. Nutr Neurosci 9(5–6):207–212
Chattopadhyaya I, Gupta S, Mohammed A, Mushtaq N, Chauhan S, Ghosh S (2015) Neuroprotective effect of Spirulina fusiform and amantadine in the 6-OHDA induced Parkinsonism in rats. BMC Complement Altern Med 15(1):296
Chu L, Huang J, Muhammad M, Deng Z, Gao J (2020) Genome mining as a biotechnological tool for the discovery of novel marine natural products. Crit Rev Biotechnol 40(5):571–589
Damsker JM, Hansen AM, Caspi RR (2010) Th1 and Th17 cells: adversaries and collaborators. Ann N Y Acad Sci 1183(1):211–221
Delpierre C, Lefèvre T (2023) Precision and personalized medicine: what their current definition says and silences about the model of health they promote. Implication for the development of personalized health. Front Sociol 8:1112159
Fakhri S, Yarmohammadi A, Yarmohammadi M, Farzaei MH, Echeverria J (2021) Marine natural products: promising candidates in the modulation of gut-brain Axis towards Neuroprotection. Mar Drugs 19(3):165
Feng CW, Hung HC, Huang SY, Chen CH, Chen YR, Chen CY et al (2016) Neuroprotective effect of the marine-derived compound 11-Dehydrosinulariolide through DJ-1-related pathway in in vitro and in vivo models of Parkinson's disease. Mar Drugs 14(10):187
Hine CM (2012) NRF2 and the phase II response in acute stress resistance induced by dietary restriction. J Clin Exp Pathol:s4. [cited 2025 Nov 25] Available from: https://www.omicsonline.org/nrf2-and-the-phase-ii-response-in-acute-stress-resistance-induced-by-dietary-restriction-2161-0681.S4-004.php?aid=7329
Hu D, Jin Y, Hou X, Zhu Y, Chen D, Tai J et al (2023) Application of marine natural products against Alzheimer's disease: past, present and future. Mar Drugs 21(1):43
Islam MN, Rauf A, Fahad FI, Emran TB, Mitra S, Olatunde A et al (2022) Superoxide dismutase: an updated review on its health benefits and industrial applications. Crit Rev Food Sci Nutr 62(26):7282–7300
Ito N, Seki S, Ueda F (2018) The protective role of Astaxanthin for UV-induced skin deterioration in healthy people—a randomized, double-blind, placebo-controlled trial. Nutrients 10(7):817
Jia C, Chai J, Zhang S, Sun Y, He L, Sang Z et al (2025) The advancements of marine natural products in the treatment of Alzheimer's disease: a study based on cell and animal experiments. Mar Drugs 23(3):91

Jiménez C (2018) Marine natural products in medicinal chemistry. ACS Med Chem Lett 9(10):959–961
Johnson JA, Johnson DA, Kraft AD, Calkins MJ, Jakel RJ, Vargas MR et al (2008) The Nrf2–ARE pathway: an indicator and modulator of oxidative stress in Neurodegeneration. Ann N Y Acad Sci 1147(1):61–69
Kaplan B, Ratner V, Haas E (2003) A-Synuclein: its biological function and role in neurodegenerative diseases. J Mol Neurosci 20(2):83–92
Keough MJ, Chernoff H (1987) Dispersal and population variation in the bryozoan Bugula Neritina. Ecology 68(1):199–210
Khalifa SAM, Elias N, Farag MA, Chen L, Saeed A, Hegazy MEF et al (2019) Marine natural products: a source of novel anticancer drugs. Mar Drugs 17(9):491
Khangura RK, Bali A, Jaggi AS, Singh N (2017) Histone acetylation and histone deacetylation in neuropathic pain: an unresolved puzzle? Eur J Pharmacol 795:36–42
Kim H, Ahn JH, Song M, Kim DW, Lee TK, Lee JC et al (2019) Pretreated fucoidan confers neuroprotection against transient global cerebral ischemic injury in the gerbil hippocampal CA1 area via reducing of glial cell activation and oxidative stress. Biomed Pharmacother 109:1718–1727
Kwak JH, Yang Z, Yoon B, He Y, Uhm S, Shin HC et al (2015) Blood-brain barrier-permeable fluorone-labeled dieckolinhibitors of s acting as neuronal ER streg inhibitors. Biomaterials 61:52–60
Lamptey RNL, Chaulagain B, Trivedi R, Gothwal A, Layek B, Singh J (2022) A review of the common neurodegenerative disorders: current therapeutic approaches and the potential role of Nanotherapeutics. Int J Mol Sci 23(3):1851
Li Q, Che H, Wang C, Zhang L, Ding L, Xue C et al (2019) Cerebrosides from sea cucumber improved a β_{1-42} -induced cognitive deficiency in a rat model of Alzheimer's disease. Mol Nutr Food Res 63(5):1800707
Liang X, Luo D, Luesch H (2019) Advances in exploring the therapeutic potential of marine natural products. Pharmacol Res 147:104373
Lin YC, Su JH, Lin SC, Chang CC, Hsia TC, Tung YT et al (2018) A soft coral-derived compound, 11-Dehydrosinulariolide, induces G2/M cell cycle arrest and apoptosis in small cell lung cancer. Mar Drugs 16(12):479
Liu L, Tong H, Sun Y, Chen X, Yang T, Zhou G et al (2023) Huntingtin interacting proteins and pathological implications. Int J Mol Sci 24(17):13060
Ma YF, Wu ZH, Gao M, Loor JJ (2018) Nuclear factor erythroid 2-related factor 2 antioxidant response element pathways protect bovine mammary epithelial cells against H2O2-induced oxidative damage in vitro. J Dairy Sci 101(6):5329–5344
Mandelkow EM, Mandelkow E (2012) Biochemistry and cell biology of tau protein in neurofibrillary degeneration. Cold Spring Harb Perspect Med 2(7):a006247
Mayer AMS, Rodríguez AD, Berlinck RGS, Fusetani N (2011) Marine pharmacology in 2007–8: marine compounds with antibacterial, anticoagulant, antifungal, anti-inflammatory, antimalarial, antiprotozoal, antituberculosis, and antiviral activities; affecting the immune and nervous system, and other miscellaneous mechanisms of action. Comp Biochem Physiol Part C Toxicol Pharmacol 153(2):191–222
Menaa F, Wijesinghe U, Thiripuranathar G, Althobaiti NA, Albalawi AE, Khan BA et al (2021) Marine algae-derived bioactive compounds: a new wave of Nanodrugs? Mar Drugs 19(9):484
Mohd Sairazi NS, Sirajudeen KNS (2020) Natural products and their bioactive compounds: neuroprotective potentials against neurodegenerative diseases. Satou T, editor. Evid Based Complement Alternat Med 2020(1):6565396
Naik B, Richa S, Bharadwaj S, Mishra S, Kumar V, Kumar V et al (2024) Utilizing marine algal metabolites to fight neurodegenerative diseases. Front Mar Sci 11:1370839
Ngo V, Duennwald ML (2022) Nrf2 and oxidative stress: a general overview of mechanisms and implications in human disease. Antioxidants. 11(12):2345
Porter AG, Jänicke RU (1999) Emerging roles of caspase-3 in apoptosis. Cell Death Differ 6(2):99–104

Provenzano F, Torazza C, Bonifacino T, Bonanno G, Milanese M (2023) The key role of astrocytes in amyotrophic lateral sclerosis and their commitment to glutamate Excitotoxicity. Int J Mol Sci 24(20):15430

Rahman MH, Bajgai J, Fadriquela A, Sharma S, Trinh TT, Akter R et al (2021) Therapeutic potential of natural products in treating neurodegenerative disorders and their future prospects and challenges. Molecules 26(17):5327

Riancho J, Delgado-Alvarado M, Andreu MD, Paz-Fajardo L, Arozamena S, Gil-Bea FJ et al (2021) Amyotrophic lateral sclerosis (ALS), cancer, autoimmunity and metabolic disorders: an unsolved tantalizing challenge. Br J Pharmacol 178(6):1269–1278

Sanghai N, Tranmer GK (2021) Hydrogen peroxide and amyotrophic lateral sclerosis: from biochemistry to pathophysiology. Antioxidants 11(1):52

Sehar U, Rawat P, Reddy AP, Kopel J, Reddy PH (2022) Amyloid Beta in aging and Alzheimer's disease. Int J Mol Sci 23(21):12924

Shan C, Zhang C, Zhang C (2024) The role of IL-6 in neurodegenerative disorders. Neurochem Res 49(4):834–846

Shimazawa M, Nakajima Y, Mashima Y, Hara H (2009) Docosahexaenoic acid (DHA) has neuroprotective effects against oxidative stress in retinal ganglion cells. Brain Res 1251:269–275

Shimizu H, Koyama T, Yamada S, Lipton SA, Satoh T (2015) Zonarol, a sesquiterpene from the brown algae Dictyopteris undulata, provides neuroprotection by activating the Nrf2/ARE pathway. Biochem Biophys Res Commun 457(4):718–722

Silva J, Alves C, Pinteus S, Mendes S, Pedrosa R (2018) Neuroprotective effects of seaweeds against 6-hydroxidopamine-induced cell death on an in vitro human neuroblastoma model. BMC Complement Altern Med 18(1):58

Silva J, Alves C, Soledade F, Martins A, Pinteus S, Gaspar H et al (2023) Marine-derived components: can they be a potential therapeutic approach to Parkinson's disease? Mar Drugs 21(8):451

Simmons TL, Andrianasolo E, McPhail K, Flatt P, Gerwick WH (2005) Marine natural products as anticancer drugs. Mol Cancer Ther 4(2):333–342

Singh R, Parihar P, Singh M, Bajguz A, Kumar J, Singh S et al (2017) Uncovering potential applications of cyanobacteria and algal metabolites in biology, agriculture and medicine: current status and future prospects. Front Microbiol 8:515

Strafella C, Caputo V, Galota MR, Zampatti S, Marella G, Mauriello S et al (2018) Application of precision medicine in neurodegenerative diseases. Front Neurol 9:701

Suk TR, Rousseaux MWC (2020) The role of TDP-43 mislocalization in amyotrophic lateral sclerosis. Mol Neurodegener 15(1):45

Sukhanova MV, Singatulina AS, Pastré D, Lavrik OI (2020) Fused in sarcoma (FUS) in DNA repair: tango with poly(ADP-ribose) polymerase 1 and compartmentalisation of damaged DNA. Int J Mol Sci 21(19):7020

Thomas N, Kim SK (2013) Beneficial effects of marine algal compounds in Cosmeceuticals. Mar Drugs 11(1):146–164

Tian Z, Lu XT, Jiang X, Tian J (2023) Bryostatin-1: a promising compound for neurological disorders. Front Pharmacol 14:1187411

Wang S, Qi X (2022) The putative role of Astaxanthin in Neuroinflammation modulation: mechanisms and therapeutic potential. Front Pharmacol 13:916653

Wang K, Song F, Fernandez-Escobar A, Luo G, Wang JH, Sun Y (2018) The properties of cytokines in multiple sclerosis: pros and cons. Am J Med Sci 356(6):552–560

Wu K, Gong Z, Zou L, Ye H, Wang C, Liu Y et al (2019) Sargassum integerrimum inhibits oestrogen deficiency and hyperlipidaemia-induced bone loss by upregulating nuclear factor (erythroid-derived 2)-like 2 in female rats. J Orthop Transl 19:106–117

Yang WN, Chen PW, Huang CY (2017) Compositional characteristics and in vitro evaluations of antioxidant and Neuroprotective properties of crude extracts of Fucoidan prepared from compressional puffing-Pretreated Sargassum crassifolium. Mar Drugs 15(6):183

Ye P, Li P, Yang W, Zhao Y, Zhao Y, Sun K et al (2019) Structure and Neuroprotective effect of polysaccharide from viscera autolysates of squid Ommastrephes bartrami. Mar Drugs 17(3):188

Yi LX, Tan EK, Zhou ZD (2024) Tyrosine hydroxylase inhibitors and dopamine receptor agonists combination therapy for Parkinson's disease. Int J Mol Sci 25(9):4643

Zhang Y, Bhavnani BR (2006) Glutamate-induced apoptosis in neuronal cells is mediated via caspase-dependent and independent mechanisms involving calpain and caspase-3 proteases as well as apoptosis inducing facequine estrogens inhibit this processuine estrogens. BMC Neurosci 7(1):49

Zhang L, Hao J, Zheng Y, Su R, Liao Y, Gong X et al (2018) Fucoidan protects dopaminergic neurons by enhancing the mitochondrial function in a rotenone-induced rat model of Parkinson's disease. Aging Dis 9(4):590

Zhang F, Zhong R-j, Cheng C, Li S, Le W-d (2021) New therapeutics beyond amyloid-β and tau for the treatment of Alzheimer's disease. Acta Pharmacol Sin 42(9):1382–1389

Zhang K, Guan X, Zhang X, Liu L, Yin R, Jiang T (2022) Protective effects of marine alkaloid Neolamellarin a derivatives against glutamate induced PC12 cell apoptosis. Mar Drugs 20(4):262

Marine-Derived Natural Bioactives for the Treatment of Stroke

12

K. Reeta Vijayarani, M. Keerthiga,
A. Theresa Thenisaimozhi, S. Anbazhagan,
R. Ramadeepa Priya, Yaman Walid Kassab, and Nishi Ann

Abstract

Stroke remains one of the leading global causes of mortality and long-term disability, with limited therapeutic options available for effective neuroprotection and neural regeneration. In recent years, marine ecosystems have emerged as a promising reservoir of bioactive compounds, distinguished by their unique molecular architectures and potent pharmacological activities. This review explores the therapeutic potential of naturally occurring bioactives derived from marine organisms—including algae, sponges, mollusks, and marine bacteria—such as carotenoids, polysaccharides, peptides, alkaloids, and phenolic compounds, aminosulfonate compounds, in the context of stroke treatment. These marine-derived agents exhibit critical biological functions, including anti-inflammatory, antioxidant, anti-apoptotic, and neuro-regenerative effects, which are essential for mitigating ischemic brain injury. The strategic integration of marine bioprospecting with contemporary drug development platforms holds considerable promise for advancing novel and efficacious interventions in stroke therapy.

K. R. Vijayarani · M. Keerthiga · S. Anbazhagan · R. R. Priya (✉)
Department of Pharmaceutics, Surya School of Pharmacy, Vikravandi, Villupuram. Affiliated to The TN Dr. M.G.R. Medical University, Chennai, Tamil Nadu, India

A. T. Thenisaimozhi
Central institute of Petrochemical Engineering and Technology, Anna University, Guindy, Chennai, Tamil Nadu, India

Y. W. Kassab
College of Pharmacy, National University of Science and Technology, Muscat, Oman

College of Pharmacy, Al-Bayan University, Baghdad, Iraq

N. Ann
Department of Zoology, Providence Women's College (Autonomous), Calicut, Kerala, India

M. Bhupathyraaj et al. (eds.), *Neuroprotective Marine Bioactives*, Nutritional Neurosciences, https://doi.org/10.1007/978-981-95-5791-2_12

Keywords

Fucoxanthin · Astaxanthin · Fucoidan · Laminarin · Phlorotannins · Xyloketal B · Manzamine A · Conantokins · Fascaplysin · Tramiprosate · NP04634

12.1 Introduction

Stroke is the world's second most common cause of death and a major contributor to disability. Ischemic stroke is the most frequent type of stroke, and its prevalence is highest in developing nations. An abrupt neurological outburst brought on by decreased blood vessel perfusion to the brain is known as a stroke (Kuriakose and Xiao 2020). The fact that only about 25% of patients have a stroke within 3 years of having a transient ischemic attack [TIA] is significant. Given that 75% of TIA patients will not experience a stroke, it is evident that any stroke prevention treatment must be somewhat safe because many patients who are not at risk will nonetheless receive treatment (Patterson Jr 1988). Although the presence of these attacks is a powerful risk for stroke occurrence, their low prevalence and the lack of clearly effective medical or surgical treatment make it unlikely that management of transient focal cerebral ischemic attacks accounts for much of the decline in the incidence or mortality of stroke (Whisnant 1983). There are significant differences in how acute transient ischemic attacks [TIAs] are managed; some hospitals admit all patients, while others conduct outpatient assessments (Johnston et al. 2000).

The exploration of marine pharmacognosy in stroke research not only expands the neurotherapeutic landscape but also exemplifies the value of interdisciplinary approaches in drug discovery. As advances in marine biotechnology, synthetic analog development, and high-throughput screening continue to evolve, the integration of marine natural products into stroke pharmacotherapy holds promise for novel, mechanism-based interventions.

Marine ecosystems have become a rich and little-studied source of physiologically powerful and structurally varied natural compounds in recent years. To live in harsh and competitive conditions, marine organisms such as sponges, tunicates, algae, and their symbiotic microorganisms have developed special chemical defense systems. A wide variety of secondary metabolites have emerged as a result of these adaptations, many of which have pharmacological properties different from those of their terrestrial counterparts.

12.2 Carotenoid

Marine ecosystems represent a vast and underexplored reservoir of bioactive compounds, among which carotenoids occupy a prominent position due to their distinctive molecular structures and potent antioxidant properties. These naturally occurring pigments are biosynthesized by a wide range of autotrophic marine organisms including microalgae, macroalgae, cyanobacteria, and certain fungi and are also found in heterotrophic marine species through dietary accumulation or metabolic modification.

12.2.1 Fucoxanthin

Fucoxanthin [C42H58O6] is a carotenoid marine xanthophyll that is found in brown seaweed such as *Undaria pinnatifida* and *Laminaria japonica* (Ferdous et al. 2024). Edible brown seaweeds are the source of fucoxanthin, the most prevalent marine carotenoid, which has been utilized as a functional component in the treatment of obesity. However, fucoxanthin's low brain bioavailability has restricted its use in the treatment of neurological conditions (Wang et al. 2022). It is a naturally occurring antioxidant pigment found in diatoms and brown macroalgae. It makes up 10% of all the carotenoids found in nature. Numerous scientific studies back up the potential advantages of using fucoxanthin in pharmaceutical and nutraceutical forms to improve human health and control illness (Mohibbullah et al. 2022). It is an orange pigment that makes up over 10% of all the carotenoids found in nature (Mohibbullah et al. 2022). It has been observed that fucoxanthin activates the nuclear factor-E2-related factor 2 [Nrf2] pathway and inhibits the nuclear factor-κB [NF-κB] pathway to generate neuroprotection against Alzheimer's disease and Parkinson's disease (Wang et al. 2022). Many Western countries currently employ fucoxanthin as a functional substance to treat obesity. Fucoxanthin, on the other hand, is rapidly digested by the body and is unstable under light. As a result, the levels of fucoxanthin that have been examined and found to have neuroprotective effects in animals can be rather high. Additionally, fucoxanthin's poor brain bioavailability has hindered its use as a medication to treat neurological disorders (Kotake-Nara et al. 2015).

12.2.2 Astaxanthin

It has been demonstrated that the natural antioxidant carotenoid astaxanthin lessens cerebral ischemia injury in animals. Nevertheless, no research has explicitly examined whether astaxanthin administered prophylactically can guard against cerebral ischemia. Examining if pretreatment with astaxanthin can shield adult rats from ischemia damage was the aim of this investigation. For 7 days, the rats received astaxanthin intragastrically once daily. One hour after the last treatment, the middle cerebral artery was occluded. Astaxanthin was discovered to lower the volume of cerebral infarction and avoid neurological impairments (Pan et al. 2017). According to tests of the pure compounds' ability to scavenge free radicals, astaxanthin is 5–15 times more powerful as an antioxidant than lutein and β-carotene, while having structural similarities (Sasaki et al. 2011). A dietary carotenoid called astaxanthin [ATX] has demonstrated potent anti-inflammatory effects in a number of tissue injury models (Zhang et al. 2014). As a xanthophyll carotenoid, astaxanthin is found in quail, shellfish, *Haematococcus pluvialis*, and certain complex plants. It has several pharmacological characteristics, such as neuroprotective, anti-inflammatory, antioxidative, and anti-apoptotic actions (Nai et al. 2018).

12.3 Polysaccharides

Marine ecosystems comprising algae, invertebrates, and microorganisms serve as a rich and underexplored reservoir of structurally diverse and biologically active polysaccharides. These high-molecular-weight macromolecules, such as alginate, fucoidan, laminarin, carrageenan, chitosan, and glycosaminoglycans, exhibit distinct monosaccharide compositions, glycosidic linkages, branching architectures, and functional substitutions that endow them with a broad spectrum of physicochemical and pharmacological properties. Their molecular complexity and bioactivity contribute to a wide array of therapeutic functions, including antioxidant, anti-inflammatory, anticoagulant, antitumor, antiviral, and immunomodulatory effects, making them valuable candidates for pharmaceutical, nutraceutical, and biomedical applications.

12.3.1 Fucoidan

Fucoidan, a sulfated polysaccharide extracted primarily from brown algae such as *Fucus vesiculosus*, *Laminaria digitata*, and *Sargassum* species, has drawn attention for its wide-ranging therapeutic properties. This review highlights the structural diversity, extraction techniques, and pharmacological activities of fucoidan, including its antidiabetic, antioxidant, anti-inflammatory, anticancer, immunomodulatory, antiviral, anticoagulant, hepatoprotective, and neuroprotective effects. The multifunctional nature of fucoidan, driven by its unique chemical structure and high sulfate content, supports its potential application in the development of novel pharmaceuticals and functional foods. Brown algae are divided into 20 groups, with Phaeophyceae accounting for nearly 1800 species and 66% of total algal consumption (Holdt and Kraan 2011).

Sargassum, Laminaria, Fucus, and Undaria are brown algae species that have been extensively researched for their potential medical benefits, which include anti-inflammatory, antidiabetic, antioxidant, and anti-obesity activities (Bringloe et al. 2020). It has been demonstrated to shield neuronal cells against oxidative stress and toxicity caused by amyloid-β, two major causes of neurodegenerative illnesses like Alzheimer's. By blocking cytochrome c release and inhibiting the activation of caspase-3 and caspase-9, fucoidan reduced apoptosis and prevented mitochondrial damage in PC12 cells exposed to amyloid-β [A β] (Elendu et al. 2023; Jiwa et al. 2010; Anand and Dhikav 2012; Kwon et al. 2023; Kim et al. 2012; Min et al. 2015; Calder 2016; Bazan 2006, 2007, 2009; Akbar et al. 2005) and D-galactose. Furthermore, fucoidan decreased reactive oxygen species [ROS] levels while increasing the expression of antioxidant enzymes such as glutathione [GSH] and superoxide dismutase [SOD]. Studies on animals also showed that it can enhance cognitive performance by lowering oxidative damage in brain tissues and restoring cholinergic activity. According to these results, fucoidan may have therapeutic advantages for the treatment or prevention of neurodegenerative diseases (Wei et al. 2017).

12.3.2 Laminarin

Laminarin, a β-glucan polysaccharide predominantly found in brown algae, has garnered attention as a marine-derived nutraceutical with multifaceted pharmacological properties. Recent studies suggest that laminarin may exert neuroprotective effects by attenuating neuroinflammatory responses implicated in ischemic brain injury (Luo et al. 2022). Seaweed-derived polysaccharides, including laminarin, are increasingly valued for their accessibility, cost-effective extraction, and diverse biological activities, encompassing antioxidant, antibacterial, anti-inflammatory, immunostimulatory, antitumor, anticoagulant, anti-obesity, antidiabetic, and wound-healing effects (Pramanik et al. 2024; Karuppusamy et al. 2022). In particular, laminarin has demonstrated potential in mitigating ischemia-induced neuronal damage. A recent investigation reported that adult gerbils pretreated with 50 mg/kg of laminarin once daily for 7 consecutive days exhibited significant neuroprotection against transient forebrain ischemia/reperfusion [IR] injury (Park et al. 2020).

Despite these promising findings, the prophylactic efficacy of laminarin in ischemic stroke remains insufficiently characterized. To address this gap, a subsequent study examined its neuroprotective effects in the cornu ammonis 1 [CA1] region of the hippocampus—a site highly susceptible to ischemia-reperfusion injury—following 5 min of transient forebrain ischemia [TFI], thereby providing further evidence of laminarin's therapeutic potential in cerebral ischemic contexts (Lee et al. 2020).

12.4 Phenolic Compound

Phenolic compounds, a diverse class of secondary metabolites found abundantly in plants and marine organisms, have garnered significant attention for their neuroprotective potential in stroke management. Phenolic compounds—including flavonoids, phenolic acids, stilbenes, and tannins—exert a wide range of biological effects through their antioxidant, anti-inflammatory, anti-apoptotic, and vascular-modulatory properties. These compounds can scavenge reactive oxygen species [ROS], inhibit pro-inflammatory cytokine production, modulate signaling pathways such as NF-κB and MAPK, and preserve mitochondrial integrity, thereby attenuating ischemia-induced neuronal injury.

The brown seaweed *Sargassum polycystum* is abundant in polyphenols, carotenoids, and polysaccharides. Additionally, the high phenolic content of *S. polycystum* is noteworthy (Catarino et al. 2023). Antioxidant activity, total flavonoid concentration, and total phenolic content were all high in *S. polycystum*.

Consequently, it seems that the extract from *S. polycystum* might have potent antioxidant qualities, most likely as a result of the phenolic compounds' great antioxidant capacities. According to earlier research, phenolic chemicals may also have strong anti-inflammatory, antioxidant, and neuroprotective properties that aid in the treatment of a number of illnesses, including memory loss (Tavan et al. 2024; Elendu et al. 2023). Therefore, using an MCAO rat model of ischemic stroke, we assessed the potential neuroprotective and antioxidant effects of *S. polycystum* extract, which has a high phenolic content. According to earlier research, cerebral ischemia brought on by the MCAO experimental model might alter blood vessels or interfere with blood flow, which can leave the brain with insufficient blood flow and impair cognitive function (Jiwa et al. 2010).

In experimental assessments of cognitive function, *Sargassum polycystum* extract demonstrated significant neuro-enhancing effects. In the Morris water maze test, administration of the extract resulted in improved spatial memory, as evidenced by prolonged retention times and reduced escape latency. Similarly, performance in the novel object recognition test indicated enhanced recognition memory. These cognitive benefits were consistently observed both prior to and following middle cerebral artery occlusion [MCAO], suggesting that *S. polycystum* may exert protective and restorative effects against ischemia-induced cognitive decline. Given the central role of the hippocampus in regulating key cognitive domains—including processing speed, working memory, executive function, and episodic memory—the observed improvements are likely attributable to hippocampal preservation and functional enhancement (Anand and Dhikav 2012). These findings support the potential of *S. polycystum* as a marine-derived therapeutic agent for mitigating post-stroke cognitive impairment.

12.5 Tannins

Tannins, a diverse group of polyphenolic compounds, are traditionally recognized for their astringent properties and are widely found in terrestrial plants. However, recent research has highlighted the marine environment as an untapped reservoir of bioactive tannins with unique structural features and potent biological activities. Marine-derived tannins, particularly phlorotannins found in brown algae, exhibit strong antioxidant, anti-inflammatory, and neuroprotective effects, making them promising candidates for stroke intervention.

12.5.1 Phlorotannins

The unique class of polyphenolic chemicals known as phenolotannins is only biosynthesized by brown sea algae, of which *Ecklonia cava* is known to be an exceptionally abundant source. Because of their high degree of polymerization and the presence of structural motifs like ether bonds, dibenzodioxin rings, and dibenzofuran connections that connect phloroglucinol units, eckol, dieckol, and phlorofucofuroeckol A [PFF-A] stand out among the others in terms of their strong antioxidant activity. Effective neutralization of reactive oxygen species is made possible by these molecular characteristics, which also confer increased redox potential and inherent stability. Both crude extracts and purified phlorotannins from brown algae have demonstrated significant antibacterial and anti-inflammatory activities in addition to their antioxidative qualities (Kwon et al. 2023).

A rat model of transient focal cerebral ischemia caused by middle cerebral artery occlusion [MCAO] was used to assess the neuroprotective effectiveness of a phlorotannin-rich extract from *E. cava*, standardized to 98.5% phloroglucinol equivalents. Doses of 0, 10, and 50 mg/kg of the extract were injected intraperitoneally 30 min before to and after surgical ischemia induction. The results showed a dose-dependent decrease in brain edema, neuronal death, and cerebral infarct volume. In particular, apoptotic cell counts in the cerebral cortex and striatum were dramatically reduced by the 10 mg/kg dose, and infarct size and brain water content were significantly reduced by the 50 mg/kg dose. Furthermore, between days two and six after ischemia, mice that received the extract showed improved neurological outcomes, underscoring the promise of phlorotannin-enriched marine extracts as a treatment for ischemic stroke and potentially other neurodegenerative diseases (Kim et al. 2012).

12.6 Phycobiliproteins

Phycobiliproteins [PBPs] are a class of highly water-soluble, pigmented proteins predominantly found in cyanobacteria, red algae [Rhodophyta], and cryptomonads, where they serve as accessory light-harvesting pigments in photosynthesis. These proteins are integral components of the phycobilisome, a supramolecular complex located on the outer surface of thylakoid membranes, which efficiently captures and transfers light energy to chlorophyll a in photosystem II, particularly in spectral regions poorly absorbed by chlorophyll [500–660 nm].

12.6.1 C-Phycocyanin [C-PC]

Green microalgae contain a biliprotein called C-phycocyanin [C-Pc], which has been shown to have strong antioxidant qualities and to provide neuroprotection in oxidative brain injury models. Although astrocytes, which make up a significant amount of the central nervous system, are essential for reducing oxidative stress, it is yet unclear how C-Pc affects astrocytic antioxidant responses. Rats undergoing middle cerebral artery occlusion [MCAO] showed notable neuroprotective effects, such as a decrease in infarct volume and an improvement in behavioral deficits, when C-Pc was administered intranasally. Without causing damage in healthy astrocytes or neurons, C-Pc therapy increased the viability and proliferation of oxidatively challenged astrocytes while also inhibiting apoptosis and the production of reactive oxygen species [ROS].

The underlying mechanisms were further clarified by creating a three-dimensional model of oxidized astrocytes. Important antioxidant enzymes like superoxide dismutase [SOD] and catalase, as well as neurotrophic factors like brain-derived neurotrophic factor [BDNF] and nerve growth factor [NGF], were discovered to be upregulated by C-Pc. At the same time, it decreased glial scar formation and inhibited the expression of pro-inflammatory cytokines IL-6 and IL-1β. Additionally, in the 3D model, C-Pc enhanced the viability of oxidized neurons. These results collectively imply that C-Pc enhances antioxidative, neurotrophic, and anti-inflammatory pathways in a synergistic manner by activating oxidized astrocytes to support neuroprotection and repair in ischemic brain tissue (Min et al. 2015).

12.7 Omega-3 Fatty Acids

Stroke remains a leading cause of global mortality and long-term neurological disability, with limited therapeutic options for effective neuroprotection and postischemic recovery. Among emerging nutritional and pharmacological interventions, omega-3 polyunsaturated fatty acids [PUFAs]—particularly eicosapentaenoic acid [EPA] and docosahexaenoic acid [DHA]—have garnered significant attention for their cerebrovascular and neuroprotective properties. These long-chain PUFAs are predominantly sourced from marine organisms, including oily fish [e.g., salmon, mackerel, sardines], microalgae, and krill, and play essential roles in maintaining neuronal membrane integrity, modulating inflammatory responses, and regulating oxidative stress.

12.7.1 Docosahexaenoic Acid

Various kinds of n-3 supplements and seafood, particularly fatty fish, contain relatively high levels of DHA (Calder 2016). In the circulation, adipose tissues, and cell membranes, DHA is esterified into complex lipids. Docosahexaenoic acid [DHA; 22:6, n-3] and other omega-3 polyunsaturated fatty acids have attracted a lot of interest among the several endogenous chemicals linked to neuroprotection. DHA is widely distributed throughout the central nervous system and is an essential structural element of phospholipids found in neuronal membranes (Bazan 2006). Cold-water fatty fish including salmon, tuna, mackerel, sardines, shellfish, and herring are the main source of this nutrient. DHA has a complex involvement in photoreceptor function, synaptic plasticity, neurodevelopment, and cognitive maintenance (Bazan 2007; Bazan 2009; Akbar et al. 2005).

NPD1 [NPD1: 10R,17S-dihydroxy docosa-4Z,7Z,11E,13E,15Z,19Z-hexaenoic acid], a bioactive docosanoid derived from DHA, has been identified as a critical modulator of neuroprotective signaling during ischemia-reperfusion and oxidative stress in recent findings (Marcheselli et al. 2003; Bazan 2005; Mukherjee et al. 2004; Lukiw et al. 2005). NPD1 has pleiotropic properties, such as promoting neuronal survival and reducing neuro inflammation (Serhan et al. 2008).

DHA treatment provides protection in models of spinal cord injury and cerebral ischemia, according to experimental research. Specifically, it has been demonstrated that low to moderate dosages of DHA enhance neurological and histological results after focal cerebral ischemia (Ward et al. 2010; Belayev et al. 2009). DHA supports neuronal survival and functional restoration by significantly reducing infarct volume and cerebral edema, improving neurobehavioral recovery, and activating NPD1 production in the ischemic penumbra when given within a therapeutic window of up to 5 h post-insult (Belayev et al. 2011).

12.8 Peptides

Marine ecosystems, encompassing a vast diversity of organisms across multiple phyla, have emerged as a prolific source of structurally unique and biologically active peptides. These low-molecular-weight compounds are biosynthesized by marine bacteria, fungi, algae, sponges, mollusks, and tunicates, often as part of adaptive responses to extreme environmental pressures such as salinity, temperature fluctuation, and microbial competition. As a result, marine-derived peptides exhibit distinctive sequence motifs, posttranslational modifications, and three-dimensional conformations that confer potent and selective bioactivity.

12.8.1 Conantokins

The venom of Conus sea snails contains a special class of peptide neurotoxins called conantokins, which are distinguished by the presence of γ-carboxyglutamate residues that are essential to their antagonistic action at N-methyl-D-aspartate receptors [NMDARs]. Among these, Conantokin-G [con-G], a gene product of *Conus geographus*, selectively binds to the GluN2B subunit and acts as a noncompetitive NMDAR antagonist. Con-G is now a viable option for neurotherapeutic intervention due to this subunit-specific inhibition. Con-G treatment dramatically reduced ischemia-induced neurological impairments and maintained neuronal architecture in the peri-infarct area in a rat model of middle cerebral artery occlusion [MCAO]. Con-G restored the ischemia-disrupted location of GluN1 and GluN2B subunits inside neuronal somata, according to histological investigations, indicating that its neuroprotective effects are mediated through NMDAR signaling regulation.

Additionally, con-G therapy decreased neuronal degeneration and calcium microdeposits, suggesting a reduction in intracellular Ca2+ overload brought on by ischemia, a major cause of excitotoxicity and cell death. These results highlight how GluN2B-targeted conantokins may be used therapeutically to reduce excitotoxic damage and aid in the healing process after an ischemic stroke (Balsara et al. 2015).

12.9 Alkaloids

Marine ecosystems, encompassing a vast diversity of organisms such as sponges, tunicates, mollusks, algae, and marine microorganisms, have emerged as prolific sources of structurally diverse and biologically potent alkaloids. These nitrogen-containing secondary metabolites exhibit a wide range of chemical scaffolds, including indole, quinoline, β-carboline, pyrrole-imidazole, and halogenated derivatives, many of which are unique to marine environments due to the ecological pressures of salinity, temperature, and microbial competition.

Marine alkaloids have demonstrated a broad spectrum of pharmacological activities, including antitumor, antiviral, antimicrobial, anti-inflammatory, neuroprotective, and cardiovascular-modulatory effects. Their mechanisms of action often involve enzyme inhibition, receptor modulation, DNA intercalation, and signal transduction interference, making them attractive candidates for drug development across multiple therapeutic domains.

12.9.1 Xyloketal B

Because xyloketal B works well in hypoxia-ischemia models, it may be used to treat hypoxic-ischemic brain injury (Xiao et al. 2014). A bioactive marine substance called xyloketal B was extracted from the south China Sea-native fungus Xylaria species, which is sourced from mangroves. Its diverse pharmacological qualities, including as strong anti-inflammatory, anti-apoptotic, and antioxidant effects, have been brought to light by mounting evidence over the last 10 years, making it a viable option for the treatment of ischemic stroke.

It controls important inflammatory and apoptotic processes, particularly by suppressing the expression of Toll-like receptor 4 [TLR4], upregulating heme oxygenase-1 [HO-1], inhibiting NADPH oxidase subunits, and increasing the Bcl-2/Bax ratio. When given within the therapeutic window of 1–2 h after the insult, xyloketal B dramatically lowers infarct volume and improves neurological outcomes, according to in vivo studies employing middle cerebral artery occlusion [MCAO] models. Reactive oxygen species [ROS] formation is attenuated, and the TLR4/NF-κB signaling cascade is inhibited, which results in decreased expression of pro-inflammatory cytokines such as IL-1β, TNF-α, and IL-6. Furthermore, xyloketal B has demonstrated effectiveness in neonatal hypoxic-ischemic models by promoting neuronal survival, lowering calcium influx, and reducing caspase-3 activation.

Furthermore, xyloketal B has demonstrated effectiveness in neonatal hypoxic-ischemic models by promoting neuronal survival, lowering calcium influx, and reducing caspase-3 activation. When taken as a whole, our results highlight the potential of xyloketal B as a novel marine-derived neuroprotectant that can target several ischemia-related pathways to reduce neuronal damage and improve recovery following a stroke (Gong et al. 2018).

12.9.2 Manzamine A

Manzamine A is a structurally intricate β-carboline alkaloid initially isolated from marine sponges belonging to the Haliclona and Xestospongia genera. It exhibits a broad spectrum of bioactivities, notably anti-inflammatory, antioxidant, and kinase-inhibitory effects, which align with key pathological mechanisms involved in ischemic stroke. While direct experimental evidence assessing its efficacy in stroke models remains limited, the compound's pharmacodynamic profile indicates promising neuroprotective potential in the context of cerebral ischemia.

The neuroprotective qualities of manzamine A, a β-carboline alkaloid that was discovered from marine sponges including Haliclona and Xestospongia species, have drawn interest due to its possible connection to ischemic stroke. Marine sponges are the source of the manzamines, a special class of alkaloids that include β-carboline and have an odd polycyclic structure (Hu et al. 2003). With their distinctive heterocyclic system, manzamine alkaloids are a special class of alkaloids that have a wide range of bioactivities, such as cytotoxicity, antimicrobial, antimalarial, antiviral, anti-inflammatory, anti-insecticidal, and proteasome inhibitory activity (Kubota et al. 2020).

These substances have been demonstrated to suppress β-site amyloid precursor protein [APP] cleaving enzyme 1 [BACE1], alter the production or function of certain neurotransmitters like glutamate and acetylcholinesterase, increase serotonin, lower oxidative stress, suppress kinases and proteases, and promote neurite growth (Alghazwi et al. 2016).

12.9.3 Fascaplysin

Because it selectively inhibits cyclin-dependent kinase 4 [CDK4], a target linked to neuronal apoptosis and cell cycle dysregulation after ischemic stroke, the marine-derived alkaloid known as "fascaplysin," which was isolated from the sponge Fascaplysinopsis sp., has attracted attention for its neuroprotective potential. Although there are not many direct clinical research on fascaplysin in stroke models, its mechanism is consistent with new neuroprotective techniques.

A significant class of secondary metabolites generated from marine sources, indole alkaloids are found in a wide range of marine sources, including sponges, tunicates, algae, worms, and microbes. Their biological activities have been thoroughly investigated.

Fascaplysin, a bis-indole alkaloid derived from sponges, demonstrated a wide range of bioactivities, including antibacterial, antifungal, antiviral, anti-HIV-1-RTase, p56 tyrosine kinase inhibition, antimalarial, antiangiogenic, and antiproliferative activity against multiple cancer cell lines, as well as specific inhibition of cyclin-dependent kinase-4 and DNA intercalator activity (Bharate et al. 2012).

12.9.4 SMTP-7 [FGFC1]

SMTP-7 [also known as FGFC1] is an isoindolone alkaloid isolated from marine-derived fungi *Stachybotrys longispora* FG216 and *Stachybotrys microspora* IFO 30018, both of which produce a range of bioactive metabolites with thrombolytic, anti-inflammatory, and antioxidative properties (Hang et al. 2022). This compound has demonstrated significant neuroprotective effects in preclinical models of ischemic stroke. Mechanistically, SMTP-7 mitigates oxidative stress and suppresses the secretion of proinflammatory cytokines, thereby reducing cerebral injury (Diao et al. 2023). In rodent models, SMTP-7 enhances plasminogen activation, facilitating thrombolysis and leading to a marked reduction in infarct volume. Its efficacy has also been validated in a nonhuman primate model involving photochemically induced thrombotic occlusion of the middle cerebral artery [MCA], where intravenous administration resulted in improved vascular recanalization, decreased infarct size, enhanced neurological outcomes, and reduced hemorrhagic transformation. Notably, while hemorrhagic risk remains a major limitation of conventional thrombolytic agents such as tissue plasminogen activator [tPA], SMTP-7 appears to exhibit a more favorable safety profile, with reduced hemorrhagic complications under certain experimental conditions (Sawada et al. 2014).

12.10 Aminosulfonate Compound

Marine ecosystems, particularly those inhabited by algae, cyanobacteria, and marine invertebrates, have yielded a diverse array of aminosulfonate compounds, characterized by the presence of both amino and sulfonic acid functional groups. These zwitterionic molecules, such as homotaurine [tramiprosate] and cysteinolic acid, exhibit unique physicochemical properties that contribute to their solubility, bioavailability, and biological activity. Their structural simplicity, combined with functional versatility, has positioned marine-derived aminosulfonates as promising candidates for therapeutic development.

12.10.1 Tramiprosate

Tramiprosate is a low-molecular-weight aminosulfonate compound naturally occurring in several species of red marine algae. Tramiprosate, also known as homotaurine, has shown promising neuroprotective effects in preclinical models of ischemic stroke. A pivotal study demonstrated that tramiprosate significantly reduced infarct volume and improved neurological outcomes in rats subjected to middle cerebral artery occlusion [MCAO], widely used model for cerebral ischemia.

It exerts neuroprotective effects by disrupting the interaction between postsynaptic density protein-95 [PSD-95] and neuronal nitric oxide synthase [nNOS], a key mediator of excitotoxic damage. Additionally, tramiprosate inhibits the translocation of nNOS from the cytosol to the plasma membrane, thereby attenuating nitric oxide-mediated neurotoxicity. This dual mechanism of action is particularly relevant in the context of ischemic stroke, where excitotoxic and oxidative pathways contribute to neuronal injury (Diao et al. 2023).

12.11 Synthetic Derivative

12.11.1 NP04634

NP04634 is a synthetic derivative of 11,19 dideoxyfistularin, a natural product of the Mediterranean sponge Aplysina cavernicola (Valero et al. 2009). It has demonstrated promising *cytoprotective effects* in preclinical models involving calcium overload and mitochondrial dysfunction—two key pathological mechanisms implicated in ischemic stroke.

In a pivotal study using *bovine chromaffin cells*, NP04634 significantly reduced cell death induced by excitotoxic stimuli such as veratridine and mitochondrial inhibitors [rotenone and oligomycin]. The compound attenuated intracellular calcium oscillations and preserved mitochondrial membrane potential, suggesting its potential to *mitigate neuronal injury* under stroke-like conditions (Valero et al. 2009).

NP04634 is a novel compound of marine origin that demonstrates significant cytoprotective activity, particularly in the context of mitochondrial dysfunction. Preclinical evidence suggests that NP04634 may offer therapeutic benefit in ischemic stroke by mitigating mitochondrial injury—a key contributor to neuronal cell death in such pathologies. Its protective effects against calcium overload and mitochondrial disruption underscore its potential as a candidate for neuroprotective intervention in stroke and related neurodegenerative conditions (Diao et al. 2023).

References

Akbar M, Calderon F, Wen Z, Kim HY (2005) Docosahexaenoic acid: a positive modulator of Akt signaling in neuronal survival. Proc Natl Acad Sci 102(31):10858–10863. https://doi.org/10.1073/pnas.0502903102

Alghazwi M, Qi Kan Y, Zhang W, Ping Gai W, Yan XX (2016) Neuroprotective activities of marine natural products from marine sponges. Curr Med Chem 23(4):360–382

Anand KS, Dhikav V (2012) Hippocampus in health and disease: an overview. Ann Indian Acad Neurol 15(4):239–246

Balsara R, Dang A, Donahue DL, Snow T, Castellino FJ (2015) Conantokin-G attenuates detrimental effects of NMDAR hyperactivity in an ischemic rat model of stroke. PLoS One 10(3):e0122840. https://doi.org/10.3390/md13010029

Bazan NG (2005) Neuroprotectin D1 (NPD1): a DHA-derived mediator that protects brain and retina against cell injury-induced oxidative stress. Brain Pathol 15(2):159–166. https://doi.org/10.1111/j.1750-3639.2005.tb00513.x

Bazan NG (2006) Cell survival matters: docosahexaenoic acid signaling, neuroprotection, and photoreceptors. Trends Neurosci 29(5):263–271. https://doi.org/10.1016/j.tins.2006.03.005

Bazan NG (2007) Homeostatic regulation of photoreceptor cell integrity: significance of the potent mediator neuroprotectin D1 biosynthesized from docosahexaenoic acid the proctor lecture. Invest Ophthalmol Vis Sci 48(11):4866–4881. https://doi.org/10.1167/iovs.07-0918

Bazan NG (2009) Neuroprotectin D1-mediated anti-inflammatory and survival signaling in stroke, retinal degeneration, and Alzheimer's disease. J Lipid Res 50:S400–S405. https://doi.org/10.1194/jlr.r800068-jlr200

Belayev L, Khoutorova L, Atkins KD, Bazan NG (2009) Robust docosahexaenoic acid–mediated neuroprotection in a rat model of transient, focal cerebral ischemia. Stroke 40(9):3121–3126. https://doi.org/10.1161/strokeaha.109.555979

Belayev L, Khoutorova L, Atkins KD, Eady TN, Hong S, Lu Y, Obenaus A, Bazan NG (2011) Docosahexaenoic acid therapy of experimental ischemic stroke. Transl Stroke Res 2(1):33–41. https://doi.org/10.1007/s12975-010-0046-0

Bharate SB, Manda S, Mupparapu N, Battini N, Vishwakarma RA (2012) Chemistry and biology of fascaplysin, a potent marine-derived CDK-4 inhibitor. Mini Rev Med Chem 12(7):650–664. https://doi.org/10.2174/138955712800626719

Bringloe TT, Starko S, Wade RM, Vieira C, Kawai H, De Clerck O, Cock JM, Coelho SM, Destombe C, Valero M, Neiva J (2020) Phylogeny and evolution of the brown algae. Crit Rev Plant Sci 39(4):281–321. https://doi.org/10.1080/07352689.2020.1787679

Calder PC (2016) Docosahexaenoic acid. Ann Nutr Metab 69(Suppl. 1):8–21. https://doi.org/10.1159/000448262

Catarino MD, Silva-Reis R, Chouh A, Silva S, Braga SS, Silva AM, Cardoso SM (2023) Applications of antioxidant secondary metabolites of Sargassum spp. Mar Drugs 21(3):172. https://doi.org/10.3390/md21030172

Diao X, Han H, Li B, Guo Z, Fu J, Wu W (2023) The rare marine bioactive compounds in neurological disorders and diseases: is the blood-brain barrier an obstacle or a target? Mar Drugs 21(7):406. https://doi.org/10.3390/md21070406

Elendu C, Amaechi DC, Elendu TC, Ibhiedu JO, Egbunu EO, Ndam AR, Ogala F, Ologunde T, Peterson JC, Boluwatife AI, Okongko AO (2023) Stroke and cognitive impairment: understanding the connection and managing symptoms. Ann Med Surg 85(12):6057–6066. http://creativecommons.org/licenses/by/4.0/

Ferdous KA, Jansen J, Amjad E, Pray E, Bloch R, Benoit A, Callahan M, Park HA (2024) Mitochondrial protective potential of fucoxanthin in brain disorders. J Nutr Sci 13:e21. https://doi.org/10.1017/jns.2024.16

Gong H, Luo Z, Chen W, Feng ZP, Wang GL, Sun HS (2018) Marine compound xyloketal B as a potential drug development target for neuroprotection. Mar Drugs 16(12):516. https://doi.org/10.3390/md16120516

Hang S, Chen H, Wu W, Wang S, Fang Y, Sheng R, Tu Q, Guo R (2022) Progress in Isoindolone alkaloid derivatives from marine microorganism: pharmacology, preparation, and mechanism. Mar Drugs 20(6):405. https://doi.org/10.3390/md20060405

Holdt SL, Kraan S (2011) Bioactive compounds in seaweed: functional food applications and legislation. J Appl Phycol 23(3):543–597. https://link.springer.com/article/10.1007/s10811-010-9632-5

Hu JF, Hamann MT, Hill R, Kelly M (2003) The manzamine alkaloids. Alkaloids Chem Biol 60:207. https://doi.org/10.1016/s0099-9598(03)60004-0

Jiwa NS, Garrard P, Hainsworth AH (2010) Experimental models of vascular dementia and vascular cognitive impairment: a systematic review. J Neurochem 115(4):814–828. https://doi.org/10.1111/j.1471-4159.2010.06958.x

Johnston SC, Gress DR, Browner WS, Sidney S (2000) Short-term prognosis after emergency department diagnosis of TIA. JAMA 284(22):2901–2906. https://doi.org/10.1001/jama.284.22.2901

Karuppusamy S, Rajauria G, Fitzpatrick S, Lyons H, McMahon H, Curtin J, Tiwari BK, O'Donnell C (2022) Biological properties and health-promoting functions of laminarin: A comprehensive review of preclinical and clinical studies. Mar Drugs 20(12):772. https://doi.org/10.3390/md20120772

Kim JH, Lee NS, Jeong YG, Lee JH, Kim EJ, Han SY (2012) Protective efficacy of an Ecklonia cava extract used to treat transient focal ischemia of the rat brain. Anat Cell Biol 45(2):103–113. https://doi.org/10.5115/acb.2012.45.2.103

Kotake-Nara E, Yonekura L, Nagao A (2015) Lysoglyceroglycolipids improve the intestinal absorption of micellar fucoxanthin by Caco-2 cells. J Oleo Sci 64(11):1207–1211. https://doi.org/10.5650/jos.ess15180

Kubota T, Kurimoto SI, Kobayashi JI (2020) The manzamine alkaloids. In: Alkaloids Chem Biol, vol 84, pp 1–24. https://doi.org/10.1016/bs.alkal.2020.03.001

Kuriakose D, Xiao Z (2020) Pathophysiology and treatment of stroke: present status and future perspectives. Int J Mol Sci 21(20):7609. https://doi.org/10.3390/ijms21207609

Kwon YJ, Kwon OI, Hwang HJ, Shin HC, Yang S (2023) Therapeutic effects of phlorotannins in the treatment of neurodegenerative disorders. Front Mol Neurosci 16:1193590. https://doi.org/10.3389/fnmol.2023.1193590

Lee TK, Ahn JH, Park CW, Kim B, Park YE, Lee JC, Park JH, Yang GE, Shin MC, Cho JH, Kang IJ (2020) Pre-treatment with laminarin protects hippocampal CA1 pyramidal neurons and attenuates reactive gliosis following transient forebrain ischemia in gerbils. Mar Drugs 18(1):52. https://doi.org/10.3390/md18010052

Lukiw WJ, Cui JG, Marcheselli VL, Bodker M, Botkjaer A, Gotlinger K, Serhan CN, Bazan NG (2005) A role for docosahexaenoic acid–derived neuroprotectin D1 in neural cell survival and Alzheimer disease. J Clin Invest 115(10):2774–2783. https://doi.org/10.1172/JCI25420

Luo J, Chen D, Qin B, Kong D (2022) Molecular mechanisms for the prevention and promoting the recovery from ischemic stroke by nutraceutical laminarin: A comparative transcriptomic approach. Front Nutr 9:999426. https://doi.org/10.3389/fnut.2022.999426

Marcheselli VL, Hong S, Lukiw WJ, Tian XH, Gronert K, Musto A, Hardy M, Gimenez JM, Chiang N, Serhan CN, Bazan NG (2003) Novel docosanoids inhibit brain ischemia-reperfusion-mediated leukocyte infiltration and pro-inflammatory gene expression. J Biol Chem 278(44):43807–43817. https://doi.org/10.1074/jbc.m305841200

Min SK, Park JS, Luo L, Kwon YS, Lee HC, Jung Shim H, Kim ID, Lee JK, Shin HS (2015) Assessment of C-phycocyanin effect on astrocytes-mediated neuroprotection against oxidative brain injury using 2D and 3D astrocyte tissue model. Sci Rep 5(1):14418

Mohibbullah M, Haque MN, Sohag AA, Hossain MT, Zahan MS, Uddin MJ, Hannan MA, Moon IS, Choi JS (2022) A systematic review on marine algae-derived fucoxanthin: An update of pharmacological insights. Mar Drugs 20(5):279. https://doi.org/10.3390/md20050279

Mukherjee PK, Marcheselli VL, Serhan CN, Bazan NG (2004) Neuroprotectin D1: a docosahexaenoic acid-derived docosatriene protects human retinal pigment epithelial cells from oxidative stress. Proc Natl Acad Sci 101(22):8491–8496. https://doi.org/10.1073/pnas.0402531101

Nai Y, Liu H, Bi X, Gao H, Ren C (2018) Protective effect of astaxanthin on acute cerebral infarction in rats. Hum Exp Toxicol 37(9):929–936. https://doi.org/10.1177/0960327117745693

Pan L, Zhou Y, Li XF, Wan QJ, Yu LH (2017) Preventive treatment of astaxanthin provides neuroprotection through suppression of reactive oxygen species and activation of antioxidant defense pathway after stroke in rats. Brain Res Bull 130:211–220. https://doi.org/10.1016/j.brainresbull.2017.01.024

Park JH, Ahn JH, Lee TK, Park CW, Kim B, Lee JC, Kim DW, Shin MC, Cho JH, Lee CH, Choi SY (2020) Laminarin pretreatment provides neuroprotection against forebrain ischemia/reperfusion injury by reducing oxidative stress and neuroinflammation in aged gerbils. Mar Drugs 18(4):213. https://doi.org/10.3390/md18040213

Patterson RH Jr (1988) Stroke: an overview. Ophthalmology 95(10):1473–1477

Pramanik S, Singh A, Abualsoud BM, Deepak A, Nainwal P, Sargsyan AS, Bellucci S (2024) From algae to advancements: laminarin in biomedicine. RSC Adv 14(5):3209–3231. https://doi.org/10.1039/D3RA08161C

Sasaki Y, Kobara N, Higashino S, Giddings JC, Yamamoto J (2011) Astaxanthin inhibits thrombosis in cerebral vessels of stroke-prone spontaneously hypertensive rats. Nutr Res 31(10):784–789. https://doi.org/10.1016/j.nutres.2011.09.010

Sawada H, Nishimura N, Suzuki E, Zhuang J, Hasegawa K, Takamatsu H, Honda K, Hasumi K (2014) SMTP-7, a novel small-molecule thrombolytic for ischemic stroke: a study in rodents and primates. J Cereb Blood Flow Metab 34(2):235–241. https://doi.org/10.1038/jcbfm.2013.191

Serhan CN, Yacoubian S, Yang R (2008) Anti-inflammatory and proresolving lipid mediators. Annu Rev Pathol 3(1):279–312. https://doi.org/10.1146/annurev.pathmechdis.3.121806.151409

Tavan M, Hanachi P, de la Luz C-GM, Segura Carretero A, Mirjalili MH (2024) Natural phenolic compounds with neuroprotective effects. Neurochem Res 49(2):306–326

Valero T, del Barrio L, Egea J, Cañas N, Martínez A, García AG, Villarroya M, López MG (2009) NP04634 prevents cell damage caused by calcium overload and mitochondrial disruption in bovine chromaffin cells. Eur J Pharmacol 607(1–3):47–53. https://doi.org/10.1016/J.EJPHAR.2009.02.021

Wang Q, Zhang D, Lu J, Zhang J, Xuan Z, Gong L, Yang M, Jin L, Le J, Zhu A, Liang H (2022) PLGA-PEG-fucoxanthin nanoparticles protect against ischemic stroke in vivo. J Funct Foods 99:105359. https://doi.org/10.1016/j.jff.2022.105359

Ward RE, Huang W, Curran OE, Priestley JV, Michael-Titus AT (2010) Docosahexaenoic acid prevents white matter damage after spinal cord injury. J Neurotrauma 27(10):1769–1780. https://doi.org/10.1089/neu.2010.1348

Wei H, Gao Z, Zheng L, Zhang C, Liu Z, Yang Y, Teng H, Hou L, Yin Y, Zou X (2017) Protective effects of fucoidan on Aβ25–35 and d-Gal-induced neurotoxicity in PC12 cells and d-Gal-induced cognitive dysfunction in mice. Mar Drugs 15(3):77. https://doi.org/10.3390/md15030077

Whisnant JP (1983) The role of the neurologist in the decline of stroke. Ann Neurol 14(1):1–7. https://doi.org/10.1111/j.1365-2044.1959.tb01656.x

Xiao AJ, Chen W, Xu B, Liu R, Turlova E, Barszczyk A, Sun CL, Liu L, Deurloo M, Wang GL, Feng ZP (2014) Marine compound xyloketal B reduces neonatal hypoxic-ischemic brain injury. Mar Drugs 13(1):29–47. https://doi.org/10.1371/journal.pone.0122840

Zhang XS, Zhang X, Wu Q, Li W, Wang CX, Xie GB, Zhou XM, Shi JX, Zhou ML (2014) Astaxanthin offers neuroprotection and reduces neuroinflammation in experimental subarachnoid hemorrhage. J Surg Res 192(1):206–213. https://doi.org/10.1016/j.jss.2014.05.029

GPSR Compliance

The European Union's (EU) General Product Safety Regulation (GPSR) is a set of rules that requires consumer products to be safe and our obligations to ensure this.

If you have any concerns about our products, you can contact us on ProductSafety@springernature.com

In case Publisher is established outside the EU, the EU authorized representative is:

Springer Nature Customer Service Center GmbH
Europaplatz 3
69115 Heidelberg, Germany

Batch number: 10731141

Printed by Printforce, the Netherlands